NANOTRIBOLOGY
Critical Assessment and Research Needs

NANOTRIBOLOGY
Critical Assessment and Research Needs

edited by

Stephen M. Hsu

and

Z. Charles Ying

National Institute of Standards and Technology
Gaithersburg, MD, U.S.A.

KLUWER ACADEMIC PUBLISHERS
Boston / Dordrecht / London

Distributors for North, Central and South America:
Kluwer Academic Publishers
101 Philip Drive
Assinippi Park
Norwell, Massachusetts 02061 USA
Telephone (781) 871-6600
Fax (781) 681-9045
E-Mail < kluwer@wkap.com>

Distributors for all other countries:
Kluwer Academic Publishers Group
Post Office Box 322
3300 AH Dordrecht, THE NETHERLANDS
Telephone 31 78 6576 000
Fax 31 78 6576 254
E-Mail < services@wkap.nl>

 Electronic Services < http://www.wkap.nl>

Library of Congress Cataloging-in-Publication Data

NANOTRIBOLOGY: Critical Assessment and Research Needs
Edited by Stephen M. Hsu and Z. Charles Ying
ISBN: 1-4020-7298-8

Copyright © 2003 by Kluwer Academic Publishers

All rights reserved. No part of this publication may be reproduced, stored in a retrieval system or transmitted in any form or by any means, mechanical, photo-copying, recording, or otherwise, without the prior written permission of the publisher, Kluwer Academic Publishers, 101 Philip Drive, Assinippi Park, Norwell, Massachusetts 02061.

Permission for books published in Europe: permissions@wkap.nl
Permissions for books published in the United States of America: permissions@wkap.com

Printed on acid-free paper.
Printed in Great Britain by IBT Global, London

Table of Contents

Preface viii

Summary of Findings 1

1. **Nanotribology and the NSF**
   Jorn Larsen-Basse 15

2. **MEMS Program at DARPA**
   William C. Tang 23

3. **Scale Effects and the Molecular Origins of Tribological Behavior**
   Gang He and Mark O. Robbins 29

4. **Nanoscale Analysis of Wear Mechanisms**
   Koji Kato 45

5. **Dependence of Frictional Properties of Hydrocarbon Chains on Tip Contact Area**
   Judith A. Harrison, Paul T. Mikulski, Steven J. Stuart and Alan B. Tutein 55

6. **Nanohydrodynamics and Coherent Structures**
   H.G.E. Hentschel and I. Tovstopyat-Nelip 63

7. **Modeling Adhesive Forces for Ultra Low Flying Head Disk Interfaces**
   Andreas A. Polycarpou 79

8. **Self-Lubricating Buckyballs and Buckytubes for Nanobearings and Gears: Science or Science F(r)iction?**
   M. N. Gardos 95

9. **Carbon Nanotubes: Objects of Well-Defined Geometry for New Studies in Nanotribology**
   Min-Feng Yu, Mark J. Dyer and Rodney S. Ruoff 109

10. **A Novel Frictional Force Microscope with 3-Dimensional Force Detection**
    M. Dienwiebel, J. A. Heimberg, T. Zijlstra, E. van der Drift,
    D. J. Spaanderman, E. de Kuyper and J. W. M. Frenken 115

11. **Quantification of Friction in Microsystem Contacts**
    Michael T. Dugger 123

12. **Nanoengineering and Tribophysics for MEMS**
    K. Komvopoulos 139

13. **Materials and Reliability Issues in MEMs and Microsystems**
    Aris Christou 165

14. **Surface Characteristics of Integrated MEMS in High-Volume Production**
    Jack Martin 177

15. **Tribological Issues in the Implementation of a MEMS-Based Torpedo Safety and Arming Device**
    Wade G. Babcock 185

16. **Challenges for Lubrication in High Speed MEMS**
    Kenneth Breuer, Frederic Ehrich, Luc Fréchette, Stuart Jacobson, C-C Lin, D.J. Orr, Edward Piekos, Nicholas Savoulides and Chee-Wei Wong 197

17. **Electrical Phenomena at the Interface of Rolling-Contact, Electrostatic Actuators**
    Cleopatra Cabuz 221

18. **The Use of Surface Tension for the Design of MEMS Actuators**
    Chang-Jin "CJ" Kim 239

19. **Mesoscale Machining Capabilities and Issues**
    Gilbert L. Benavides, David P. Adams and Pin Yang 247

20. An Overview of Nano-Micro-Meso Scale Manufacturing at the NIST

Edward Amatucci, Nicholas Dagalakis, Bradley Damazo, Matthew Davies, John Evans, John Song, Clayton Teague and Theodore Vorburger 259

21. LIGA-Based Micromechanical Systems and Ceramic Nanocomposite Surface Coating

W. J. Meng, L. S. Stephens and K. W. Kelly 271

22. Nanotribology of Thin Film Magnetic Recording Media

T. E. Karis 291

23. Nanolubrication: Concept and Design

Stephen M. Hsu 327

24. Molecularly Assembled Interfaces for Nanomachines

Vladimir V. Tsukruk 347

25. STM-QCM Studies of Vapor Phase Lubricants

B. Borovsky, M. Abdelmaksoud and J. Krim 361

26. Micro-Hardness and Micro-Wear Measurements on Magnetic Heads

V. Prabhakaran, F. E. Talke and J. T. Wyrobek 377

27. Bonding and Interparticle Interactions of Silica Nanoparticles

James D. Batteas, Marcus K. Weldon and Krishnan Raghavachari 387

28. Macroscale Insight from Nanoscale Testing

Steven R. Schmid and Louis G. Hector, Jr. 399

29. Surface Forces and Friction Between Cellulose Surfaces in Aqueous Media

Stefan Zauscher and Daniel J. Klingenberg 411

Index 441

PREFACE

This book is based on the Nanotribology Workshop held at the National Institute of Standards and Technology in Gaithersburg, Maryland, on March 13–15, 2000. Nearly 100 leading researchers from the academia, government, and industry in nanotribology, microelectromechanical systems (MEMS), nanotechnology and meso-manufacturing attended the Workshop. The purpose of the Workshop was to critically assess the current state-of-the-art of nanotribology within the context of MEMS, meso-manufacturing, nanotechnology and microsystems, to identify gaps in current knowledge and barriers to applications, and to recommend research areas that need to be addressed to enable the rapid development of these technologies.

The trend to miniaturize mechanical, electronic and optical components has been ongoing in the last several decades and the pace has accelerated in recent years. Advances in semiconductor processing, sensors, actuators, and microprocessors have made possible the development of revolutionary technologies such as MEMS, smart structures, micro-engines, smart controllers, lab-on-a-chip devices, and micro-biomedical devices, which have abilities to sense/detect, compute, and activate/control in real time. Coupled with neural network and artificial intelligence, manufacturing processes will enter a new era of intelligent manufacture unthinkable a decade ago.

Integration of sensing and computing to actuation is currently in its embryonic stage. Rapid actuation requires fast moving contacting surfaces. Frictional heating associated with moving parts requires innovative materials as well as creative design. Adhesion and stiction become the dominant issues for durability and reliability of MEMS devices.

Nanotribology deals with friction, wear, and lubrication of interacting surfaces in relative motion at the nanometer scale. Current advances in magnetic recording, MEMS, nanotechnology, meso-manufacturing, and microsystems technology all involve the basic issues of friction, wear, adhesion, and lubrication. Micromachines and normal miniaturization demands also are pushing the scale of investigation of nanotribology smaller and smaller. Many phenomena at small scales present new challenges in concept as well as practice. In the last several years, some of the important issues in nanotribology such as scaling laws, measurement and handling problems, contamination issues, and modeling have begun to be addressed. While these issues have often been dealt with at macroscale, nanoscale phenomena have only been dealt with at the molecular dynamics and atomically flat surface levels. Much is unknown at this time. Measurement of friction and wear at nanometer scale, for example, is problematic at best. Atomic force microscope and its modifications are often used, but the issues are many; statically sampling, localized deformation, tip-substrate interactions, and environmental effects (*e.g.*, water condensation and the resultant meniscus) often make the results difficult to reproduce. "Simple" nanomechanical property measurement of materials is equally difficult.

Nanotribology is a new field and current research efforts in nanotribology are scattered through the communities in physics, electronics, MEMS, microprocessors, surface sciences, materials, mechanical engineering, and tribology. The issues in each of the technology areas have different technical communities working on them; communication and exchange of ideas have been problematic. A public forum to exchange information across a wide spectrum of disciplines is needed to provide cooperation and mutual support. It was the purpose of this Workshop to invite these communities on a worldwide basis to come together to share common technical issues, provide assessments, and together identify areas that have the maximum impact across the technologies.

In addition to oral and poster presentations, the Workshop organized extensive discussion sessions that covered areas of nanotechnology, MEMS, mesomanufacturing, and nanolubrication. The discussions were focused on theoretical needs, measurement challenges, materials requirements, and barrier to applications. The findings of the discussion sessions are summarized in Chapter 1 of this book.

Chapters 2 and 3 present overviews of the nanotribology programs supported by the National Science Foundation and the MEMS programs supported by the Defense Advanced Research Projects Agency, respectively.

Theoretical studies on nanotribology are presented in Chapters 4 to 8. Chapter 4 discusses breakdown of continuum behavior at nanometer scales and, using the molecular dynamics (MD) simulation technique, explores the origin of friction at the molecular level. Also using the MD technique, Chapter 5 investigates three abrasive wear mechanisms: cutting, wedging and plowing. Chapter 6 reports MD studies of frictional forces between hydrocarbon chains and a nanotube or a diamond tip, especially dependence of frictional properties on tip contact area. Hydrodynamics or flow of lubricant at nanometer scale is studied numerically in Chapter 7. The last chapter on theory, Chapter 8, models the adhesive force at the interface between a magnetic storage disk and an ultralow flying head.

The next two chapters explore applications of fullerenes and nanotubes for nanotribology, particularly as self-lubricating nanobearings and nanogears (Chapter 9) and as probe tips for atomic force microscopy (Chapter 10).

Novel instrumentation for quantitative nanotribology measurements is urgently needed. Design, fabrication, and performance of a "tribolever," which allows quantitative measurements of frictional forces without the shortcomings associated with the cantilever probes commonly used in friction force measurements, are described in Chapter 11. Chapter 12 presents a MEMS-based device that permits quantitative measurements of frictional forces at surface of micromachined sidewall contacts and study of environmental effects using this device.

Nanotribology research is intimately associated with research and development in MEMS, meso-scale machining, and magnetic recording industry. Chapters 13 and 14 present overviews of nanotribology, materials, and reliability issues associated with MEMS. The following four chapters discuss development of MEMS-based accelerometer (Chapter 15), safety and arming devices for weapons especially torpedoes (Chapter 16), high-speed rotating MEMS devices (Chapter 17), and electrostatic actuators (Chapter 18), plus technology challenges to the nanotribology community that are associated with these MEMS devices. Chapter 19 presents several novel designs of MEMS actuators, including a microinjector, a bubble pump, a micromechanical switch, and a liquid micrometer, that fully utilize surface tension at nanometer scale.

The next two chapters describe meso-scale machining capabilities developed at the Sandia National Laboratories (Chapter 20) and at the National Institute of Standards and Technology (Chapter 21) plus the technical issues associated with meso-scale machining. As a specific example, Chapter 22 presents fabrication and testing of prototype face seals with embedded micro-heat-exchangers based on X-ray microfabrication technique.

Chapter 23 describes lubricants used on magnetic recording media, the properties and testing methods used in the magnetic recording industry.

Development of lubricants and study of lubrication phenomena at the nanometer scale are of current focus in nanotribology. Chapter 24 discusses concepts and designs of nanolubrication. Chapter 25 presents current development of molecularly assembled interfaces with emphasis on future needs and possible fabrication methods of these interfaces for nanomachines. Experimental research of gas-phase lubricants using a combination of scanning tunneling microscopy and quartz crystal microbalance in vacuum is presented in Chapter 26.

The final four chapters of the book present research on wear at the nanometer scale. Chapter 27 describes study of wear of diamond-like-carbon overcoats on magnetic-recording heads. Experimental research using atomic force microscopy on adhesion at silica nanoparticle surfaces and on single-asperity plowing of aluminum surface is presented in Chapters 28 and 29, respectively. Chapter 30 describes study normal forces and sliding friction forces between cellelose surfaces in aqueous solution using scanning probe microscopy.

The Nanotribology Workshop was sponsored by the National Science Foundation, Defense Advanced Research Projects Agency, University of Maryland, National Institute of Standards and Technology, and Sandia National Laboratories. Their financial support of the Workshop is acknowledged.

We express appreciation to the session chairs for the assistance they provided during the Workshop. Special thanks go to the chairs of the breakout discussion sessions, who not only organized the discussion sessions and also provided summary reports, which are the basis for Chapter 1 of this book.

Finally, we would like to thank the editorial and production staff at Kluwer Academic Publishers, especially Editor Jennifer Evans, Editor Rob Zeller, and Ms Anne Murray for their support and patience.

Stephen M. Hsu
Z. Charles Ying
Gaithersburg, Maryland

SUMMARY OF FINDINGS

Four breakout discussion sessions were held during the Workshop, which covered areas of nanotechnology, microelectromechanical systems (MEMS), mesomanufacturing, and nanolubrication. For each of the areas, topics of theory, measurement, materials, and barriers to applications were discussed. This chapter summarizes the findings from the discussions.

The discussions on theory were led by Professor Don Brenner at North Carolina State University, and participated by Thomas Conry, Martin Edelstein, Feyeydoon Family, Bill Goddard, Peter Gordon, Judith Harrison, George Hentchel, Paul Mikulski, Tim Power, Mark Robbins, Lawrence Romana, Michael Rozman, and Igor Tovstopyat-Nelip.

The participants of the discussions on measurement were James Battreas, Kenny Breuer, Robert Carpick, Stephen Hsu, Andrew Jackson, Koji Kato, Kyriakos Komvopoulos, Jack Martin, Martin Munz, Ronald Ott, Andreas Polycarpou, Helena Ronkainen, Rodney Ruoff, Steven Schmid, and Stefan Setz. The discussions were led by Professor Komvopoulos at Berkeley (on nanotechnology and MEMS) and by Dr. Hsu at the NIST (on mesomanufacturing and nanolubrication).

Dr. Mike Gardos at Raytheon led discussions on materials. The participants included Stephanie Broak, Charles Drain, Christine Grant, Tom Karis, Gary Katulka, Peter Kotvis, Jorn Larson-Basse, Mike McNallan, Wen-Jin Meng, Steve Patton, Paul Sutor, Paul Trulove, Vladimir Tsukruk, Simon Tung, and Zhize Zhu.

The discussions on barriers to applications were participated by Phillip Abel, Wade Babcock, Gil Benavides, Bharat Bhushan, Gordon Brown, Cleo Cabuz, Martin Edelstein, Jeff Hillesheim, Stephen Hsu, Koji Kato, Kevin Kelly, James Kiely, C. J. Kim, Jackie Krim, Howard Last, Jack Martin, Gouri Radhakrishnan, Helena Ronkainen, Frank Talke, Min-Feng Yu, and Jane Wang. Dr. Mike Dugger at Sandia National Laboratories led the discussions.

The following is a summary of the findings generated during the discussions for the areas of nanotechnology, MEMS, mesomanufacturing, and nanolubrication:

1. NANOTECHNOLOGY

1.1 Theory

Reliable experimental data: The group identified a strong need for good, reliable experimental data on the nanoscale with which theory and modeling efforts can be verified and/or calibrated. This includes well-defined surface geometry and chemistry.

Time scale models: A proper treatment of multi-time scales is currently needed. Classical molecular dynamics (CMD) appears appropriate for describing much of the short-time dynamics, including quantum effects in thermal conductivity as demonstrated by Professor Goddard's results using linear response theory. Monte Carlo modeling appears appropriate for equilibrium long-term dynamics. For non-equilibrium dynamics (kinetics) at time scales beyond those accessible to CMD (beyond the nanosecond scale), a satisfactory scheme akin to CMD is needed to address many issues appropriate to tribology at the nanoscale.

Length scale: An appropriate treatment of phenomena at the mesoscale (roughly between 1-1000 nm) is needed, although the need is presently less critical than time scales due to recent progress in this area. The discussion focussed on the fact that the required length scale depends on the phenomena being modeled, and a need to identify appropriate scales for given phenomena was stressed. Generating data at the atomic scale that can be used as input to mesoscale modeling (*e.g.*, boundary conditions, constitutive relations) was identified as important. Overlap between scales and modeling methods was also discussed as being critical to effectively bridging length scales.

Surface effects: The discussion stressed the critical role played by surfaces at the nanoscale and the need for good treatments of properties such as surface energies was identified.

Surface roughness: The need to address non-ideal interfaces corresponding to experimentally realizable systems was identified, specifically surface roughness. A key issue is incorporating the long-range nature of elastic effects.

Defect (failure) initiation at the nanoscale: An emphasis on connecting dynamics at the nanoscale to larger-scale defect formation and ultimately materials failure at the macroscopic scale was identified as a critical direction for theory.

Nanorheology: The need to understand fluid dynamics at the nanoscale, including ordering in confined fluids, was identified as an important area to address in theory/modeling.

Quantum-mechanical based force fields: Using data bases derived solely from quantum mechanics to parameterize force fields was identified as a path to building quantum mechanics into larger-scale CMD simulations. In open discussion between breakout groups it was suggested that ideal force fields will likely never be developed that are capable of addressing forces in all situations for a given system, and that a certain degree of *ad hoc* potential development is still needed. In particular, potentials for dissimilar materials at tribological interfaces are needed, including modeling charge transfer, tribochemistry (*e.g.*, bond breaking and making), surface charges, effective double layers, *etc.*

1.2 Measurement

The theme of the discussion was what kind of measurements and techniques could be used/developed to advance our state of knowledge in tribology at the nanoscale. An ultimate goal is to facilitate bridging from the atomic scale analytical studies (molecular dynamics or MD, Monte Carlo simulations) to the nanometer continuum description. The discussion of the group led to the following main issues:

What kind of measurements and techniques are available (or need to be developed) for atomic-scale tribology?

What are the critical parameters that must be determined experimentally and used in the theory/simulations to make the simulations more "usable?"

What scales are "critical" to real applications vs. scales in simulations? An attempt should be made to move both theoretical and experimental scale closer to those of the problem of interest.

To approach the scales in the simulation studies, sharper tips and more sensitive probes are needed. Possible approach includes growth of nanotubes on existing AFM/STM tips, which has problems of (i) controlled growth, (ii) adhesion of nanotube to tip, and (iii) other materials (*e.g.*, nanorods) and advantages of (i) high resolution, (ii) nanoscale, and (iii) absence of capillary effects. The actual material properties of the ultra-sharp tip need to be investigated.

Nanoengineered materials for good nanowear resistance can only be developed provided we can reliably determine the nanomechanical properties. A challenging task in nanocomposites is the objectives evaluation of the properties of laminates (multilayered composites) and the interfacial strength (nanotube-reinforced composites). What is the critical force to pull out a nanotube from its matrix? How do we measure it?

Simulations should be extended to incorporate realistic features with scales up to ~10 nm to approach scale of measurements.

Adhesion at the atomic level. What is it? How do we measure it?

Can frictionless material removal at the atomic level provide insight into the material bonding strength?

Environmental effects on measurements include adsorbates, temperature, and time. How to identify the adsorbate is a challenge.

What can we learn from other model systems that exhibit high uniformity and linearity?

What is the significance of the measuring device's stiffness? Simulations assume the device is infinitely rigid. Is this justifiable at the atomic scale?

It is important to identify those physical properties that are measurable at length and time scales that are of the same order with those in the simulations and vice visa.

In addition to the above issues mostly pertaining to the determination of the material properties at the atomic scale, it is necessary to obtain information about the surface chemical state. This is especially critical to the simulations, since it affects surface interatomic forces and, in turn, adhesion (tip wetting) and deformation [surface potential function, spectroscopic techniques, especially for sliding surfaces (limitation: one surface must be transparent)].

Nano-patterned surfaces: How do we evaluate such surfaces? Advantages could include less adhesion, lube reservoir, lower friction, less wear, *etc.*

Acoustic emission measurements at the nanoscale.

Characterization of wear debris at nanoscale. How do we detect and collect it? Is this feasible?

Perhaps we can study the dynamic characteristics of the measuring devices to extract information about friction. This concept has been used at larger scale, but not at the AFM level.

Currently, it is extremely difficult to quantify the properties of the very soft, viscoelatic materials (*e.g.*, tissues, skin, arteries), especially at the atomic scale. Creep effects at the nanoscale are not well understood. Temperature and time effects play an even greater role in such materials. Advances in atomic-scale bio-tribology and biomaterials characterization can only be achieved with ultra-sensitive, ultra-light-load force instruments. A polymeric AFM/STM cantilever might be the right tool in this case.

Flow in highly constraint media (*e.g.*, nanochannels, nanotubes), which has direct implications in pharmaceutical field.

What should be the next generation nano-stress/strain apparatus? To interface atomic scale results with results obtained at scales an order of magnitude greater, it is imperative to measure and/or determine analytically stress and strains at the atomic scale and integrate such constitutive models into continuum models.

1.3 Materials

The first day's sessions on materials clearly brought out the fact that there is a dearth of viable bearing materials and shapes for nanotribological applications. An unavoidable result of this is the inordinate amount of attention paid to computer-based modeling compared to developing the necessary materials/shape synthesis and the applicable tribometric test techniques on the nanoscale. The wide availability of AFM/FFM/SFA apparatus should improve this incomplete process. The order of the day should be: "If you can't measure it (well), don't model it (wrong)!"

Therefore, first identify those organic/inorganic materials, which can be synthesized in the necessary nanomechanism shapes, as we imagine them according to the currently accepted Drexlerian concepts. Then, reconfigure available AFM/FFM/SFA to measure interfacial attraction or repulsion as a function of realistic environments and temperatures, which can be handled by the apparatus. This materials class could include nanocomposites, which exhibit extraordinarily desirable physical properties (*e.g.*, high strength and modulus) compared to conventional tribomaterials, but cannot yet be shaped to useful nanomechanical configurations. In such cases, the specimens representing the best approximation of nanotribological interfaces could be prepared by avail-able CVD-PVD coating techniques, even if the interfaces were made planar for convenience's sake (*i.e.*, ignoring nanoscale shape factors).

Second, some nanocrystalline materials exhibit super plasticity at the other end of the property spectrum. There are presently no tribometric techniques at any scale that can be employed to characterize them. Clearly, the type of materials needed cannot be separated from the measurement techniques used to evaluate them, nor can either of these separated from the modeling methodologies used to approximate nano-tribological interactions.

1.4 Barriers to Applications

The discussion identified many barriers to applications, which are prioritized below:

Group 1 (highest priority): Surface energetic (bonding) tailoring; high-volume manufacturability; nanometer scale filters - lack of process control, inhibits fabrication; VLSI for nano devices and their packaging requirements; and lack of biocompatibility.

Group 2: Use friction in manufacturing/assembly/transport; metrology and verification/calibration tools; need R&D on a single nanosystem or device to demonstrate system viability (only building blocks of nanodevices have been demonstrated); active control of properties; and metrology and verification tools.

Group 3: Environment of manufacture vs. operation; Does ENV constraint prohibit manufacture? Manufacturability control of stochastic process; high-volume manufacturability; lack of tools from simulation-based design; We are stuck in a paradigm-actual; Nanotechnological devices may look very different than what have been imagined; reproducibility/process control-quality control; Is this a solution looking for a problem? Bioengineering? How to power nano devices? Operational reliability as a function of environment (high temperatures, shock, vibration, and irradiation); lack of command/control method and tools (telemetry?); packaging; diffusion/migration which causes changing device and surrounding over time; nanometer scale filters—lack of process control; active control of properties after fab.; "VLSI" for nano devices (packaging impact); and lack of biocompatibility.

2. MEMS

2.1 Theory

What theory and modeling should strive to accomplish in the MEMS community?

Predict behavior of devices prior to building them: provide specific MEMS designs from theory and modeling. Critical concepts include performance, instabilities, *e.g.*, stiction, component wear rates and lifetimes, plus component failure modes and mechanisms.

Rational materials design needs to identify general properties required for specific MEMS operating and processing conditions and to suggest candidate materials for specific device conditions.

Rational lubricant design should predict lifetimes, and thermal/chemical/mechanical stabilities of candidate lubricants and predict adhesive strengths for specific lubricant/materials combinations.

What do theory and modeling need to accomplish the above goals? The needs include (i) good quantitative experimental data, *e.g.*, adhesive strengths for specific materials/boundary lubricant combinations; (ii) new computational tools (multi-scale, multi-grid) in order to attack appropriate time scales and appropriate length scales; (iii) governing constitutive equations; and (iv) well-defined direction from experimentalists.

2.2 Measurement

The focus of this breakout session was focused on tribological measurements for the microscopic scales, typical to those of MEMS devices. The main issues raised are:

There is a need for funding on basic and applied research at the sub-micrometer level, with focus on the advancement of the present state of understanding occurring at micromachine levels.

It is critical to develop testing standards at micrometer scale which will allow reliable and robust evaluation of microtribological and micromechanical properties such as: friction, adhesion, surface roughness, hardness, toughness, wear, and reliability (fatigue). This might spark the interest from the development of new instrument that will allow performing such measurements. It is desirable to identify simple test structures for property measurements.

Surface engineering should continue focusing on the following areas: (i) chemical surface treatments to control surface termination; (ii) texturing for surface roughness modification; (iii) development of SAMs that are compatible with microfabrication and packaging; and (iv) new coatings (both soft and hard) for lowering the surface energy (hydrophobic surfaces). All the above treatments are targeting anti-stiction, low-wear surface characteristics.

There are needs for new testing methods involving conditions resembling those of MEMS devices.

Rheology in confined channels is expected to be impotent in biofluids-surface tension, roughness of inner wall surfaces, and understanding of bubble dynamics.

Although most tests performed with MEMS-like structures may achieve operating conditions closely resembling those of field conditions, there is still a need for accelerated tests.

Interfacing of bio-MEMS-proper electronics is a challenge.

Determination of polymers used in biological applications is essential since decoupling time rate effects of polymeric parts of the device (*e.g.*, polymer cantilevers used in AFM imaging of tissue and arteries) from those of the test specimen should be achieved in order to measure the properties of the biomaterial reliably.

2.3 Materials

Materials suitable for MEMS moving mechanical assembly (MEMS-MMA) applications are only one part of a particular tribosystem. Simultaneous considerations must include antiwear and lubricating films applied to the MMAs, the configurational design of the nanoscaled power generation or transfer mechanism and its operational environment. Depending on a given system, single-crystal, nanocrystalline or amorphous versions of metals, ceramics, polymers and a variety of nanocomposites may be considered for MEMS-MMA materials of construction.

Metals include those that can be plated and otherwise shaped by LIGA techniques such as Ni, Ni-Fe, Ti-Ni shape memory alloy, Al, and W. However, bare-metallic MEMS-MMAs have the same friction and wear problems their unlubricated macro-versions do.

Ceramics: The most often used ceramics include, in an order of increasingly advantageous properties, $Si < SiC \sim Si_3N_4 <$ nanocrystalline and amorphous diamond. Unfortunately, the list is disadvantageously reversed in a decreasing ability to fabricate them into MEMS-MMA shapes. However, a recent Sandia announcement and a poster from the Argonne National Laboratory at this Workshop showed that versions of the fabrication and self-assembly techniques used with Si MEMS-MMAs can also be employed with amorphous and ultrananocrystalline diamond films and self-standing structures. However, as with metals, none of the ceramic materials perform well without some microscale lubrication.

Polymers, which can be LIGA-fabricated into MEMS-MMAs include POM and PMMA. European tribologists (France, Poland) are paying more attention to other polymeric MEMS-MMA materials capable of self-lubrication than we do here in the U.S.

Nanocomposites may be promising due to the ability to tailor structural integrity, but are seldom fabricated to component shapes by techniques that lend themselves readily to easy assembly.

As to wear-resistant coatings, the MEMS community is slowly beginning to realize the shortcomings of silicon for MEMS-MMAs (especially for extreme environment applications) and has begun to gradually shift to coating the tribocomponents with more wear-resistant CVD SiC. Other coatings such as diamond-like-carbon, silicon nitride or thermally grown SiO_2 did not perform nearly as well. In particular, Case-Western University researchers proposed the application of SiC, both as a coating on silicon and as a self-standing alternative for MEMS-MMAs.

Researchers from the Aerospace Corporation have also developed an *in-situ* TiC coating process by pulsed laser deposition, which could coat fully assembled silicon MEMS-MMAs with a thin, wear-resistant film (see one of the poster papers at this Workshop). Unfortunately, the magnetic field-guided, energetic constituent atoms of TiC delivered to the slightly biased parts or the energetic neutrals of the SiC CVD process precursors surrounding the MEMS-MMAs during deposition at low partial pressures may not reach the insides of narrow and tortuous bearing clearances.

If conditions for any deposition technique were kept in the *molecular flow* regime, they must provide mean free paths for the gaseous constituents *long enough* to penetrate the narrow gaps of the already assembled silicon MEMS pin/hub interfaces. If the reacting vapor-phase species were held at pressures *high enough* for *slip- or viscous- flow* conditions, they could reach the frictional surfaces even better. Ideally, these constituents should reassemble deep in the narrow crevices generating the coatings, provided they did not preferentially react at the entrance. Attempts to solid lubricate fully assembled MEMS-MMAs with a Teflon-like CVD coating, or even film coverage of deep cavities also failed due to the same fundamental flow condition limitations. Notwithstanding any applicable deposition method or the wear-resistance of any likely hardcoats such as SiC and TiC, these are still high friction materials without additional lubrication.

2.4 Barriers to Applications

In MEMS applications, devices can be divided into two classifications: those operate statically and those operate under dynamic sliding conditions. These two classes have different research issues and needs.

Non-sliding. The issues include adhesion; lack of conformal surface treatments (compatible with fab packaging); metrology for surface characterization (electrical properties, "bulk" mechanical properties, surface flaws, high spatial resolution, surface analysis, and failure analysis, "Shell Game" or "NTF"); long term reliability (how to accelerate aging?); surface inhomogeneity (environmental effects on materials and crystalline plane orientation, and stress corrosion cracking measurements and data); effect of BEOL on surfaces (both operational and long term); and standard (relevant) test structures.

Sliding: The issues are "flash" temperature; quantifying friction (use and age); wear (change in morphology and chemistry with wear); contamination (particles); effect of wear debris; lubricant delivery and replacement; lubricant durability; friction instability at micron scale limited set of materials; lack of data on adhesion energy for material pairs; lack of simulation tools for surface interactions; include surface int. in numerical design tools; need accelerated testing with correlations to field results; and long term storage stability.

3. MESOMANUFACTURING

3.1 Theory

Theoretical studies are needed to model current processing techniques, which include (i) laser ablation (*e.g.*, heat removal, heat transfer, debris formation, thermal vs. chemical mechanisms); (ii) chemical etching (*e.g.*, masking and patterning, mechanisms, directionality); (iii) electrode discharge machining; and (iv) ion-beam milling (*e.g.*, masking and patterning, mechanisms, directionality, products, energy transfer).

Statistical properties of reliability (*e.g.*, yields) and self-assembly are other theoretical issues associated with nanomanufacturing.

3.2 Measurement

The discussion identified following measurement needs for mesomanufacturing: metrology capabilities and standards; ability to measure the dimensional parameters at nanometer scale with accuracy and precision; ability to measure the mechanical properties of materials at nanometer scale to millimeter scale with a self-consistent test methods and procedures; measurement capability of mass at nanometer scale; manipulation tools at meso to nanometer scales to handle parts of different dimensions; remote sensing/measuring tools to measure product dimensions and quality control; materials property data spanning from nanometer to millimeter scales; standards and reference materials for calibration of instruments; standards in optical, mechanical, electrical and magnetic measurement techniques for dimensional, materials properties and mechanical properties; and interface standards for manufacturing using surface micromachining and conventional mechanical machining.

3.3 Materials

"Mesoscale" is in the eye of the beholder. On one hand, physical and electronic phenomena studied under the aegis of mesoscopic physics involve hardware in the 1–1000 nm range. In our parlance, nano- and micro-mechanisms fall more into this category. A search of the literature shows that each discipline, ranging from electronic to dislocation science and molecular self-assembly, has its own definition of "mesoscopic" within this rather wide size regime.

On the other hand, the current DARPA/DSO program on mesoscale electromechanical devices deals with equipment in the sugar cube-to-fist size domain. Components of these devices may actually be smaller than 1 mm; they are mesoscale mechanisms, possibly containing a few MEMS parts here and there. Taking the lead from DARPA, we considered only the sugar-cube-fist size boundaries, along with any smaller component which may fit within these limits.

Accordingly, conventional materials of construction could be used in the fabrication of mesoscale mechanisms. Certain composites, however, where the size of the reinforcement (*e.g.*, particles, chopped fibers, fiber weave geometries) are on the scale of the Hertzian contacts aren't acceptable, because their desired composite nature cannot manifest itself in such small contact zones. This is where new nano- or ultranano-composites will have to come into play, for both MEMS and mesoscale tribocomponent materials structured on the smallest possible scale.

There is also a need for more innovative machining of conventional (and a few unconventional) materials to suit the mesoscale. We must develop (i) nanocermets as machine tool materials, along with tech-niques to shape them into microcutters, drills, endmills, gear hobs and other tool varieties that can provide the necessary high surface finishes and tolerances in a single machining step; (ii) ultrathin and ultratough hardcoats to reinforce the tool tips and edges for maximum wear life; (iii) machining-cooling fluids capable of providing bearing contacts with higher surface finishes and more controlled surface chemistries; (iv) laser and spark discharge machining to complement mechanical stock removal; (v) microinjection molding techniques to inexpensively shape polymers (and polymeric nanocomposites) into the necessary micro- and meso-shapes; and (vi) CAD-CAM methods for all these techniques, suitable for the various size regimes.

3.4 Barriers to Applications

Because of the multidisciplinary nature, one needs teams with different expertise; systematic databases on materials, design principles, and tools; dimensional control/metrology; reliability and durability criteria for product production; system integration and design software; lubrication and ease of dimensional control; and sensors and quality control standards.

4 NANOLUBRICATION

The need for nanolubrication exists in nanotechnology, MEMS, and mesomanufacturing, so many of the research needs identified previously are also needed here. At the same time, there are specific issues that are particularly important to nanolubrication.

4.1 Theory

Theoretical investigations are needed in order to predict structure-property relations for specific lubricants to optimize performance. Effects of surface roughness and additives need to be studied.

There is a need for basic theory and force field estimation for surface and molecular interactions. In particular, theory on adsorption, desorption, and reaction from first principles is needed. There are needs for solution of (i) molecular mixtures and aggregation phenomena prediction, (ii) surface conformity of molecules and stacking tendency prediction, and (iii) molecule to molecule interaction patterns in the presence of surfaces and surface defects.

In nanorheology, effects of electrostatic forces and surface forces on fluid flow should be investigated. There is a need for surface wetting prediction based on molecular structures and molecular sizes.

In addition, one needs air-bearing theory and model at high speed, high temperature, and low mass region.

4.2 Measurement

There are a number of needs that are particularly important to nanolubrication measurements. The discussion session identified the following: accurate measurement of force down to nano Newton range; method and instrumentation to measure nanomechanical properties of nanometer films; model to account for the surface deformation caused by the tips; method and instrumentation to measure stiction, under both static and dynamic conditions; method and instrumentation to measure friction, surface forces and their effects on adhesion, charge transfer and charge density measurement, effect of environment on lubricant degradation, lubricant deposition into crevices and three dimensional structures, lubricant flow in microfluidic devices, influence of electrostatic forces on molecular attachment and orientations, and effects of high speed or high frequency motion on lubrication requirements; ability to measure the effect of a monolayer on friction without substrate effect; ability to measure molecular wear loss of materials; ability to measure surface forces of realistic surfaces such as silicon, polysilicon, nickel, silicon nitride, and nanocomposites formed under CMOS conditions; atomic level measurement of lubrication effectiveness in nanotechnology; and device level testing for lubricant effectiveness assessment.

4.3 Materials

The "lubrication" of nanoscale MMAs is highly dependent on the chemical bonding-controlled interaction of the molecular-sized components, as solvated by the medium of operation. To reduce the attractive interaction (and the drive power needed) in the presence of gases or fluids, one must find the right kind of lubricating gas or liquid medium within which the magnitude of friction remains acceptable. The concept of friction is more relevant than wear, because nanoscale wear means introducing life-threatening flaws or destroying the entire molecular MMA component.

As to MEMS, since silicon micromachining is such a well-established technology, the most logical (and cost-effective) solution would be is to forget about any other likely material of construction and retain silicon and its fabrication methodology, but devise methods to reduce the friction and wear of the MEMS-MMAs by surface engineering. Pre-hydrogenation of all silicon surface bonds by HF rinses and non-replenished self-assembled monolayers solvent-introduced into fully assembled MEMS-MMAs have been suggested to improve their tribological behavior. Such one-time applications depend on capillary forces delivering the "lubricant" to the functional surfaces through tortuous paths with narrow bearing clearances, and achieving lifetime lubrication with the residual, sacrificial layers that form after the evaporation of the volatile solvent. Wear-induced destruction of the initially hydrogenated silicon surfaces and tribothermal-catalytic degradation of incomplete lubricant films render these methods marginal. Effective lubrication is especially problematic in reactive environments (*e.g.*, dry and humid air, in-vivo and in other corrosive process fluids) kept at elevated temperatures.

The available lubricants do not limit lubrication at the mesoscale. It is the conventional lubricant re-supply methods that cannot be employed, because lube reservoir and the delivery systems are usually far larger than the meso-device itself. The resultant packaging problems are untenable. Furthermore, assuring lubricant-provided low surface energies is insufficient, because unwanted surface charging must also be eliminated.

The effectiveness of any lubricant mitigating the wear of bearing components of any size must be modeled (predicted) by wear equations. For example, the tribochemical kinetics of MEMS/mesoscale-MMA wear in any environment, which result in chemical alteration of the bearing surfaces into an *in-situ* generated solid lubricant/wear-resistant coating, may be modeled by some variant of the Arrhenius equation, $k = A \exp(-E_{act}/RT)$, where A is the frequency factor, E_{act} is the activation energy, R is the gas constant and T is the absolute temperature. These parameters associated with say the *static* oxidation mode can be obtained by isothermal oxidation data usually generated by thermo gravimetric analysis (TGA). The TGA-determined oxidation kinetics of a large number of materials can be defined by the simple equation of $m^n = kt$, where m is the change in the mass of the sample due to oxidation, t is the time of oxidation, and n is the order of reaction. It is implicit in this methodology that the flow rate or partial pressure of the oxidizing agent (*e.g.*, air, oxygen) is high enough around the sample. Gas delivery to the reacting interface is impeded ($n = 2$ or higher) or unimpeded ($n = 1$ or less) by the respective protective-nonprotective nature of the tribothermally reacted surface layer determining the order of reaction n, and not by the insufficiency of the supplied gas pressure.

Variations of the Arrhenius equation must still be formulated to deal with *dynamic* tribochemical wear. Although the specific form good enough for MEMS/meso-MMA wear predictions is yet to be developed, it is known that the E_{act} on *dynamic* tribo-oxidation is less than that associated with *static* oxidation. The equation should take into account any thermal-atmospheric environment-induced property degradation of the bulk bearing material (and its surfaces) as a function of storage time.

There was a consensus that discussing nano/micro/meso-materials properties cannot be separated from talking about their synthesis and test methodology rationalized by *ab-initio*, molecular dynamics and continuum-mechanical models. A holistic approach is more appropriate, as described in Horst Czichos' 1978 Elsevier book entitled "Tribology - a systems approach to the science and technology of friction, lubrication and wear," along the lines of the now-famous space-time continuum chain-link chart of overlapping size-temporal domains shown Prof. Goddard's paper in this book. Complete evaluation always comes in three simultaneously performed and closely iterated steps: chemical modeling of the material structure and chemistry, preparation of the theoretically most promising and realistically attainable materials into the appropriate MMA shapes, as well as testing the tiny mechanisms of any size for ease of assembly (low cost) and performance.

For now, producing winning mechanisms in our size domains of interest usually depends on the insight and finesse of the very few. Ideally, user-friendly design tools of all types should be developed and made available to any reasonably competent scientist and engineer. The tribologist's role in all this remains the same now as it has ever been, for any MMA size regime: provide the *technological push* in the form of low stiction/friction and the longest possible wear life mechanisms. However, before our efforts are truly appreciated, we must first define those early nano-to-meso "killer applications," which provide the *technological pull* from the market place. Unfortunately, industry has yet to be factored into the national nanotechnology initiative. Since without industry product lines will not appear on the scene, this oversight must be remedied soon.

4.4 Barriers to Applications

In order to overcome barriers to applications, on needs design guideline for what molecule for what application; large database on lubricant used for MEMS and other applications; a central warehouse of information; aging effect assessment on lubricants; data and test methodology of environmental effects on lubricant degradation and ways to predict lubricant useful life; teaming arrangement drawing expertise from universities, government labs and industries due to the multidisciplinary in nature of this area; figures of merit comparison of lubricants and coatings under standardized conditions; and measurement standards and calibration materials.

Chapter 1

NANOTRIBOLOGY AND THE NSF

Jorn Larsen-Basse
National Science Foundation
Arlington, VA 22230

Abstract Nanotribology has been an active field of research for well over a decade. In evolved in response to the search for better understanding of some of the many elusive tribological phenomena, such as friction and solid lubrication. It evolved rapidly, in constant step with the development of new instrument capabilities, which enabled closer and closer observation of surface features and events. Nanotribology also became an important development tool for hard disk technology and MEMS. The National Science Foundation has supported a modest amount of basic research in this area for a number of years. A small sampling of projects funded during the past decade is briefly outlined to give a flavor of the program.
Recently, the concept of nanotechnology – the building of devices from the individual atom and up – has captured the policymakers' imagination, promising a new industrial revolution. This has resulted in a proposed new Federal Nanotechnology Initiative. Many barriers must be overcome before nanotechnology becomes commonplace and a substantial research effort is therefore envisioned. The proposed funding directions for this new initiative are outlined, with emphasis on areas where research in nanotribology and nano-surface engineering promises to make major contributions. Since many of the research problems have strong relevance to tribology and the active engineering of surfaces, it is expected that nanotribology will serve as an important enabler for much nanotechnology. Eventually, nanotribology will need to evolve from the laboratory and become integrated into the engineering design and processing of nano- and meso-scale devices, including tribologically functional design and engineering of surface material and features, atom-by-atom if necessary.

1 INTRODUCTION

Most of the research supported by the NSF develops from unsolicited research proposals. The assumption is that the research community, rather than the Federal bureaucracy, has the innovative ideas and is at the forefront of knowledge. Proposals are almost all peer reviewed and those considered to be the very best are selected for funding, or "investment" as the current nomenclature has it. The early investment in nanotribology began in the mid-1980s, partly as an aid in development of hard disk tribology, partly as an extension of our knowledge of tribological events in general and partly as a curiosity because new observations or computations suddenly became possible. The materials scientist has been in a constant search for better and better resolution of the structure of matter, ever since an optical microscope was focused on an etched metal surface and it was discovered that metals have microstructure. Tribologists have shared in that quest but also had to deal with problems specific to observation of tribological surfaces, namely the lack of depth of focus. This was a very severe obstacle until the SEM arrived on the scene in the 1960s. Much more detail of microstructure became observable with the high resolution TEM and STEM and surfaces could be observed at the atomic scale with the arrival of the STM on the scene in the early-to-mid 1980s, followed shortly by the AFM and a host of microanalytical techniques. In parallel with this development there was a rapid growth in computer-based control and modeling capabilities.

2 PRESENT AND RECENT NSF PROJECTS

A small sample of projects in nanotribology (and some in nanomechanics and in MEMS tribology) supported by the NSF during the past decade, many of them still current, is given below. The grouping fairly well follows the general themes in the evolution of nanotribology. First, the curiosity/ capability-driven studies, where the availability of new instruments and modeling capabilities allowed a closer look at parts of the tribological world hitherto unknowable. Second, the technology-driven studies, trying to develop a basis for micro- and nano-scale technologies, such as hard disk technology and chemo-mechanical polishing of wafers, and related mechanics efforts in electronic packaging and mechanical properties of nanostructures. And third, the possibility-driven attempts to utilize some of the new nano-structures for tribological benefit, such as buckytubes, carbon nanotubes, nano-particulate reinforcement, self-assembled boundary lubricant layers, etc.

2.1 Curiosity/Capability Driven Projects

- Molecular dynamics simulations of asperity contact and friction (Landman, Georgia Tech)
- Surface force microscopy of mono- and bi-layer lubricants (Granick, U. Illinois U/C)

- AFM generation of single molecule tribochemical reaction product and subsequent AFM imaging of the same molecule (Krim, NC State U)
- Elementary particle emissions in tribological contacts; nano-step tribochemical events (Dickinson, Washington State U)
- Interdisciplinary mechanics for MEMS (Boyd, U Illinois Chicago)
- Strain fields at the nanometer scale (Chang, AFOSR)
- Tribomechanics of nanostructured materials (Espinosa, Purdue)
- Workshop on research needs in nano and micromechanics of solids for emerging science and technology, Oct. 1999 (Kim, Brown U) (1)

2.2 Technology Driven Projects

- Fundamentals of stiction in MEMS manufacture and application (Komvopoulos, UC Berkeley)
- AFM modification for simultaneous determination of surface roughness and dielectric properties (Bhushan, Ohio State U)
- Nano-layered superhard coatings by design, using carbon nitride and TiC nanolayers (Keer and Chung, Northwestern U)
- Head-disk dynamics, coatings, lubricants (multiple projects during the past decade, for example by Bogy, UC Berkeley; Streator, Georgia Tech; Gellman, CMU; Lin, U Michigan and Shen, U Washington; Komvopoulos, UC Berkeley)
- Visualizing nanoparticle interactions with silicon wafer during chemo-mechanical polishing (Yoda, Georgia Tech)
- Thermomechanical reliability of interconnects in microelectronics (Shen, U New Mexico)

2.3 Possibility Driven Projects

- Tribochemical wear of ceramics as a nano-polishing process (Fischer, Stevens)
- Tribochemical generation and self-replenishment of boundary lubricant films (Furey, Virginia Tech)
- Buckyballs as a last-resort disk/head crash prevention "bearing" (Bogy, UC Berkeley)
- Carbon nanotubes as reservoir for boundary lubricant. Unsuccessful but an interesting nanotube "grass" was grown which can have strongly directional friction properties (Shaw, SUNY Buffalo)

- Use of nanoparticulate fillers in polymers in attempt to strengthen transfer layer (Bahadur, Iowa State U)
- Self-assembled monolayers of polymers on surfaces as potential boundary lubricants (Tsukruk, W. Michigan U and Iowa State U, and D. Grainger, Colorado State U)
- Technique for depositing wear resistant coating on as-fabricated MEMS devices (Radhakrishnan, Aerospace Corp.)

3 THE NEAR FUTURE: THE NANOTECHNOLOGY INITIATIVE

Nanotechnology has attracted the attention of policymakers in recent years. The first Senate hearing was called by then-senator Gore in 1992. A worldwide study (WTEC) of the state-of-the-art in the US as compared with Europe and Japan was carried out in 1996-98 (2,3) and a number of workshops were held to define agency-specific interests. The issue reached Congress when the President's Assistant for Science and Technology and OSTP Director (and former NSF Director) Dr. Neal Lane stated in Congressional testimony in April of 1998:

> "If I were asked for an area of science and engineering that will most likely produce the breakthroughs of tomorrow, I would point to nanoscale science and engineering" (4).

Since 1998 an interagency committee has been hard at work trying to come to grips with what role the government should play in development of this new initiative. The Interagency Working Group on Nanotechnology (IWGN) reports to the President's National Science and Technology Council (NSTC). It has developed a funding plan for the next five years, held forums for community input, and prepared a plan for a national initiative entitled "National Nanotechnology Initiative - Leading to the Next Industrial Revolution" (February 2000) (5). President Clinton drew the public's attention to the topic area in a speech at Caltech on January 21, 2000:

> "My budget supports a major new National Nanotechnology Initiative, worth $ 500 million ... the ability to manipulate matter at the atomic and molecular level. Imagine the possibilities: materials with ten times the strength of steel and only a small fraction of the weight -- shrinking all the information housed at the library of congress into a device the size of a sugar cube -- detecting cancerous tumors when they are only a few cells in size. Some of our research goals may take 20 or more years to achieve, but that is precisely why there is an important role for the Federal government." (5).

This is quite remarkable exposure for a science and technology issue and illustrates how the concept of nanotechnology has captured the imagination. The Administration's proposed budget for FY '01 contains $

495 M for the Nanotechnology Initiative, with a major portion ($217 M) targeted for NSF, $ 110 M for DOD, $ 94 M for DOE, $ 36 M for NIST and about $ 20 M each for NIST and NASA. It is well to remember that at this stage this is a proposed budget and that not all the funds are "new" money. It still has to be approved by Congress.

The funding themes outlined in the IWGN report (5) are:
- fundamental research ($ 170 M)
- grand challenges - interdisciplinary teams in research and education on nanomaterials, nanoelectronics, bio-nanodevices, and nanotechnology applications in healthcare, environment, energy, microspacecraft, transportation, and national security ($ 140 M)
- centers and networks of excellence ($ 77 M)
- research infrastructure ($ 80 M)
- societal implications ($ 28 M)

The total proposed budget of $ 495 M represents an 83% increase above the estimated expenditures for nanotechnology and related research in FY 2000.

4 POTENTIAL ROLE OF NANOTRIBOLOGY

Each of the funding themes contains a role for tribology and surface science and engineering. Some of those are outlined below. I should stress that this is a very subjective and personal view and that the NSF takes no position in the matter.
- <u>Fundamental research:</u> how to control and manipulate matter at the nanoscale. The hard disk community is already pretty much there and the coatings area is making great strides. But we need to learn to take advantage of new electronic and atomic interactions which become possible only at the nanoscale; also to take advantage of new properties, such as unusual magnetic properties, mechanical properties dominated by defect free crystallites and grain boundary material and/or large surface area. In this area we also need scaling laws, synthesis, self-assembly, processing of nanometer-scale building blocks, etc.
- <u>Nanostructured materials and surfaces by design:</u> how to design and process nano-scale networks, composites, coatings, porous materials, smart materials with built-in condition-based maintenance and self-repair, self-cleaning surfaces with reduced and controlled friction, wear and corrosion.

- **Advanced healthcare:** surface modifications to create novel structures that control interactions between materials and biological systems, for example bioactive particle coatings for implants.
- **Efficient energy conversion and storage:** wear resistant composites, nanoscaled carbide coatings for cutting tools, self-assembled layers for friction control, information on nano-mechanics and materials performance at nano- and MEMS scales as function of aging. Also, interfacial force microscopy and surface manipulation at the nano scale for better adhesion, friction, wear and corrosion. Finally, diagnostic techniques for degradation, and for in-situ lubrication study and control.
- **Microcraft space exploration and industrialization:** materials and systems for ultra-long mission performance under microcraft conditions, such as ultralight weight, ultrastrong materials and materials with unique piezoelectric and optical properties from utilization of the special properties at the nano scale. Also self-repairing materials and self-replicating, biomimetic materials, and last but not least: nanoscale devices which can sustain any needed movement of sliding surfaces for extended periods under very severe conditions - switches, positioning systems, etc.

In addition to its role in support of the evolving nanotechnology one can expect nanotribology to continue to play important roles as part of the discovery process for fundamental phenomena of interacting surfaces and as an invaluable support for the continuing development of technologies. These include hard disk systems with ever-smaller flying heights and processing of IC wafers to ever-greater accuracy as Moore's "law" drives the line width below 100 nm in less than a decade. In the area of nano-scale surface engineering there will be needs for surface treatment of MEMS machines, for smart surfaces and for surfaces with some embedded logic, and for surfaces with instantly changeable properties (6).

5 A NOTE OF CAUTION

A few notes of caution are in order. First, a proposed budget does not always pass Congress in exactly the form it was proposed. The actually available funding may well be significantly less than envisioned and earmarks could encumber some of it.

Second, the promise of new technologies is easily oversold and the public's interest may quickly cool, with consequent decrease in public funding for research. Examples from the past include high temperature superconductivity and the concept of MEMS machines. The latter was actually presented as having some of the same potential promises that are now used for nanotechnology. And, since nothing happened at Y2K it is easy for an impatient public to think that the technical community had oversold the problem for its own benefit. Therefore, some general caution is in order. Also, as demonstrated by M. Gardos at this meeting (7) nano-scale devices are not developed by simply scaling down from the macroscopic scale. He uses as example the proposed use of two individual carbon nanotubes, one grown inside the other, as a nano-scale journal bearing. He demonstrates that secondary bonding forces must be reckoned with and would make this concept totally unworkable, even though it looks attractive in a conceptual drawing.

Third, from a more general and long-term point of view the research community may some day face some substantial ethical questions. Examples are outlined by Bill Joy, chief scientist and co-founder of Sun Microsoft and quoted in a recent Washington Post article (8). His point of caution is that nanotechnology in combination with biotechnology can not only bring amazing benefits to mankind, the combination can also pose a severe threat because it creates the ability to unleash self-replicating, mutating, mechanical or biological plagues which could attack the physical world and would be difficult to stop, even by those who first unleash it. At present we face threats from "rogue states" that with great difficulty have mastered the technology needed to make nuclear weapons. In the future we may have to deal with "rogue individuals" who fairly easily could have mastered the new technologies and may pose a potentially much worse threat. Whether this is a real concern and if anything can be done at this stage remain unanswered questions.

6 CONCLUSIONS

Nanotribology is a rapidly expanding and growing part of tribology. Its future role has two components: it will continue as a tool to expand our knowledge of microscopic and mesoscopic tribological phenomena and it will be an important enabling technology for much of the emerging nanotechnology. It will be necessary for the tribology community to be proactive in developing the field, in its broadest sense, to the point where nanotribological phenomena are included in the original design and where optimal surface structures and compositions for tribological performance are designed in, atom-by-atom as needed.

REFERENCES

1. Workshop report will be available at <http://en732c.engin.brown.edu/NSF-workshop.html>
2. Siegel RW, Hu E, Roco MC (eds.), *Nanostructure Science and Technology*. Kluwer Academic Publishers, 1999.
3. WTEC Workshop Report on *R&D Status and Trends in Nanoparticles, Nanostructures Materials, and Nanodevices in the United States,* Proceedings of the May 8-9, 1997 Workshop, International Technology Institute, Jan. 1998.
4. Amato I, *Nanotechnology - Shaping the World Atom by Atom,* (NSTC Report, Brochure for the public), 1999.
5. *National Nanotechnology Initiative - Leading to the Next Industrial Revolution,* Supplement to the President's FY 2001 Budget, National Science and Technology Council, Committee on Technology, Iteragency Working Group on Nanoscience, Engineering and Technology.
6. Chong KP, Larsen-Basse J, Some Research Activities and Opportunities in Smart Materials and Structures, in *Adaptive Structures and Materials Systems,* Sirkis J and Washington G (eds.), ASME 1998, AD-Vol. 57/MD-Vol. 83, 1-7.
7. Gardos M, Self-lubricating Buckyballs and Buckytubes for Nanobearings and Gears: Science or Science F®iction? (this workshop).
8. Garreau J, From Internet Scientist, a Preview of Extinction, The Washington Post, March 12, 2000, A 15.

Note: Opinions expressed in this paper are those of the author alone. The National Science Foundation takes no position in the matter.

Chapter 2

MEMS PROGRAM AT DARPA

William C. Tang
Microsystems Technology Office
Defense Advanced Research Projects Agency
Arlington, Virginia 22203
wtang@darpa.mil

Microelectromechanical systems (MEMS) are one of the three core enabling technologies within the Microsystems Technology Office (MTO) of the Defense Advanced Research Projects Agency (DARPA). Together with Photonics and Electronics, MEMS forms the foundation for a broad variety of advanced research projects sponsored by MTO. MEMS technology merges the functions of compute, communicate and power together with sense, actuate and control to change completely the way people and machines interact with the physical world. Using an ever-expanding set of fabrication processes and materials, MEMS will provide the advantages of small size, low-power, low-mass, low-cost and high-functionality to integrated electromechanical systems both on the micro as well as on the macro scales. Further, demands for increased performance, reliability, robustness, lifetime, maintainability and capability of military equipment of all kinds can be met by the integration of MEMS into macro devices and systems.

MEMS is based on a manufacturing technology that has had roots in microelectronics, but MEMS has gone beyond this initial set of processes as it became more intimately integrated into macro devices and systems. MEMS will be successful in all applications where size, weight and power must decrease simultaneously with functionality increases, and all while done under extreme cost pressure. Typical applications include, but are not limited to:

- Inertial measurement units for munitions, military platforms and personal navigation;
- Electromechanical signal processing;
- Distributed control of aerodynamic and hydrodynamic systems;
- Distributed sensors both for condition-based maintenance and for structural health and monitoring;

- Distributed unattended sensors both for asset tracking and for environmental/security surveillance;
- Atomic resolution data storage devices;
- Miniature analytical instruments;
- Non-invasive biomedical sensors; and
- Optical fiber components and networks.

The long-term goal of the DARPA MEMS program is to merge information processing with sensing and actuation to realize new systems and strategies to bring co-located perception and control to the physical, biological, and chemical environment. Short-term goals include: demonstration of key devices, processes and prototype systems using MEMS technologies; development and insertion of MEMS products into commercial and defense systems; and lowering the barriers to access and commercialization by catalyzing an infrastructure that can support shared, multi-user design, fabrication and testing. Accordingly, three major research thrusts comprise the DARPA MEMS program: 1) advanced device and process concepts, 2) microsystems development and demonstration and 3) support and access technology.

1 ADVANCED DEVICE AND PROCESS CONCEPTS

The advanced device and process concepts thrust enables the integration and co-location of actuators, sensors, electronics, and power supplies to achieve new functionality, increased sensitivity, wider dynamic range, programmable characteristics, designed-in reliability, self-testing, and autonomy. New device concepts include but are not limited to: the integration of micro devices with communication, control, computation and power components, miniature electromechanical signal processing elements (tuning elements, antennas, filters, mixers), miniature opto-electromechanical devices (cross-bar switches, fiber-optic interconnects and aligners, deformable gratings, and tunable interferometers), force/motion balanced accelerometers and pressure sensors, atomic-resolution data storage, electromechanical signal processing, process control (HVAC equipment, mass flow controllers), simultaneous, multi-parameter sensing with monolithic sensor clusters, and biochemical identification and manipulation.

Examples of innovative processes that enable integration include but are not limited to: surface micromachining processes integrated with all-CMOS microelectronics for low-cost, monolithic motion detection (fuzing, safing and arming functions, tamper detection, impact sensors, velocimeters); deep reactive-ion etching (DRIE) for integrating bulk-micromachined sensors with CMOS; wafer-level encapsulation of micromechanical devices with on-chip electronics; silicon carbide deposition and machining for extreme and hazardous environment sensing; non-silicon micromachining for fluid valves and regulators appropriate for hydraulic and pneumatic systems pressures and flows; and machining and fabrication techniques at the nanometer scale to achieve high-capacity data storage as well as molecular identification and manipulation.

2 MICROSYSTEMS DEVELOPMENT AND DEMONSTRATION

The microsystems development and demonstration thrust focuses both on high-density arrays and on demonstrations of MEMS devices (either in single instances or in dense arrays) in applications of DoD relevance.

The batch fabrication inherent in photolithography-based processing makes it possible to fabricate thousands or millions of components and their interconnections as easily and at the same time that it takes to fabricate one component. This multiplicity allows additional flexibility in the design of electromechanical systems to solve problems. Rather than designing components, the emphasis can shift to designing the pattern and form of interconnections between thousands or millions of components. The diversity and complexity of function in ICs is a direct result of the diversity and complexity of the interconnections and it is the differences in these patterns that differentiate a microprocessor from a memory chip. MEMS technology makes available a distributed approach to design for solving problems in electromechanical systems design.

The incipient maturity of some MEMS devices augers for the demonstration of these devices in applications of DoD relevance. Future MEMS research requires an increasing level of knowledge about the uses and environments MEMS will encounter. It appears prudent to initiate the process of determining this information by sponsoring MEMS demonstrations. These demonstrations will assist in determining the limits of applicability of MEMS technology and foster the adoption of this technology by developers of commercial and DoD systems into which MEMS will be incorporated.

3 SUPPORT AND ACCESS TECHNOLOGY

The support and access technology thrust provides the services, tools, processes and equipment that will accelerate the affordability and manufacturability of MEMS devices and systems. While MEMS is a new way to make electromechanical systems that leverage microelectronics fabrication, significant differences in MEMS devices and fabrication processes, (particularly at end-stage processes and interfaces) require new processing and packaging approaches. Conventional electronics packaging and interconnects seek to provide an appropriate electrical, thermal and mechanical environment for networks of electronic devices. Such packaging often also aims to shrink large quantities of electronics into a small volume, with attendant improvements in performance and reliability. The interface/package manufacturing part of this thrust acknowledges that in many cases the electronics and interconnections need to be not only small, but conform to, rather than dictate the system form factor. Further, many applications require that MEMS devices be integrated intimately into a macro device such as arrays of actuators into airfoils for aerodynamic control. In these cases, traditional packaging and interconnect are inadequate and new technologies must be developed. Examples are the skin of a hypervelocity missile, an unattended sensor, or the knee joint of an exoskeleton. In addition to providing appropriate isolation or protection from some of the environment, packages for MEMS components and systems need to also allow controlled access to selected physical parameters that are either being sensed or controlled. Examples include deformable gratings that need access to light but need to operate in vacuum, MEMS-based aerodynamic control, and silicon carbide sensors that need to access pressure and temperature inside engines.

Sufficient access to MEMS fabrication is necessary so that the rate of technological advance remains high. Thus, it is essential that the MEMS program assist the support and development of affordable, distributed fabrication services for users in industry, academia and government. Previous efforts of the MEMS program have resulted in the MUMPS process. Current efforts are the establishment of a national distributed MEMS fabrication service, the MEMS Exchange. The access to MEMS technology is accelerating both innovation and commercialization of MEMS products. These fabrication facilities are used by other federal and service agencies for education and as a training program in MEMS.

4 DEFENSE AND TECHNOLOGY IMPACT

Experiences in recent conflicts and the evolving role of the US military in rapid response to new kinds of missions have demonstrated the compelling advantage of securing accurate and timely information. Coupled with smart weapons systems, the resulting combination of awareness and lethality is key to increasing and projecting military capability in the 21st

century at reduced cost. MEMS embedded into weapons systems, ranging from competent munitions and sensor networks to high-maneuverability aircraft and identify-friend-or-foe systems, will bring to the military new levels of situational awareness, information to the warrior, precision strike capability, and weapons performance/reliability. These heightened capabilities will translate directly into tactical and strategic military advantage, reduced casualties and reduced material loss. Further, the ability of individual war fighters of the future will be greatly enhanced with wearable and portable instruments that are enabled with MEMS technology. These highly capable and rugged instruments not only are lightweight and small, but also consume far less power than conventional counterparts, and therefore significantly reducing the battery weight. Some of the ultra-low-power MEMS devices will have on-board MEMS power sources based on liquid fuel, which has orders-of-magnitude higher energy density than batteries. These devices, coupled with wireless communication capabilities, will form truly autonomous, standalone, remote sensors that can be deployed en mass for situation awareness and battlefield surveillance.

MEMS will make high-end functionality affordable to low-end military systems and extend the operational performance and lifetimes of existing weapons platforms. Because devices and systems that will be produced or enabled by this program can be expected to be deployed in significant numbers, affordability and manufacturing issues are key to DoD use. The manufacturing processes resulting from this program will be capable of producing a diversity of components and systems without retooling and of realizing near-equivalent unit costs from either prototype or full-scale production quantities. Procurement costs for new components, systems and spare parts will decrease, inventory/storage costs will diminish and manufacturing capability/facilities will consolidate. Because of these advantages, the development, acceptance and acquisition of MEMS devices and technologies will accelerate.

The ability of MEMS to gather and process information, decide on a course of action and control the environment through actuators increases the affordability, functionality and the number of smart systems. Dynamic devices and systems will merge with communication, power, control and processing components; three-dimensional micromachining technologies; and flexible packaging techniques. The range of current design and simulation tools for electronic devices will be extended with three-dimensional mechanical and electric field modeling modules and process-related material property modules to improve the design, integration and operation of dynamic devices.

MEMS devices will see wide use in both military and commercial arenas, with applications ranging from automobiles and fighter aircraft to printers and munitions. While MEMS devices will be a relatively small fraction of the cost, size and weight of these systems, MEMS will be critical to their operation, reliability and affordability. MEMS devices, and the products they enable, will increasingly be the performance differentiation for both defense and commercial systems.

Chapter 3

SCALE EFFECTS AND THE MOLECULAR ORIGINS OF TRIBOLOGICAL BEHAVIOR

Gang He and Mark O. Robbins
Department of Physics and Astronomy
Johns Hopkins University
Baltimore, MD 21218
mr@pha.jhu.edu

1 INTRODUCTION

There has been a great deal of progress in probing the molecular origins of friction in recent years. New experimental tools such as the surface force apparatus,[1-3] quartz microbalance,[4] scanning probe microscopy,[5-6] and other methods,[7-9] allow measurements with controlled chemistry and, in some cases, geometry. At the same time, tremendous increases in computer power have allowed increasingly sophisticated models of these ideal systems.[10-11]

These nanotribological studies reveal behavior that can often be quite different than that observed at macroscopic scales. For example, crystalline monolayers exhibit no static friction on incommensurate substrates, and may even slide more easily than the fluid phase of the same monolayer![4,12] In contrast, most lubricants begin to exhibit solid-like behavior when confined in nanoscale contacts.[1,3,10,13-15] Such dramatic changes with contact size have tremendous implications for the developing area of nanotechnology. They also pose questions of great fundamental interest about the molecular origins of conventional macroscopic behavior.

In the following sections we outline some of our group's work on nanoscale contacts. We begin by describing work that examines the limits of continuum approaches, culminating in a discussion of confinement induced glass transitions. Then the static and kinetic friction of glassy films are explored and shown to provide a molecular-scale explanation for Amontons' laws.

2 COMPUTATIONAL METHODOLOGY

Our focus has been on exploring general phenomena using molecular dynamics (MD) simulations with relatively simple interaction potentials.

These allow us to quickly span a wide range of shear rates, system sizes, geometries and interactions. They also allow treatment of longer time and length scales than more detailed potentials.

In all of the work described below, the two sliding solids contain discrete atoms. To minimize the number of atoms, the elastic interactions within the solids are treated in an Einstein model. Each atom is coupled to its equilibrium position by a spring of stiffness κ. These equilibrium positions can form a crystalline or disordered surface. The coordinate system is chosen so that the surfaces lie in the x-y plane and are normal to the z-axis.

One important variable is the relative registry and orientation of the two surfaces. Identical, aligned crystals can easily lock together to produce static friction, while analytic studies indicate that incommensurate surfaces (those with no common periodicity) should slide with zero static friction.[10,12,16-18] Our simulations use periodic boundary conditions in the plane of the surfaces, and so can never be strictly incommensurate. However, we find that surfaces behave as if they were incommensurate once the common period becomes longer than a few lattice constants.[18]

In most of our simulations, a layer of "fluid" molecules is placed between the two solid surfaces. Some work considers spherical molecules that interact with a truncated Lennard-Jones potential:[19]

$$V_{LJ}(r) = 4\varepsilon\left[(\sigma/r)^{12} - (\sigma/r)^{6}\right], \quad r < r_c, \qquad (1)$$

where r is the distance between molecules and the potential is zero for $r > r_c$. The parameters ε and σ are characteristic energy and length scales, respectively. The characteristic time is $t_{LJ} \equiv (\sigma^2/m\varepsilon)^{1/2}$ where m is the molecular mass. These characteristic scales are used to normalize other quantities. Values that would be representative of hydrocarbons are[20] $\varepsilon \sim 30$ meV, $\sigma \sim 0.5$ nm and $t_{LJ} \sim 3$ ps.

Other simulations use a simple bead-spring model for oligomers.[20] Each molecule contains n spherical molecules of mass m. All monomers interact through a truncated Lennard-Jones potential, and adjacent monomers along the chain are coupled by an additional potential that prevents chains from crossing:

$$V_{CH}(r) = -(1/2)kR_0^2 \ln\left[1 - (r/R_0)^2\right], \qquad (2)$$

where $R_0 = 1.5\sigma$ and $k = 30\varepsilon/\sigma^2$. Previous studies have shown that this bead-spring potential yields realistic dynamics for polymer melts,[20] and shown how to map between it and detailed chemical models of polymers.[21] These detailed models take orders of magnitude more computer time than the bead-spring model.

Wall atoms interact with fluid molecules or monomers through a Lennard-Jones potential with modified parameters ε_w and σ_w. This allows us to increase or decrease the amount of adhesion and the effective surface corrugation of the walls.

All of our runs are done in a constant temperature ensemble. In most cases a Langevin heat bath is coupled to the equations of motion of wall atoms.[14] In the case of infinitely rigid walls $\kappa \to \infty$ the thermostat is instead coupled to the velocity components of the fluid molecules that are orthogonal to any imposed shear. We have performed extensive tests on the effects of thermostats.[22] At the shear rates discussed below, their only effect is to prevent a gradual rise in temperature due to energy dissipated by friction. At higher shear rates, the structure and other properties of the film may be affected.[23-24]

It is well known that experimental measurements of friction are influenced by the mechanical properties of the measuring device. The same is true in simulations, and we have attempted to mimic typical surface force apparatus or atomic-force microscope experiments. For example, we generally apply a constant normal pressure, and allow the separation between surfaces to equilibrate.[14] We find that using fixed wall separation can lead to very different behavior, especially when the walls are commensurate. The bottom wall is held fixed and the top wall is moved laterally by applying a constant velocity, a constant force, or by coupling the wall to a constant velocity stage with a spring. In most simulations the walls are also allowed to move freely in the direction normal to z and to the applied shear.

3 BREAKDOWN OF CONTINUUM BEHAVIOR AT NANOSCALES

A large number of studies have addressed the breakdown of continuum equations as length scales approach atomic dimensions.[10] In the following we will focus on examples that our group has been involved in. The surprising conclusion is that continuum descriptions of elasticity and surface tension describe stresses down to a few lattice constants. Continuum viscoelastic equations can also apply to this scale. However, confinement in such small pores may change the phase of a fluid film, so that the appropriate value for the viscosity may differ dramatically from the bulk value.

3.1 Elasticity

Most analyses of the stress distribution near a contact are based on continuum elasticity. One might wonder whether such analyses are relevant in nanoscale contacts. Landman and coworkers[25] have examined this issue for tips with a diameter of about 10 atoms and find a strong qualitative correspondence between the observed stresses and continuum calculations. However, one aspect of continuum calculations that cannot be correct is the prediction of infinite stresses at the edge of a contact. Such singularities are predicted by continuum theory whenever materials of sufficiently dissimilar elastic constants intersect at a sharp corner.[26] These mathematical singularities are unphysical and must be cut off in some manner at the atomic scale. The nature of this cutoff is important in determining the ultimate shear strength of a contact.

Vafek and Robbins have performed a detailed atomistic study of the stress near such corners.[27] At distances greater than a few lattice constants from the corner, the MD results agree quantitatively with continuum theory. When a perfectly harmonic interaction potential is used between the atoms, the singular behavior is cut off by the atomic size. For Lennard-Jones potentials, large stresses lead to anharmonic behavior within a few lattice constants from the corner, and eventually to plastic flow. One of the most intriguing results is that the maximum stress at the corner increases with system size. This can be understood from scaling analysis of the continuum equations.[26-27]

The observation that the peak stress at the edge of a contact decreases with contact diameter has potential consequences for nanodevices. In some models the large stresses at the edges of contacts facilitate sliding by nucleating defects.[28] If the stresses decrease in magnitude with the size of the contacts, the mechanism of sliding may change.

3.2 Surface Tension

As length scales decrease, capillary forces become increasingly important. Capillary forces are a major component of the normal load in many AFM and SFA experiments. Their magnitude is usually calculated with continuum theory and bulk surface tensions even though the radii of curvature of the menisci may be only a few nanometers or less.

Thompson *et al.*[29] have calculated the Laplace pressure and contact angle of menisci with radii of curvature as small as a few molecular diameters. They found that the results were in excellent agreement with continuum equations for all systems considered. Earlier work showed that even the non-equilibrium dynamic contact angle in systems of similar size could be described by continuum theory.[30]

Figure 1 replots some of Thompson *et al.*'s data. Continuum theory predicts that the surface tension γ divided by the pressure drop across the interface ΔP should equal the radius of curvature, ρ. As illustrated in the figure, values of $\gamma/\Delta P$ are consistent with directly measured radii of curvature down to about 7 times a typical atomic diameter. It would be interesting to test this relation to even smaller radii in future work.

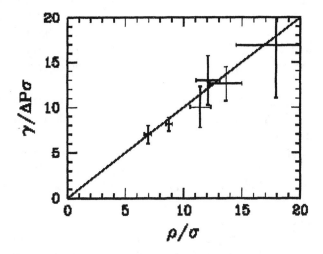

Figure 1. The radius of curvature ρ calculated from the pressure drop ΔP using the continuum relation $\rho = \gamma/\Delta P$ is plotted against the value of ρ measured directly from the meniscus geometry. The points should lie on the solid line if continuum theory applies.

3.3 Flow Boundary Conditions

Continuum theories of lubrication need boundary conditions to describe the dynamics at solid surfaces. The usual assumption is that there is no slip, i.e. that the two phases have the same velocity at the interface. However, a number of theoretical and experimental studies have shown that deviations from the no-slip boundary condition are common. They can be characterized by a slip length S that is normally of atomic dimensions,[31-32] but can be much larger.[32-35] Typical examples of slip for simple spherical molecules are shown in Fig. 2(b).

Calculations based on the no-slip assumption become inadequate when S is of the same order as the geometrical separation h between two interfaces. There are many cases where S is comparable to the minimum separation between lubricated surfaces in conventional applications. More dramatic effects can be expected at the ever shrinking average separations between hard disks and read heads, between the components of microelectromechanical machines, or in the tiny pipes being constructed with lithography.

3.4 Viscosity and Phase Changes

Many recent experimental[1-3,13,36,37] and theoretical[14,38-40] studies have examined the effect of confining walls on the viscosity of fluid films. Most fluids show bulk behavior when the separation h between the walls is more than ten molecular diameters. In this limit, their lubricating properties could be described by continuum equations with an appropriate flow boundary condition.

As the film thickness drops below a few to ten molecular diameters, the viscosity of most fluids diverges. Sharp divergences that are indicative of a first order phase transition are seen in some systems.[13,14,38,39] Most systems exhibit a continuous divergence in the viscosity that is reminiscent of a bulk glass transition,[2,14,38,39] but at temperatures and pressures that are far from the glass region in the bulk phase diagram. In fact, the molecules may not readily form glasses in the bulk, and the nature of this fluid/solid transition has been of considerable interest.

The onset of solid-like behavior in confined films has important implications for nanodevices and for boundary lubrication, where the separation between asperities on opposing surfaces decreases to molecular scales.[41] For example, confinement-induced solidification would prevent lubricant from being squeezed out of contacts between static or moving surfaces. The remaining film would be able to accommodate shear, and thus help to lower both friction and wear.

3.4.1 Structural Changes

Even before films solidify there are dramatic changes in their structure. The most widely studied is ordering of atoms into layers that are parallel to the wall. This layering is induced by the sharp cutoff in fluid density at the wall and the bulk pair correlation function.[42-44] A fluid layer forms at the preferred wall-fluid spacing. Additional fluid molecules then tend to lie in a second layer at the preferred fluid-fluid spacing, and so on.

Figure 2(a) illustrates layering of simple spherical molecules. The magnitude of the first peak depends strongly on the interaction between wall and fluid atoms ε_w, while the rate of decay of peaks into the fluid does not. This decay rate is determined by the decay of oscillations in the pair correlation function of the pure fluid.

Walls also induce epitaxial in-plane order in the fluid.[32,45-47] This lateral registry is crucial to the transfer of shear stress and is inversely related to the amount of slip at the interface.[32] It is illustrated in Fig. 3 for the three layers nearest to the (100) face of an fcc crystal. The range of in-plane order is similar to the range of layering, and is also related to the pair correlation function.

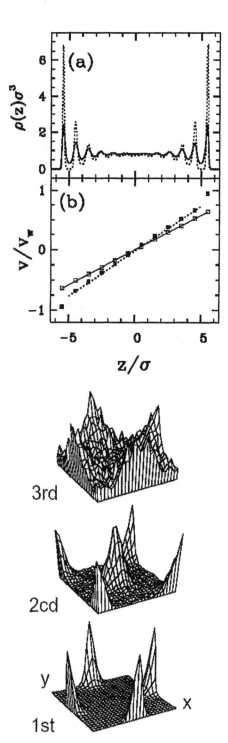

Figure 2. (a) Density as a function of distance between two walls for $\varepsilon_w/\varepsilon = 0.4$ (solid line) and 4 (dotted line). The wall atoms are centered on the left and right boundaries of the plots. Increasing the attraction to the wall increases the magnitude of the density oscillations, but not the decay length. (b) Velocity divided by the wall velocity v_w as a function of distance between walls for $\varepsilon_w/\varepsilon = 0.4$ (open squares) and 4 (closed squares). This data is from the low wall velocity regime where the curve is independent of v_w. Points show the average velocity in each layer, and linear fits to each set of points are shown. Note that the velocity extrapolates to a value less than v_w, indicating slip.

Figure 3. Probability of finding a fluid atom at a given lateral position in the 1st through third layers above a (100) surface of an fcc solid. Only one periodic unit cell of the surface is shown. Wall atoms lie at the center and corners of the cell. Atoms in the first layer lie centered between these wall atoms on the edges of the unit cell. Atoms in the second layer are centered between those on the first layer and so on. (Reproduced from Ref. 32.)

Layering and in-plane order are present near any solid/fluid interface. When the fluid film is thicker than the range of structural change, its main effect is to alter the flow boundary condition. However, when the order spans the entire thickness, the film may begin to exhibit solid-like behavior.

For the spherical molecules of Figs. 2 and 3 the order must be surprisingly large before solidification occurs. This can be seen from Fig. 2(b) where any change in viscosity would produce a change in the velocity gradient (since the stress is constant). All but the last solid square near each wall fall onto a single straight line, implying that the viscosity retains its bulk value down to the first layer. This is true even though the density changes by up to a factor of seven near the second and third layers.[48] Studies of chain molecules show much smaller density modulations before solidification occurs.

3.4.2 Glass Transition

Figure 4 shows how the viscosity changes in the limit of extreme confinement. The spacing between the walls was decreased by (a) decreasing the number of layers at fixed pressure, or (b) increasing pressure at a fixed number of layers. To compare to experiments, an effective shear rate $\dot{\gamma}_{\text{eff}}$ was defined as the velocity difference between two parallel walls divided by their separation. The effective viscosity μ_{eff} is then the shear stress divided by $\dot{\gamma}_{\text{eff}}$.

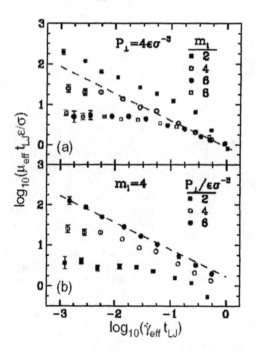

Figure 4. Plots of the effective viscosity μ_{eff} vs. effective shear rate $\dot{\gamma}_{\text{eff}}$ at (a) fixed normal pressure $P_\perp = 4\varepsilon\sigma^{-3}$ and varying numbers of layers m_l, and (b) fixed $m_l = 4$ and varying $P_\perp$. Dashed lines have slope $-2/3$. (Reproduced from Ref. 14.)

For thick films and low pressures, the viscosity shows a broad Newtonian region where μ_{eff} is constant. The shear rate where the viscosity begins to drop is the inverse of the characteristic relaxation time of the film. As the degree of confinement increases, both the low shear rate viscosity and the relaxation time rise dramatically. These changes are typical of a fluid approaching a glass transition.

Demirel and Granick[2] showed that the frequency-dependent complex elastic moduli measured at different film thicknesses could be collapsed onto a single universal curve using a generalization of the time-temperature scaling that describes bulk glasses.[49,50] Baljon and Robbins examined the relationship to bulk glass transitions in more detail.[15,51] The shear-rate dependent viscosity was calculated for films thick enough to exhibit bulk behavior and the temperature was varied through the glass transition.[52] Fig. 5 shows that these results can be collapsed onto a universal curve using time-temperature scaling.[15,51] More importantly, viscosity curves for thin films that were brought through a glass transition by either increasing pressure at fixed number of fluid layers or by decreasing the number of layers at fixed pressure could be collapsed onto the same universal curve. This indicates that the same glass transition occurs whether thickness, normal pressure, or temperature is varied. The effect of walls is similar to that of a change in thermodynamic variables.

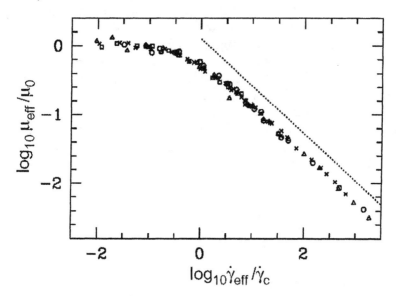

Figure 5. Universal response function for bulk and confined films obtained by scaling the effective viscosity and shear rate. Circles show results for decreasing temperature in bulk systems, squares are results from Fig. 4(a), crosses are from similar runs with different wall interactions, and triangles are results from Fig. 4(b). (Reproduced from Ref. 15.)

4 MOLECULAR ORIGINS OF FRICTION

4.1 How Do Solids Lock Together?

One of the outstanding puzzles in studies of friction has been why macroscopic bodies normally exhibit static friction.[41] That is, they do not move relative to each other until a finite threshold force F_s, or static friction, is exceeded. This implies that the two bodies have locked together into a local energy minimum, and F_s is the force needed to break them free from this minimum.

The first difficulty is to understand how almost any pair of macroscopic bodies manages to lock together.[10,41,51] Analytic[53] and numerical work[12,16,18,54,55] shows that two crystals should almost always slide over each other without static friction. Static friction only occurs if the alignment and lattice constants are exactly tuned to produce a common periodicity (commensurability), or if the interactions between the two surfaces are strong compared to the interactions within each surface. Surfaces are extremely unlikely to be commensurate, and the interactions between incommensurate surfaces are unlikely to be strong enough to produce locking. Indeed, even calculations for two clean incommensurate surfaces of the same metal show no static friction.[16,55] One might expect that roughness or other disorder on the surface produces locking, as in charge-density waves and other systems. However scaling analysis shows that the pinning force between three dimensional objects is exponentially weak.[56-58]

Smith and Robbins[41] noted that most surfaces are coated with a layer of hydrocarbon molecules and other debris that settles out from the air and suggested that glassy behavior of these thin layers might be responsible for static friction. Müser and Robbins considered commensurate, incommensurate and atomically flat amorphous surfaces, that were either bare or had a layer of molecules separating them.[18-19] Bare rigid surfaces only exhibited static friction in the thermodynamic limit if they were commensurate. Adding elasticity only produced locking between bare incommensurate surfaces when the interaction between surfaces was an order of magnitude larger than the internal interactions. In contrast, all surfaces pinned together readily when a monolayer or submonolayer of molecules was placed in between them. Indeed, static friction was observed between rigid commensurate and incommensurate surfaces even when the molecules separating them formed a freely diffusing fluid layer.[18]

4.2 Molecular Origins of Amontons' Laws

The above results indicate that the airborne molecules present between any ambient surfaces can naturally lead to widespread observation of static friction. The next question is whether the resulting friction is consistent with experiment. Recent work by He *et al.*[17] shows that glassy adsorbed layers provide a simple molecular explanation for Amontons' 300-year-old laws for friction. These state that friction is proportional to the normal load L and independent of the apparent geometric area of the surfaces A_{app}.

He *et al.* built on a phenomenological model developed by Bowden and Tabor.[60] These authors noted that A_{app} is usually much larger than the area of intimate molecular contact A_{real}. They pointed out that both Amontons' laws and many exceptions to them could be explained *if* the local shear stress, τ, in the contacts increased linearly with the local pressure:

$$\tau = \tau_0 + \alpha P. \quad (3)$$

Summing over A_{real} and dividing by load gives a friction coefficient

$$\mu \equiv F/L = \alpha + \tau_0/P. \quad (4)$$

This is independent of load if P is constant or $\tau_0/P \ll \alpha$. At high loads, both elastic[58,61] and plastic[60] models of multiple contacts yield a constant P. At low loads, P decreases, explaining why many systems show an increase in μ in this limit.[60,62]

Molecular dynamics simulations with commensurate, incommensurate and disordered walls show that adsorbed molecular layers naturally lead to a linear relation between τ and P like that assumed by Bowden and Tabor.[17,59] This linear relation extends into the gigapascal range that is typical of real contacts. Moreover, τ_s is insensitive to parameters that are not controlled in experiments, including the relative crystallographic orientation of the surfaces, sliding direction, chain length, chain density, *etc*. These results can be understood from simple geometric arguments[17] and a microscopic theory.[59]

Figure 6 illustrates some of the walls that were studied. In each case the bottom wall is a (111) surface of an fcc crystal with nearest neighbor spacing $d = 1.2\sigma$. In (A-C) the top wall is identical, but is rotated by 0, 8.2 or 90°, respectively. In (D), the lattice constant of the top wall is slightly smaller than that of the bottom wall. Only (A) is commensurate.

4.2.1 Results for Static Friction

The static shear stress τ_s for each system and pressure was determined by increasing the force until the system began to slide steadily. The static friction was zero when no atoms were placed between the surfaces, except for the commensurate case (Fig. 6(A)).

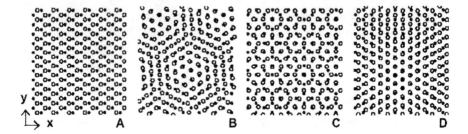

Figure 6. Projections of atoms from the bottom (filled circles) and top (open circles) surfaces into the plane of the walls. In A-C the two walls have the same structure and lattice constant but the top wall has been rotated by 0°, 8.2° or 90°, respectively. In D the walls are aligned, but the lattice constant of the top wall has been reduced by 12/13. Note that the atoms can only achieve perfect registry in the commensurate case A. The simulation cell was at least four times the area shown here. (Reproduced from Ref. 17.)

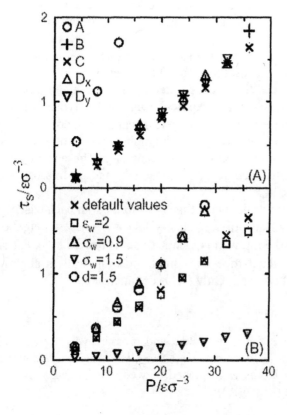

Figure 7. Variation of the yield stress τ_s with pressure P for (a) coverage equal to 1/8 for the different walls shown in Fig. 6. The results for incommensurate surfaces are independent of sliding direction as illustrated for sliding along x (up triangles) and y (down triangles). Panel (b) shows results for different potential parameters. The default values are $n = 6$, $\varepsilon_w = \varepsilon$, $\sigma_w = \sigma$, $d = 1.2\sigma$, $r_c = 2^{1/6}\sigma$, and $k_B T = 0.7\varepsilon$. (Reproduced from Ref. 17.)

Figure 7(a) compares τ_s for the different surfaces shown in Fig. 6 at an adsorbate coverage of 1/8, defined as the ratio of monomers to wall atoms on each surface before they are brought into contact. For each case the data fall onto the straight line implied by Equation 3 up to $P=36\varepsilon/\sigma^3$ or about 1.5 GPa. The incommensurate cases [Fig. 6(B)-(D)] all give very similar values of τ. Changing the direction of sliding also has little effect on τ, as indicated by the comparison between sliding along x or y for the surfaces of Fig. 6(D) [up and down triangles in Fig. 7(a)]. Only the commensurate case [Fig. 6(A)] produces different values of τ_s. As expected, the friction is enhanced because of the uniform registry between the two surfaces. However, the ratio between commensurate and incommensurate values of α has been reduced from infinity to a factor of three due to the presence of an adsorbed layer.

A variety of interaction potentials and geometries have been considered to determine which factors influence α and τ_0 and which leave them unchanged. The alignment of the surfaces and the direction of sliding always leave τ_s nearly unchanged, except in the unlikely case where the surfaces are commensurate. Other parameters that are not controlled in experiments have little effect on the friction. For example, decreasing the chain length n from 6 to 3 to 1 produces no change in τ_s, within statistical fluctuations. Increasing the coverage on each surface up to one or more monolayers on separated surfaces also produces little change in τ_s.

Since different materials do have different experimental friction coefficients, some potential parameters must change τ_s. Fig. 7(b) shows the effect of changing one parameter at a time from our standard set. Increasing $\varepsilon_w/\varepsilon$ from one to two produces almost no change in τ_s. In contrast, decreasing σ_w/σ from 1 to 0.9 increases α by 50% and increasing this ratio to 1.5 decreases α by a factor of 6. The opposite trend is seen for d/σ, where an increase from 1.2 to 1.5 increases α to 0.073.

These trends can be understood from a simple geometric argument. At the large pressures of interest, the repulsive interactions between wall atoms and monomers are dominant. Atoms and monomers cannot be closer than an effective hard sphere diameter of order σ_w. This is very insensitive to pressure because the repulsive force rises as $(\sigma_w/r)^{-13}$ as r decreases. For the same reason, changing σ_w by a factor of two has little effect on the system. In contrast, decreasing σ_w/d allows monomers to penetrate more deeply into the wells between wall atoms. Tracing out the height of closest approach as a function of lateral displacement produces a ramped surface. The value of α corresponds to the effective slope of the ramps, as envisioned in the early models of friction.[63] As σ_w/d decreases, the ramp becomes steeper, resulting in the increase in α seen in Fig. 7(b). We have confirmed that the surfaces do indeed move apart as the yield stress is approached and that this displacement increases with α.

4.2.2 Kinetic Friction

In recent studies we have examined the influence of adsorbed layers on kinetic friction.[64] We find that the friction rises logarithmically with sliding velocity v as observed in many experiments.[65] As for static friction, $\tau(v)$ rises linearly with P over the experimentally relevant range of pressures. Moreover the kinetic and static friction follow the same trends with ε_w, σ_w and other parameters. At the lowest velocities accessible to our simulations ~cm/s, the kinetic friction is usually 70 to 90% of the static friction. Such ratios are typical of many experiments with dry surfaces.

5 CONCLUSIONS

The field of nanotechnology offers great opportunities for fundamental science as well as industrial impact. Experimental systems continue to become better characterized and smaller in scale, while simulations increase in scale and complexity. There is already considerable overlap between the two, and promise for closer interplay between them in the future.

The work presented here emphasizes the important role that surface species can play. Two general factors that have yet to be addressed in detailed simulations are surface roughness and long-range elastic deformations. These effects and detailed models of molecular structure and bonding will be needed to improve comparisons of theory and experiment.

ACKNOWLEDGMENTS

The authors wish to thank Drs. A. R. C. Baljon, M. Cieplak, G. S. Grest, M. H. Müser, P. M. McGuiggan and P. A. Thompson for useful discussions. Support from the National Science Foundation Grant DMR-9634131 is gratefully acknowledged.

REFERENCES

1. M. L. Gee, P. M. McGuiggan, J. N. Israelachvili, and A. M. Homola, J. Chem. Phys. **93**, 1895 (1990).
2. A. L. Demirel and S. Granick, Phys. Rev. Lett. **77**, 2261 (1996).
3. H.-W. Hu, G. A. Carson, and S. Granick, Phys. Rev. Lett. **66**, 2758 (1991).
4. J. Krim, E. T. Watts, and J. Digel, J. Vac. Sci. Technol. A **8**, 3417 (1990); J. Krim, D. H. Solina, and R. Chiarello, Phys. Rev. Lett. **66**, 181 (1991).
5. C. M. Mate, G. M. McClelland, R. Erlandsson, and S. Chiang, Phys. Rev. Lett. **59**, 1942 (1987).
6. R. W. Carpick and M. Salmeron, Chem. Rev. **97**, 1163 (1997).
7. P. Berthoud and T. Baumberger, Proc. R. Soc. Lond A **454**, 1615 (1998).

8   A. J. Gellman, J. Vac. Sci. Technol. A 10, 180 (1992); C. F. McFadden and A. J. Gellman, Surf. Sci. 391, 287 (1997).
9   J. H. Dieterich and B. D. Kilgore, Tectonophysics 256, 219 (1996).
10 M. O. Robbins and M. H. Müser, in *Modern Tribology Handbook*, edited by B. Bhushan (CRC Press, Boca Raton, 2001) pp. 717-765 and cond-mat/0001056.
11 J. A. Harrison, S. J. Stuart, and D. W. Brenner, in *Handbook of Micro/Nanotribology*, edited by B. Bhushan (CRC Press, Boca Raton, 1999), pp. 525–594.
12 M. Cieplak, E. D. Smith, and M. O. Robbins, Science 265, 1209 (1994); E. D. Smith, M. Cieplak, and M. O. Robbins, Phys. Rev. B. 54, 8252 (1996).
13 J. Klein and E. Kumacheva, Science 269, 816 (1995).
14 P. A. Thompson, G. S. Grest, and M. O. Robbins, Phys. Rev. Lett. 68, 3448 (1992); P. A. Thompson, M. O. Robbins, and G. S. Grest, Israel J. of Chem. 35, 93 (1995).
15 M. O. Robbins and A. R. C. Baljon, in *Microstructure and Microtribology of Polymer Surfaces*, edited by V. V. Tsukruk and K. J. Wahl (American Chemical Society, Washington DC, 2000), pp. 91–117.
16 M. Hirano and K. Shinjo, Phys. Rev. B 41, 11837 (1990); K. Shinjo and M. Hirano, Surface Science 283, 473 (1993).
17 G. He, M. H. Müser, and M. O. Robbins, Science 284, 1650 (1999), and G. He and M. O. Robbins, Phys. Rev. B64, 035413 (2001).
18 M. H. Müser and M. O. Robbins, Phys. Rev. B 64, 2335 (2000).
19 M. P. Allen and D. J. Tildesley, *Computer Simulation of Liquids* (Clarendon Press, Oxford, 1987).
20 K. Kremer and G. S. Grest, J. Chem. Phys. 92, 5057 (1990).
21 W. Tschöp, K. Kremer, J. Batoulis, T. Bürger, and O. Hahn, Acta Polym. 49, 61 (1998); 75 (1998).
22 M. J. Stevens and M. O. Robbins, Phys. Rev. E 48, 3778 (1993).
23 E. Manias, G. Hadziioannou, and G. T. Brinke, J. Chem. Phys. 101, 1721 (1994).
24 R. Khare, J. J. de Pablo, and A. Yethiraj, Macromolecules 29, 7910 (1996).
25 U. Landman, W. D. Luedtke, and J. Gao, Langmuir 12, 4514 (1996).
26 E. D. Reedy, Jr., Engineering Fracture Mech. 36, 575 (1990).
27 O. Vafek and M. O. Robbins, Phys. Rev. B 60, 12002 (1999).
28 J. A. Hurtado and K. S. Kim, Proc. R. Soc. Ser. A. (London) 455, 3363 (1999).
29 P. A. Thompson, W. B. Brinckerhoff, and M. O. Robbins, J. Adhesion Sci. Technol. 7, 535 (1993).
30 P. A. Thompson and M. O. Robbins, Phys. Rev. Lett. 63, 766 (1989).
31 D. Y. C. Chan and R. G. Horn, J. Chem. Phys. 83, 5311 (1985); J. N. Israelachvili, J. Colloid Interface Sci. 110, 263 (1986).
32 P. A. Thompson and M. O. Robbins, Phys. Rev. A 41, 6830 (1990).
33 P. A. Thompson and S. M. Troian, Nature 389, 360 (1997).
34 J.-L. Barrat and L. Bocquet, Phys. Rev. Lett. 82, 4671 (1999).
35 P. G. de Gennes, C. R. Acad. Sci. Ser. B 288, 219 (1979).
36 J. M. Georges, S. Millot, J. L. Loubet, A. Touck and D. Mazuyer, in *Thin Films in Tribology*, edited by D. Dowson, C. M. Taylor, T. H. C. Childs, M. Godet, and G. Dalmaz (Elsevier, Amsterdam, 1993), pp. 443–452.
37 J. N. Israelachvili, P. M. McGuiggan, and A. M. Homola, Science 240, 189 (1988).
38 E. Manias, I. Bitsanis, G. Hadziioannou, and G. T. Brinke, Europhys. Lett. 33, 371 (1996).
39 J. Gao, W. D. Luedtke, and U. Landman, Phys. Rev. Lett. 79, 705 (1997); J. Chem. Phys. 106, 4309 (1997); J. Phys. Chem. B 101, 4013 (1997).
40 I. Bitsanis, S. A. Somers, H. T. Davis, and M. Tirrell, J. Chem. Phys. 93, 3427 (1990); I. Bitsanis and C. Pan, J. Chem. Phys. 99, 5520 (1993).
41 M. O. Robbins and E. D. Smith, Langmuir 12, 4543 (1996).

42. F. F. Abraham, J. Chem. Phys. 68, 3713 (1978).
43. S. Toxvaerd, J. Chem. Phys. 74, 1998 (1981).
44. I. K. Snook and W. van Megen, J. Chem. Phys. 72, 2907 (1980).
45. P. A. Thompson and M. O. Robbins, Science 250, 792 (1990).
46. U. Landman, W. D. Luedtke, and M. W. Ribarsky, J. Vac. Sci. Technol. A 7, 2829 (1989).
47. M. Schoen, C. L. Rhykerd, D. J. Diestler, and J. H. Cushman, J. Chem. Phys. 87, 5464 (1987); Science 245, 1223 (1989); M. Schoen, J. H. Cushman, D. J. Diestler, and C. L. Rhykerd, J. Chem. Phys. 88, 1394 (1988).
48. Note that if one defines the velocity gradient on a finer mesh, there are fluctuations in slope. The concept of a bulk viscosity is meaningless at such small scales.
49. J. D. Ferry, *Viscoelastic Properties of Polymers*, 3rd Ed. (Wiley, New York, 1980).
50. W. Götze and L. Sjögren, Rep. Prog. Phys. 55, 241 (1992).
51. M. O. Robbins, in *Jamming and Rheology: Constrained dynamics on microscopic and macroscopic scales*, edited by A. J. Liu and S. R. Nagel (Taylor and Francis, London, 2000) and cond-mat/9912337.
52. A. R. C. Baljon and M. O. Robbins, Mat. Res. Soc. Bull. 22(1), 22 (1997); and in *Micro/Nanotribology and Its Applications*, edited by B. Bhushan (Kluwer, Dordrecht, 1997), pp. 533–553.
53. S. Aubry, in *Solitons and Condensed Matter Physics*, edited by A. R. Bishop and T. Schneider (Springer-Verlag, Berlin, 1979), pp. 264–290. P. Bak, Rep. Prog. Phys. 45, 587 (1982).
54. B. N. J. Persson, Phys. Rev. B 48, 18140 (1993).
55. M. R. Sørensen, K. W. Jacobsen, and P. Stoltze, Phys Rev. B 53, 2101 (1996).
56. B. N. J. Persson and E. Tosatti, in *Physics of Sliding Friction*, edited by B. N. J. Persson and E. Tosatti (Kluwer, Dordrecht, 1996), pp. 179–189.
57. C. Caroli and P. Nozieres, in *Physics of Sliding Friction*, edited by B. N. J. Persson and E. Tosatti (Kluwer, Dordrecht, 1996), pp. 27–49.
58. A. Volmer and T. Natterman, Z. Phys. B 104, 363 (1997).
59. M. H. Müser, L. Wenning, and M. O. Robbins, Phys. Rev. Lett. 86, 1295 (2001) and cond-mat/0004494.
60. F. P. Bowden and D. Tabor, *The Friction and Lubrication of Solids* (Clarendon Press, Oxford, 1986).
61. J. A. Greenwood and J. B. P. Williamson, Proc. Roy. Soc. A 295, 300 (1966).
62. E. Rabinowicz, *Friction and Wear of Materials* (Wiley, New York, 1965).
63. D. Dowson, *History of Tribology* (Longman Inc., New York, 1979).
64. G. He and M. O. Robbins, Tribo. Lett. 10, 7 (2001) and cond-mat/0008196.
65. J. H. Dieterich, J. Geophys. Res. 84, 2169 (1979).

Chapter 4

NANOSCALE ANALYSES OF WEAR MECHANISMS

Koji Kato
School of Mechanical Engineering
Tohoku University
Sendai, Japan

Abstract This paper compares the usefulness of continuum mechanics with that of MD simulation for the analysis of abrasive wear mechanism in nanometer scale. The wear in repeated abrasive contact is also analyzed experimentally in nanoscale. In abrasive wear, there are three abrasive wear modes; cutting mode, wedge-forming mode and ploughing mode. Wear particles are generated by one pass of sliding in cutting mode or wedge forming mode, but no wear particle in ploughing mode. The generation mechanisms of these three abrasive wear modes have been well observed with SEM in micrometer scale and theoretically analyzed with continuum mechanics. Abrasive wear mode map has been established with these results. The predictions of abrasive wear modes by this map are compared in this paper with those by the MD simulations calculated for nanometer scale machining. The comparison shows that the prediction of nanoscale wear mode by the continuum mechanics agrees well with that by the MD simulation. In the case of ploughing mode, wear particles are generated only after repeated friction, which has been well observed experimentally in micrometer scale. This wear process is experimentally observed on carbon nitride coating, in this paper, in nanometer scale and its wear mechanism is confirmed as low cycle fatigue.

1 INTRODUCTION

When the size of a machine or a device becomes small from meter size to millimeter size and then to micrometer size, the effects of adhesion, friction and wear on the performance of the machine or device becomes relatively large. Tribological properties known with macrosystems, on the other hand, are not always available for such MEMS.

For example, adhesive force or pull-off force is negligibly small in macrosystems, but it becomes relatively larger than external load in nanonewton range and generates problems in MEMS. Friction coefficient in unlubricated sliding between metals or ceramics can generally be supposed below unit for macrosystems, but it must be supposed to increase to above five with the decrease of contact load in the range of nanonewton. Oil film at the sliding interface is supposed to reduce friction in its micrometer scale thickness for macrosystems, but its shear strength increases to much higher values as the film thickness is reduced down to mono-layer.

The existence of adhesion or pull-off force and its dependency on surface roughness and material properties have been shown experimentally by Tabor et al. (1,2). The contact mechanics under the effect of adhesion by long distance atomic force was theoretically analyzed by Johnson et al., and the JKR model introduced with continuum mechanics has been shown effective to explain experimental results (3). The study of Landman et al. (4) on atomic contact and pull-off force with MD simulation showed its usefulness for the predictions of contact resistance and pull-off force in nanoscale. It also showed the good agreement between predictions of stress distributions by continuum mechanics and MD simulations. The atomic friction properties were observed by Mate et al. (5) and Mori et al. (6) with FFM on ideal atomic surfaces and the experimental results were well reproduced by Sasaki et al. with MD simulations (7).

The lubricating properties including shear strength of a lubricant film in the nanometer range of thickness were shown experimentally by Israelachvili et al. (8) and simulated by Miyamoto et al. (9) with MD. The predictions of traction coefficients showed reasonable values to accept for practice. The lubricating effect of hydrogen on carbon coating which has been used in practice, was shown by Harrison et al. (10) with MD simulation. The super low friction in sliding of solid krypton surface on crystalline gold surface was observed by Krim et al. (11) and explained by Robbins with MD simulation (12).

Similar extremely low friction between unaligned MoS_2 crystals was observed by Martin (13) in ultrahigh vacuum, but friction rises rapidly with exposure to air. The atomic model for such low friction was proposed by Hirano et al. (14).

In comparison with above introduced recent experimental and simulational understandings on adhesion, friction and lubrication in nanoscale, understanding on wear in the same scale stays at its initial stage.

Well used wear testers are AFM and FFM, which have low stiffness and operate at very small speed in a limited area. It is difficult to find wear scars with present microscopes and is difficult to collect or pick up wear particles for analysis. It is also difficult in nanoscale to make real time observation of the wear process with any present microscope. Setting up a piece of equipment for nanoscale wear test is also not common because of the difficulty of controlling all related parameters in nanoscale.

Therefore scratch test or abrasive wear test with AFM or FFM has been commonly carried in recent wear researches. Groove depth formed by scratching with a diamond or ceramic tip is the measure to evaluate the wear property of the surface. The experimental results obtained by Miyake et al.,

Bhushan, Komvopoulus et al., and Kato et al. showed that the nanoscale wear rate of hard coatings such as DLC and diamond coatings varied between 10^{-6} and 10^{-4} mm^3/Nm in air without lubricants. Khurshudov et al. showed that the wear rate of diamond pin in wear test with AFM was in the order of 10^{-7} mm^3/Nm. These experimental wear results including wear modes and values are not yet explained with MD simulations.

In this paper, therefore, the wear mode prediction by the abrasive wear theory based on the traditional continuum mechanisms is compared with that by the MD simulations to confirm the degree of agreement in both predictions as the first step of the understandings of wear mechanisms in nanoscale. For the further step of understanding of wear, an experimental result of low cycle fatigue wear in nanoscale is introduced and its mechanism is analyzed.

2 ABRASIVE WEAR MODE MAP

Three wear modes are observed in abrasive wear of metals: cutting mode, wedge forming mode and ploughing mode. These three modes were confirmed experimentally with the SEM-tribosystem at micrometer scale and abrasive wear mode map was proposed by Hokkirigawa and Kato (15). Their mechanisms were explained with the solutions of slip line field theory by Challen and Oxley (16).

Figure 1 shows the proposed abrasive wear mode map where f is the shear strength τ of the contact interface normalized by the shear strength κ of the bulk material and D_p is defined by the following equation,

$$D_p = \frac{h \text{ (Indentation depth)}}{a \text{ (Contact radius)}} = R\sqrt{\frac{\pi H}{W}} - R\sqrt{\frac{\pi H}{2W} - 1}, \quad (1)$$

where W is the load, H hardness and R tip radius. D_p is the measure of severity of contact and is called as degree of penetration.

It is related to the representative plastic strain ε_R in the plastic deformation zone generated by indentation of a spherical indenter. ε_R is given by the following equation with the radius R of the indenter and the contact radius r (17),

$$\varepsilon_R = \alpha \frac{r}{R}, \quad (2)$$

where α takes 0.2 by D. Tabor and 0.17–0.19 for materials of work hardening property by J.R. Mathews.

On the other hand, geometrical relationship gives the following equation for D_p,

$$D_p \cong \frac{a}{2R}. \quad (3)$$

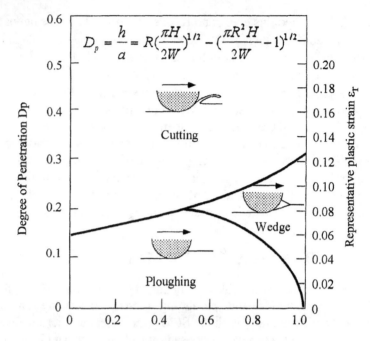

Figure 1. Abrasive wear mode map described with degree of penetration D_p and the relative shear strength f, which is defined by the shear strength of the contact interface divided by the bulk shear strength. ε_R is the representative plastic strain (15).

By supposing the linear relationship between a and r as follows

$$a = \gamma \times r \quad (1 \leq \gamma). \tag{4}$$

D_p and ε_R are related with Eqs. (2), (3) and (4) as follows,

$$\varepsilon_R = \frac{2\alpha}{\gamma} D_p. \tag{5}$$

The values of ε_R of this meaning are also shown in Fig. 1 for the case of $\gamma=1$.

3 MD SIMULATIONS OF CUTTING MODE

For the MD analysis of cutting mode in abrasive wear, the simulation results in Fig. 2 for cutting of copper with a diamond tool by S. Shimada *et al.* (18) are available. Morse potential is postulated between two atoms.

In the figure, the uncut chip thickness is similar to the indentation depth h and the tool edge radius is similar to a in Eq. (1). Therefore, Fig. 2 tells that ideal cutting is performed at $h/a > 0.2$ and becomes difficult at $h/a < 0.2$ where specific cutting energy stays at high values over 4.8×10^4 N/mm².

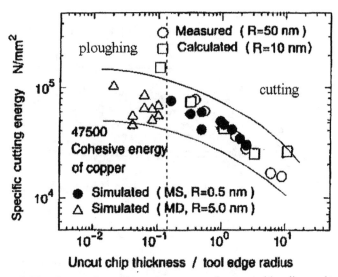

Figure 2. The change of specific cutting energy of copper with a diamond tool in relation to the uncut chip thickness which is similar to cutting depth. (18).

The high value of specific cutting energy means that the larger part of the energy is used only for plastic deformation in ploughing mode, and its low value at high value of h/a means that efficient cutting is performed when the tool edge is sharp.

Although the values of f in the simulation is not known, we may suppose that f can not be over 0.5 at the contact interface between diamond and copper by considering practical experiences. Therefore, we may confirm with Fig. 2 that MD simulation predicts the cutting mode of copper at $f < 0.5$ and $h/a > 0.2$ on the abrasive wear mode map shown in Fig. 1.

4 MD SIMULATIONS OF WEDGE FORMING AND PLOUGHING MODES

For the MD analyses of wedge forming mode and ploughing mode in abrasive wear, the simulation results for the sliding of a diamond slider on a copper flat by Ohmura *et al.* (19) are available. The normal and horizontal reactive forces are shown in Figs. 3 and 4, where Morse potential is supposed between two atoms, and a Maxwell distribution of velocities of atoms is confirmed (18).

The tangential force F_x in Fig. 3 changes in cyclic way with the interval of about 0.25 nm, which is the distance between neighboring two atoms. It means that F_x shows the atomic structure of the contact surfaces and the contact stress is within the elastic limit at the indentation depth of 0.1 nm.

Wear particles are not generated in this sliding as the result. Therefore, Fig. 3 is understood as the simulation of ploughing mode in abrasive sliding.

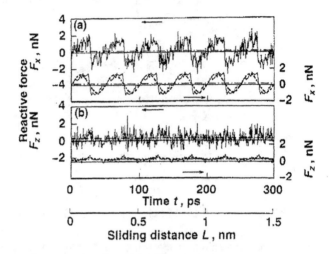

Figure 3. Variation of tangential (F_x) and normal (F_z) forces acting on the diamond slider in sliding against the copper flat. The broken lines and solid lines in the lower part of (a) or (b) show approximations with sine functions and Fourier expansion, respectively (19). Slider diameter: 10.4 nm, indentation depth: 0.1 nm, sliding speed: 5 m/s.

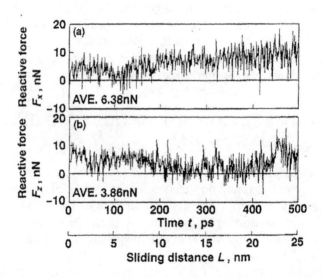

Figure 4. Variation of tangential (F_x) and normal (F_z) forces acting on the diamond slider in sliding against the copper flat. (19). Slider diameter: 10.4 nm, indentation depth: 0.4 nm, sliding speed: 50 m/s.

By considering the geometry of contact for Eq. (3), Dp is described by

$$D_p = \sqrt{\frac{h}{2R}} \qquad (6)$$

The value of for D_p for $2R = 10.4$ nm and $h = 0.1$ nm in Fig. 3 is calculated as $D_p \cong 0.10$. It is the value below which ploughing mode takes place over the wide range of f in the abrasive wear mode map shown in Fig. 1. It would be reasonable from the past experimental results to suppose the value of f below 0.5 for the contact between the diamond slider and the copper flat surface. Therefore, the simulation of Fig. 3 including the confirmation of no wear predicts the generation of ploughing mode in abrasive sliding, and this prediction agrees with that of abrasive wear mode map in Fig. 1 for the same contact condition.

The tangential force F_x in Fig. 4 does not change in cyclic way at the indentation depth of 0.4 nm compared with that in Fig 3. For this indentation depth, a typical wedge is formed in MD simulation and relative sliding takes place in the bulk of copper, as shown in Fig. 5 (19).

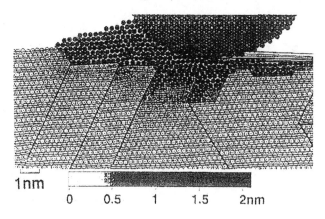

Figure 5. Snapshot of atomic array in wedge forming process with a diamond slider on a copper flat surface. A gray scale shows the distance moved from the initial array (19). Slider diameter: 10.4 nm, indentation depth: 0.4 nm, sliding speed: 50 m/s, sliding distance: 25 nm.

The value of Dp for this contact condition is calculated as $D_p = 0.20$ and f should be supposed as $f = 1$. The set of these values of D_p and f take place in the region of wedge forming mode on the abrasive wear mode map in Fig. 1. Therefore, we may say that the prediction of wedge forming mode by MD simulation agrees well with that by the abrasive wear map.

5 LOW CYCLE FATIGUE WEAR IN PLOUGHING MODE

As already discussed in Fig.3, abrasive wear mode map and MD simulations predict ploughing mode at D_p below 0.10 over the wide range of f values. In order to understand wear generated in the mode by the repeated sliding contact, carbon nitride coating was rubbed by an AFM diamond pin at around $D_p = 0.05$ or below it.

Fig. 6 shows the observed relationship between the depth of wear scar on the carbon nitride coating and the number of friction cycles in the repeated abrasive sliding of a diamond spherical pin. At the sudden increase in the depth of wear scar, the delamination of a thin filmy wear particle was observed (20).

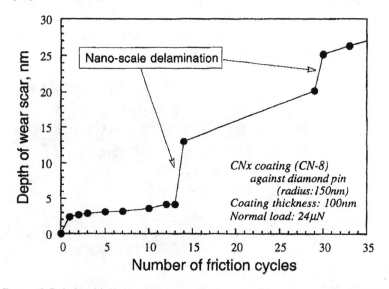

Figure 6. Relationship between the depth of wear scar on the carbon nitride coating and the number of friction cycles in repeated abrasive sliding of a diamond spherical pin. At the sudden increase in depth of wear scar, the delamination of a thin filmy wear particle was observed (20).

Similar observations were made under different contact loads, and the relationship between the plastic strain amplitude $\Delta\varepsilon_p$ in the contact region and the critical number of friction cycles N_d for the initiation of coating delamination was obtained as shown in Fig. 7.

The experimental equation, $\Delta\varepsilon_p \times N_d^{0.73} = 0.079$, in Fig. 7, tells that the delamination mechanism is understood as low cycle fatigue described by Coffin-Manson's equation.

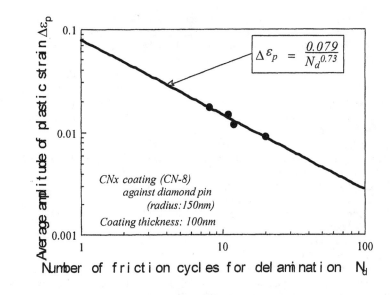

Figure 7. Relationship between the plastic strain amplitude at the contact region in carbon nitride coating and the critical number of friction cycles for the coating delamination in repeated abrasive sliding of a diamond spherical pin under the loads of 14, 24, 35 and 45 μN. (20)

6 CONCLUSION

The abrasive wear mode map formed by the continuum mechanics and confirmed by the micron scale experiments predicts abrasive wear modes simulated by MD in nanoscale. This confirms the usefulness of the continuum mechanics for the analyses of wear mechanisms in nanoscale.

Carbon nitride coating on silicon wafer shows low cycle fatigue wear of Coffin-Manson's relationship in nanoscale abrasive contact of ploughing mode.

ACKNOWLEDGEMENT

The author wishes to express his thanks to Dr. Lin Zhou, Dr. Ming-wu Bai, Dr. Dong-fang Wang and Dr. Unchung Cho for their discussions and material preparations.

REFERENCE

(1) J. S. McFarlane and D. Tabor, Proc. R. Soc. Lond. (1950) 224.
(2) N. Gane, P. F. Pfaelzer and D. Tabor, Proc. R. Soc. Lond. A 340 (1974) 495.

(3) K. L. Johnson, K. Kendall and A. D. Roberts, Proc. R. Soc. Lond. A 324 (1971), 301.
(4) U. Landman, W. D. Luedtke and R. J. Cilton, Science 248 (1990) 454.
(5) C. M. Mate, G. M. McClelland, R. Erlandsson and S. Chiang, Phys. Rev. Lett. 59 (1987) 1942.
(6) S. Fujisawa, Y. Sugawara and S. Morita, Phil. Mag. A 74 (1996) 1329.
(7) N. Sasaki, M. Tsukada, S. Fujisawa. Y. Sugawara, S. Morita and K. Koboyashi, Phys. Rev. B 57 (1998) 3785.
(8) J. N. Israelachvili, in Fundamentals of Friction, edited by I. L. Singer and H. M. Pollock, Kluwer, Dordrecht, (1992).
(9) A. Miyamoto et al, Proc. Annual Meeting, JSI, 1999.
(10) A. Harrison, C. White, R. J. Colton and D. W. Brenner, Phys. Rev. B 46 (1992) 9700.
(11) J. Krim, D. H. Solina and R. Chiarello, Phys. Rev. Lett. 66 (1991) 181.
(12) M. Cieplak, E. D. Smith and M. O. Robbins, Science 265 (1994) 1209.
(13) J. M. Martin, Phys. Rev. B 48 (1993) 10583.
(14) M. Hirano, K. Shinjyo, R. Kaneko and Y. Murata, Phys. Rev. Lett. 78 (1997) 1448.
(15) K. Hokkirigawa and K. Kato, Tribology International 21 (1988) 51.
(16) J. M. Challen and P. L. P. Oxley, Wear 53 (1979) 229.
(17) K. L. Johnson, Contact Mechanics, Oxford Press, (1985).
(18) S. Shimada, N. Ikawa, H. Tanaka, G. Ohmori and J. Uchikoshi, J. JSPE 12 (1993) 2015.
(19) E. Ohmura, J. Shimizu and H. Eda, Proc. Int. Trib. Conf., Yokohama, (1995) 103.
(20) K. Kato, H. Koide and N. Umehara, Wear (2000) in press.

Chapter 5

DEPENDENCE OF FRICTIONAL PROPERTIES OF HYDROCARBON CHAINS ON TIP CONTACT AREA

Judith A. Harrison,[1] Paul T. Mikulski,[1] Steven J. Stuart[2] and Alan B. Tutein[1]
[1]*Department of Chemistry, United States Naval Academy, Annapolis, Maryland 21402*
[2]*Department of Chemistry, Clemson University, Clemson, South Carolina 29634*

1 INTRODUCTION

A number of groups have examined the frictional properties of alkanethiols on Au and alkylsilane monolayers on various substrates using the atomic force microscope (AFM).[1-6] Such kinds of monolayers may be useful in the construction of MEMS with low friction, reduced stiction, and the desired wear properties. In addition, the frictional properties of related systems, such as Langmuir-Blodgett films, have also been investigated.[7,8] Complementary work described here uses classical molecular dynamics (MD) simulations to study monolayers of n-alkanes that are chemically bound to diamond substrates. Other MD simulations have examined the structure[9-12] and compression of n-alkanethiols on Au.[13,14] In this work, two kinds of probes with different contact areas are utilized to study the effects of chain length on mechanical and frictional properties. The probes are a nanotube, which penetrates into the monolayers (Figure 1, left), and a countersurface, which compresses the monolayers (Figure 1, right).

2 METHODS AND PROCEDURES

To examine the effect of contact area on friction, two kinds of probe tips are placed in contact with the n-alkane monolayers. The first is a capped single-wall, [10,10] carbon nanotube that is 5.6 nm long and has a diameter of approximately 1.4 nm. (The nanotube contains 870 atoms.) This tube was chosen because its mechanical response upon interaction with a diamond substrate and indentation into hydrocarbon monolayers is well known.[15,16] The second probe is a hydrogen-terminated diamond surface containing 1152 atoms, which has been described previously.[17] While the nanotube's

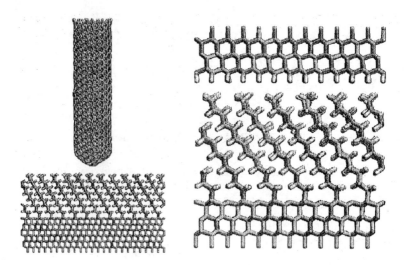

Figure 1. Starting configurations for indentation of a carbon nanotube (left) and for compression of a diamond counterface (right) into a monolayer of hydrocarbon chains attached to the (111) face of a diamond substrate. The depicted chains each contain 13 carbon atoms. The total number of atoms is 11910 and 3600 for the systems on the right and left, respectively.

diameter permits it to contact only a few chains during indentation, the countersurface contacts all of the alkane chains in the simulation. In fact, due to the periodic boundary conditions imposed, the countersurface effectively has an infinite contact area.

Each of the samples to be probed consists of a monolayer of hydrocarbon chains chemically bound to the (111) face of a diamond substrate. A monolayer is comprised of saturated, straight-chain hydrocarbons: $CH_3(CH_2)_{n-1}$ with $n = 8$, 13, or 22 (referred to as C_8, C_{13}, and C_{22} respectively). The chains are attached to the diamond substrate in a (2x2) arrangement with all dihedral angles of the carbon backbone in the anti-configuration. (This yields a surface density of 0.224 nm^2 / molecule.) This packing density and *cant* is similar to alkanethiols on Au(111).[1,18] Periodic boundary conditions are imposed in the plane of the monolayer.

The bulk of the diamond substrate and the attached hydrocarbon chains are free to move according to classical dynamics. The bottom two layers of the diamond substrate (those farthest removed from the chains in Figure 1) are held rigid while the next two layers are maintained at a temperature of 300 K using Langevin dynamics to simulate coupling to an external heat bath.[19] The equations of motion for nonrigid atoms are integrated with the velocity Verlet algorithm using a step size of 0.25 fs.[20]

The diamond counterface is constructed in an analogous fashion with the top two layers held rigid and the next two coupled to a second external heat bath. Indenting (moving the probe closer to the diamond substrate) and sliding (moving across at a fixed distance from the substrate) are conducted by moving all rigid atoms in the probe with a constant velocity. Limits on computation time require indentation (and sliding) speeds on the order of 100 m/s. This is orders of magnitude faster than characteristic speeds in nanoindentation experiments[5,21] (10^{-8} m/s) and in AFM tapping mode (10^{-1} m/s). Simulations conducted at both 100 m/s and 50 m/s show similar results for most system properties, though some quantities (such as the rate of defect formation) are markedly different.[16,17] In the case of the single-walled carbon nanotube, all atoms within the tube are held rigid.[22]

The potential governing atomic motions is the recently developed adaptive intermolecular reactive empirical bond-order (AIREBO) potential.[23] Long-range interactions (via a Lennard-Jones 12-6 potential) and torsional interactions were used to enhance the Brenner reactive empirical bond-order (REBO) potential[24,25]. A novel adaptive algorithm accomplishes this without compromising the unique ability of the REBO potential to model chemical reactions, i.e. changes in hybridization.

3 RESULTS

3.1 Probing with the Nanotube

Figure 2 shows the frictional force as a function of load for the case of the nanotube sliding through a monolayer. For a given load, the C_{22} chains have higher friction than the shorter C_{13} chains. For both chain types, higher loads result in higher friction. Salmeron and co-workers have suggested that gauche defects form during AFM indentation and sliding experiments in self

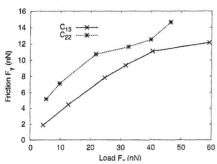

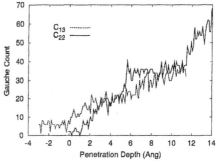

Figure 2. Sliding of the nanotube: average friction as a function of average load. Lines are drawn to aid the eye.

Figure 3. Indentation of the nanotube: number of gauche defects as a function of tip penetration distance.

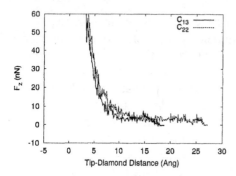

Figure 4. Indentation of the nanotube: load on the tip as a function of distance from the diamond substrate.

assembled monolayer films.[21] The formation of these defects has been proposed to be a significant source of energy dissipation in these systems.[21] If the formation of gauche defects is a significant channel for the dissipation of energy, this process should be strongly correlated with frictional properties. With that in mind, particular attention has been focused on the formation of these defects during the simulations reported here. For the purposes of this analysis, a gauche defect is defined as an intrachain C-C-C-C dihedral angle that is greater than 270° or less than 90°.

During indentation with the nanotube, gauche defects form underneath the tip and along the sides of the tip.[16] Over the load range examined, the formation of gauche defects is governed by the depth of penetration of the nanotube, but insensitive to the tube's distance from the diamond substrate. This is illustrated in Figure 3, which shows that the number of defects versus distance into the monolayer is not sensitive to chain length (monolayer thickness).

In contrast, the load on the nanotube tip is largely determined by the tube's distance from the diamond substrate (Figure 4). Therefore, to achieve a given load, the nanotube must penetrate deeper into the C_{22} monolayer than into the C_{13} monolayer. Thus, C_{22} has a larger number of defects prior

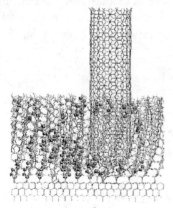

Figure 5. Snapshot of the nanotube sliding through the C_{22} monolayer. The load is approximately 40 nN. Gauche defects are shown as gray spheres.

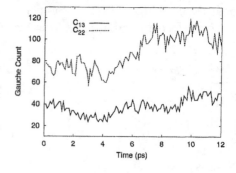

Figure 6. Sliding of the nanotube: number of gauche defects within the monolayer as a function of time. The load on the tube for both slides is approximately 40nN.

to sliding because the defects congregate not only underneath the tip but also along the sides of the tip.

During sliding, additional defects are generated around the nanotube as it slices through the monolayer. A wake of defects is formed behind the tube which extends from the bottom of the tip up to the top of the monolayer surface (Figure 5). Thus, the number of defects formed in the wake of the tube is typically higher if the tube is deeper into the monolayer. This is evident in Figure 6 where the overall rise in the number of gauche defects is larger for the C_{22} slide where the nanotube is deeper into the monolayer than for the C_{13} slide. Assuming a larger number of gauche defects is connected with higher friction, deeper penetration into the monolayer leads to higher friction. This explains both trends in Figure 2: higher friction is achieved by fixing the load and lengthening the chain or by fixing the chain length and increasing the load.

3.2 Probing with the Countersurface

Figure 7 shows the frictional force as a function of load for the case of the counterface sliding over an *n*-alkane monolayer. As with the case of the nanotube, the friction increases with load. As the load increases the monolayer becomes increasingly compressed. This leads to a rising number of gauche defects that form at the ends of the chains.[17] Given the tendency of the defects to form at the ends of the chains, one might expect little difference in frictional properties with chain length; however under similar loads, C_8 generally gives a higher number of defects and thus higher friction than C_{13}.

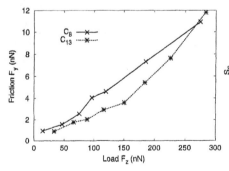

Figure 7. Sliding of the counterface: average friction force as a function of average load. Lines are drawn to aid the eye.

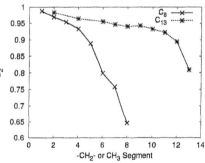

Figure 8. Two-dimensional structure factor S_2 as a function of chain segment (-CH_2- or -CH_3) for equilibrated monolayers.

One difference between the C_8 and C_{13} monolayers, which may be responsible for their differing frictional properties, is that C_8 is less ordered than C_{13}. This is quantified by calculating the two-dimensional structure factor based on the (2×2) arrangement of the attachment sites of the chains to the diamond substrate.[16] This quantity for a layer of chain segments ($-CH_2-$ or $-CH_3$) equals one if the ordering is that of the attachment sites and averages about zero if the chain segments do not possess that ordering.

Figure 8 shows the structure factor for various chain segments in the C_8 and C_{13} monolayers after equilibration. As expected, the structure factor is close to unity for the segment closest to the diamond substrate. Comparing the topmost segment in each chain type, it is evident that at the surface, C_8 is much less ordered than C_{13}. This lack of order at the surface may be responsible for the formation of the greater number of gauche defects upon compression. This hypothesis is supported by a similar comparison between C_{13} and C_{22}. Our simulation results show that both the ordering at the surface and frictional properties are more similar for C_{13} and C_{22} relative to the comparison between C_8 and C_{13}. This suggests that longer chain lengths give rise to lower friction only up to a certain point. That is, beyond some critical length, all chains give similar frictional properties. Evidence for this trend has also been found experimentally.[2]

4 FUTURE WORK

These simulation results show that the process of gauche defect formation may be closely linked with frictional properties of self-assembled monolayers. In addition, the locations of the defects within a monolayer depend strongly on the geometry of the probing tip. The AFM experiments that have examined these types of monolayers give results that are similar to those produced by the countersurface. This is not surprising considering the AFM tips that have been used have radii of hundreds of angstroms, a dimension much larger than the spacing between hydrocarbon chains. This suggests that the frictional properties observed in AFM experiments are largely due to the response of the compressed chains underneath the tip rather than the response of the chains that lie near the sides of the probe.

In the case of a countersurface probe, it seems that the degree of order of the monolayer is related to defect formation and frictional properties. All the simulations reported here utilize chains that are attached in a perfect (2×2) arrangement. The importance of ordering at the monolayer surface suggests that the presence of structural defects could give rise to significantly higher friction. In addition to structural defects, a lower packing density could also give rise to disorder at the monolayer surface. In both cases, the potential for disorder is far greater than in the case of the tightly packed C_8 monolayers. This suggests that friction would be greater than the tightly packed C_8 regardless of the chain length used to construct the monolayer if significant defects are present or if the packing density were less. The increased freedom of the chains associated with these scenarios also suggests that other mechanisms beside gauche defect formation (such as entanglement of chains) may provide significant channels for energy dissipation. Simulations that examine both these possibilities are in progress.

ACKNOWLEDGMENTS

This work was supported by the Office of Naval Research under contract N00014-WR-20186 and by the Air Force Office of Scientific Research under contract NMIPR-00-5203503.

REFERENCES

(1) Lio, A.; Charych, D.H.; Salmeron, M. *J. Phys. Chem. B* **1997**, *101*, 3800-305.
(2) Lio, A.; Morant, C.; Ogletree, D.F.; Salmeron, M. *J. Phys. Chem. B* **1997**, *101*, 4767-4773.
(3) Kim, H.I.; Grapue, M.; Oloba, O.; Doini, T.; Imaduddin, S.; Lee, T.R.; Perry, S.S. *Langmuir* **1999**, *15*, 3179-3185.
(4) Xiao, X.; Hu, J; Charych, D.H.; Salmeron, M. *Langmuir* **1996**, *12*, 235-237.
(5) Harrison, J.A.; Perry, S.S. *MRS Bull.* **1998**, *23*, 27-31.
(6) Barrena, E.; Kopta, S.; Ogletree, D.F.; Charych, D.H.; Salmeron, M. *Phys. Rev. Lett.* **1999**, *82*, 2880-2883.
(7) Liley, M.; Gourdon, D.; Stamou, D.; Meseth, U.; Fisher, T.M.; Lautz, C.; Stahlberg, H.; Vogel, H.; Burnham, N.A.; Duschl, C. *Science* **1998**, *280*, 273-275.
(8) Burns, A. R.; Houston, J. E.; Carpick, R. W.; Michalske, T. A.. *Langmuir.* **1999**, *15*, 2922-2930.
(9) Mar, W.; Klein, M.L. *Langmuir* **1994**, *10*, 188-196.
(10) Bhatia, R; Garrison, B.J. *Langmuir* **1997**, *13*, 765-769.
(11) Bhatia, R; Garrison, B.J. *Langmuir* **1997**, *13*, 4038-4043.
(12) Luedtke, W.D.; Landman, U. *J. Phys. Chem.* **1998**, *102*, 6566-6572.
(13) Tupper, K.J.; Colton, R.J.; Brenner, D.W. *Langmuir* **1994**, *10*, 2041-2043.
(14) Tupper, K.J.; Brenner, D.W. *Langmuir* **1994**, *10*, 2335-2338.
(15) Harrison, J.A.; Stuart, S.J.; Robertson, D.H.; White, C.T. *J. Phys. Chem. B* **1997**, *101*, 9682-9685.
(16) Tutein, A.B.; Stuart, S.J.; Harrison, J.A. *J. Phys. Chem. B* **1999**, *103*, 11357-11365.
(17) Tutein, A.B.; Stuart, S.J.; Harrison, J.A. *J. Langmuir* **2000**, *16*, 291-296.
(18) Fenter, P.; Eberhardt, A.; Eisenberger, P. *Science* **1994**, *266*, 1216-1218.
(19) Adelman, S.A.; Doll, J.D. *J. Chem. Phys.* **1976**, *64*, 2375-2388.
(20) Swope, W.C.; Anderson, H.C.; Berens, P.H.; Wilson, K.R. *J. Chem. Phys.* **1982**, *76*, 637-649.
(21) Carpick, R.W.; Salmeron, M. *Chem. Rev.* **1997**, *97*, 1163-1194.
(22) For a comparison of the flexible and rigid nanotube behavior during sliding see Harrison, J.A.; Stuart, S.J.; Tutein, A.B. in Interfacial Properties on the Submicron

Scale, eds. J. E. Frommer and R. Overney, ACS Press, "A New, Reactive Potential Energy Function to Study Indentation and Friction of C_{13} *n*-Alkane Monolayers", in press.
(23) Stuart, S.J.; Tutein, A.B.; Harrison, J.A. *J. Chem. Phys.* **2000**, *112*, 6472-6486.
(24) Brenner, D.W. *Phys. Rev. B* **1990**, *42*, 9458-9471.
(25) Brenner, D.W.; Harrison, J.A.; Colton, R.J.; White, C.T. *Thin Solid Films* **1991** *206*, 220-223.

Chapter 6

NANOHYDRODYNAMICS AND COHERENT STRUCTURES

H.G.E. Hentschel and I. Tovstopyat-Nelip
Department of Physics
Emory University
Atlanta, GA 30322

1 INTRODUCTION

What types of new flow phenomena may be expected at the nanoscale, and how will such flow phenomena affect lubrication and transport? The study of hydrodynamic flows at a macroscopic scale has revealed an astonishing variety of phenomena including creeping flows, interfacial patterns, laminar flows, flows in channels, Couette flows, vortex structures, turbulence, free boundary flows, convections, waves and shocks (see for example, Van Dyke, 1982). With the development of MEMS devices such as sensors and motors on the micron lengthscale and more recently with advances in nanolithography (Xia & Whitesides, 1998), the question arises as to what types of new flow phenomena may appear at these scales. Flows in microchannels, microcavities, and nanotubes at periodic, structured, and rough surfaces all need to be studied.

We are entering a new world as regards the lengthscales and timescales required to describe phenomena of relevance to nanotechnology. With relevant lengthscales ranging from about 10 nm to 10 μm which are well above the microscopic scale ruled by quantum phenomena and atomic potentials, yet well below the macroscopic world so well described by bulk hydrodynamics and elasticity new phenomena and approaches are required. Consider, for example, a thin film of dimensions 10 nm by 1 μm by 1 μm which consists of about 10^7 atoms and whose viscoelastic behavior we would like to follow for about 1 second. Clearly such a task is well beyond the capabilities of molecular dynamics, yet not a question easily resolved using macroscopic continuum theories.

In fact it is clear that we are dealing with a multiscale problem, and many approaches including molecular dynamics, kinetic equations and continuum theories will be required to resolve it. Thus by using molecular dynamics approaches Koplik and Banavar (Koplik *et al.*, 1989, Koplik and

Banavar, 1995) were able to show that in the presence of Couette and Poiselle flows the stick boundary conditions usually employed in macroscopic hydrodynamics held down to microscopic dimensions, while apparent singularities due to such stick boundary conditions, for example, at flow past sharp edges, and at moving contact lines during wetting, could be resolved by molecular dynamics contributing information on more realistic boundary conditions.

At the mesoscale, however, stick boundary conditions are often observed to fail, with slip occurring at the interface (Ho & Tai, 1998). Thus the question of what types of flows, mass transport and heat transfer occur in the remarkable variety of new MEMS devices with complex geometries now being developed (Xia & Whitesides, 1998) is both of fundamental theoretical interest and of great importance technologically. Some of these MEMS devices will require lubrication, others may involve gas flows through constricted domains for their operation. Whatever the application, it is clear that system geometry and surface characterization will be key variables in controlling the fluid dynamics

The main difference from bulk flows as system size is reduced is the ever-increasing importance of surfaces in controlling the fluid dynamics. Consider a small volume of fluid of linear dimensions L. Then the body forces specifically viscous effects will scale as L^3, while surface forces will scale as L^2. Clearly surface forces will become dominant at small scales, but what may be surprising is that as far as MEMS are concerned even at scales of 10^2-10^3 microns surface effects dominate (Ho and Tai, 1998). Effects connected to stiction, slip at surfaces, the influence of charge distributions at surfaces, and roughness all involve flows at interfaces.

Another question that arises as system size is reduced concerns the breakdown of continuum assumptions incorporated into classical hydrodynamics and elasticity. Fluctuation phenomena start to become significant, and mean free paths l, especially of gases, can begin to approach the system size L leading to large Knudsen numbers $K_n = l/L$ and slip at boundaries. Yet it is obvious that in order to understand transport and lubrication at the mesoscale such phenomena must be understood.

Thus in mesoscopic fluid layers of thickness around 100 nm, where it is often supposed that bulk hydrodynamics is sufficient to describe the flow, noise, large mean free paths or boundary effects associated with rough or charged surfaces might seriously perturb bulk behavior. If such surface effects are significant they may dominate fluid flowing through MEMS devices and in biomedical applications where fluid transport is required for chemical analysis or in drug delivery. Even for unlubricated surfaces in MEMS devices, deviations from classical behavior are often observed. For example, during the development of the micromotor it was found that the frictional force between the rotor and substrate is a function of the contact area, a behavior at odds with classical notions of dry friction. How lubricating such surfaces would change this frictional behavior is open question.

2 RECENT DEVELOPMENTS IN NANOSCALE FLUID FLOW

The last decade (Bhushan *et al.*, 1995; Krim, 1998; Ho & Tai, 1998) has seen tremendous growth in our understanding of nanotribology and mesoscopic fluid flows. We cannot give a comprehensive account of all these developments here. We will, however, try to point the interested reader to some of the directions in which new developments have occurred, and to some of the references useful in further exploring this exciting topic.

New experimental techniques to grow and study interfaces with well-defined geometries have been developed creating the foundations of a new technology. Using methods ranging from molecular beam epitaxy (MBE) to nanolithography it is now possible to create devices of well defined geometry made from materials with well-characterized surfaces. The structures can vary from distributions of clusters of atoms in submonolayer growth, interfaces that are flat at an atomic scale, rough surfaces with well defined self-affine statistical properties, and moundlike structures which can self-assemble on atomic surfaces (Family & Vicsek, 1991, Barabasi & Stanley, 1995). Studies of the dynamics and frictional behavior of individual atoms, molecules, polymer chains, nanoclusters, and microparticles such as fullerenes and nanotubes have been carried out.

It is due to the development of a series of powerful experimental techniques allowing an unparalleled resolution of nanoscale phenomena that these advances have been made. The development of the scanning tunneling microscope (STM) (Binnig & Rohrer, 1982; 1983), and its extensions – the atomic force microscope (AFM), and the friction force microscope (FFM) – have allowed nanoscale measurements of forces both parallel and transverse to an interface and the imaging of both topographical and frictional maps of interfaces.

The development of the quartz-crystal microbalance (QCM) by Krim and Widom (1988) have allowed studies on atomic sliding on periodic and rough surfaces to be carried out by the direct measurement of slip times as well as the mass and dissipation of a thin film sliding on a surface (Daly & Krim, 1996; Krim *et al.*, 1991). Such studies are very versatile and have allowed information on the velocity dependence (Mak & Krim, 1998a) and the effect of disorder on slip times (Mak & Krim, 1998b). The surface force apparatus (SFA) has allowed studies of nanoscale lubrication by shearing thin layers of fluid between atomically flat layers of mica (Israelachvili,

1988; 1991; Luengo *et al.*, 1996) in the process uncovering the complex viscoelastic properties of thin films (Alsten & Granick, 1988; Gee *et al.*, 1990; Klein & Kumacheva, 1998a,b).

Concomitant with these experiments, large scale molecular dynamics simulations of surface structure and nanotribological events in both dry friction and lubrication were carried out (Baljon & Robbins, 1997; Gao *et al.*, 1997; Landmann *et al.*, 1991; Landmann *et al.*, 1996), revealing mechanisms for wear, adhesion, density layering and flow at the atomic scale in agreement with experiment and possibly relevant at the micron length scale determining the tribological properties of MEMS.

Both experiments (Israelichvili et al., 1988; Gee *et al.*, 1990; Yoshizawa *et al.*, 1993; Klein & Kumacheva, 1998; Hu & Granick, 1998) and molecular dynamics (Landman *et al.*, 1996; Gao *et al.*, 1997) simulations have uncovered a number of phenomena that will need to be considered in the development of a detailed understanding of fluid flows at scales between about 100nm and 10 microns – the scales relevant to MEMS. Thus for thin films, as the number of molecular layers is decreased by changing the applied normal load on the films, a sudden confinement induced solidifications in which the layers are ordered occurs (Klein and Kumacheva, 1995; Reiter *et al.*, 1994; Gao *et al.*, 1997). This transition typically occurs at around 10 molecular layers. On further increasing in the normal loading molecular layers can be squeezed out, and in this manner an exquisite degree of control applied to the film. The degree of in plane ordering is, however, less well understood, and several possibilities exist including ordered phases, a two-dimensional liquid-like structure, or glassy phases. In the presence of applied shear viscoelastic behavior is exhibited by the thin film. Initial elastic behavior is observed; then at a well defined yield stress, shear melting occurs followed by ductile flow. Stick-slip dynamics by the thin film is typically observed. Finally on relaxing the applied stresses glassy properties such as slow relaxation and creep back to an ordered elastic phase occurs.

The observed stick-slip behavior is not only due to the ordered state of the thin film of fluid but also to the presence of velocity weakening boundary conditions at the mesoscale, which are a direct result of the strong intermolecular forces that exist between a sliding thin film and the surface at the atomic scale (Cieplak *et al.*, 1994; Smith *et al.*, 1996; Thompson & Robbins, 1990a,b; Thompson et al., 1992). Stick forces on ordered arrays are much larger when the array is interacting with a commensurate substrate than an incommensurate one (Cieplak, Smith, and Robbins; 1994,1996), and it is possible to analyze very general equations of motion for arrays of particles interacting with periodic interfaces, such as the Frenkel-Kontorova model and its generalizations, that lead to the conclusion that such velocity weakening boundary conditions are generic (Persson, 1993,1997; Braiman *et al.*; 1996, 1997; Rozman *et al.*, 1996a,b; Batista & Carlson, 1996, 1998), as is periodic and chaotic stick-slip dynamics depending on the relative strength of the substrate-array interactions and the material properties of the array.

In fact one can go further. Molecular dynamics simulations indicate that the velocity-weakening stresses at a boundary typically have a universal material-independent velocity-weakening form when properly scaled (Thompson & Troian, 1997). While universal bifurcations between different types of stick-slip boundary conditions may also exist if macroscopic investigations of dry friction are any guide. For in such investigations of dry friction at low velocities the dynamics was also usually characterized by several stick-slip regimes (Baumberger *et al.*, 1994) including: a supercritical Hopf bifurcation between stick-slip and steady sliding; and hysteresis due to subcritical Hopf bifurcations. Such bifurcations in boundary conditions might significantly influence the lubricating properties of thin films.

3 MESOSCALE FLUCTUATIONS AND LUBRICATION

One of the most exciting recent developments in nanotribology, has been the realization that long-lived fluid temporal fluctuations in driven thin films may be associated with the existence of coherent structures at the nanoscale. Thus using a SFA, Demirel and Granick (1996) observed a power law distribution in the frictional forces generated by molecularly thin films of squalane between mica. Let us estimate the order of magnitude of the scales of length L and time T that describe their experiment. The film thickness in this experiment was 1.8 nm, which constituted several molecular layers of squalane, the diameter of the contact interface in their SFA was about 45 μm, and the value of the external oscillatory driving force was about 10^2 μN. Thus the typical accelerations applied to the film during the experiment were in the order of $a \sim 10^{10}-10^{11}$ m/s^2, while the speed of sound in such compressible films is about $c \sim 10-10^2$ m/s. Thus a typical lengthscale in the experiment of $L = c^2/a \sim 10^{-2}-10^0$ μm and timescale $T = c/a \sim 10^{-9}-10^{-8}$ s describe the size and duration of the fluctuations. These length and timescales are much larger than atomic length scales (~1 nm) and phonon time scales (~10^{-12} s).

An important question in nanohydrodynamics lies in the appearance, properties and consequences of such mesoscale fluctuations for nanotribology. The boundary conditions affecting mesoscale flows are often competitive and cannot be simultaneously satisfied at several surfaces. Consider first stick-slip dynamics due to the presence of velocity-weakening stresses at a driven interface. Stick at one surface implies slip at another nearby surface, and such frustration can lead to the development of mesoscale coherent structures in the lubricating fluid. Consider a second example in which one surface is hydrophobic and another is hydrophilic. Again mesoscale fluctuations may appear in a fluid separating these surfaces as it try to accommodate itself to the presence of competing interactions. Clearly such surface effects scale as L^2 while bulk forces scale as L^3, and therefore their relative importance diverges as L^{-1}. Similarly, in both these problems and more generally at the nanoscale we may expect thermal noise to have a significant impact of the resulting fluid behavior. This is because thermal fluctuations scale as $L^{3/2}$ and therefore the relative level of noise fluctuations diverge as $L^{-3/2}$.

Our approach to such questions is to combine numerical integration of the relevant hydrodynamic equations, analyze the resulting fluid dynamics and make comparisons and predictions for real fluids under such nonequilibrium conditions. Let us consider in more detail the case of a thin fluid film driven between two surfaces. We can describe the driven thin film as a viscoelastic compressible fluid, in which strong velocity weakening stresses exist at the interface and thermal noise due to coupling with the interface is a significant factor determining the fluid dynamics.

$$\rho \partial v_\alpha / \partial t + \rho \vec{v} \cdot \vec{\nabla} v_\alpha = -\partial p / \partial x_\alpha - \partial \tau_{\alpha\beta} / \partial x_\beta + \rho G_\alpha(\vec{x},t),$$
$$\partial \rho / \partial t + \vec{\nabla} \cdot (\rho \vec{v}) = 0, \qquad (1)$$

where $\tau_{\alpha\beta}$ is the stress tensor in the thin film, while the term G_α on the right hand side represents the thermal noise in the fluid which is taken here to have Gaussian statistics.

$$< G_\alpha(\vec{x}_1,t_1) G_\beta(\vec{x}_2,t_2) >= 2\Gamma \delta_{\alpha\beta} \delta(\vec{x}_1 - \vec{x}_2) \delta(t_1 - t_2). \qquad (2)$$

The stress tensor will have strong viscoelastic characteristics resulting in a relationship between the rate of strain $\partial v_\alpha/\partial x_\beta$ and the stress $\tau_{\alpha\beta}$ of the form

$$\partial v_\alpha / \partial x_\beta = (1+\sigma)/E d\tau_{\alpha\beta}/dt + (\tau_0/\eta) F_{\alpha\beta}(\tau/\tau_0), \qquad (3)$$

where σ is Poisson's ratio, E is Young's modulus, η is the dynamic viscosity of the film, and τ_0 is a stress scale such that for $\tau << \tau_0$ the film behaves as a viscous fluid.

The basic stick boundary conditions for bulk flows need to be replaced by nonlinear velocity-weakening shear stresses $\tau_{xz}(v_x)$, $\tau_{yz}(v_y)$ at the boundaries of the driven thin film

$$\eta \partial v_\parallel(x,y,z=0,t)/\partial z = \tau(v_\parallel(x,y,z=0,t))$$
$$\eta \partial v_\parallel(x,y,z=H,t)/\partial z = \tau(v_\parallel(x,y,z=H,t)-V) \qquad (4)$$

where V is the velocity of the upper plate separated from the lower plate by a distance H. As the speed of sound c is the relevant velocity scale for a compressible film, we can parametrize the stress at the interface as

$$\tau(v) = \tau_{\max} f(v/c), \qquad (5)$$

where the stress has a velocity-weakening form (for example $f(u) = 1/(bu + \text{sign } u) + Ku$ is a possible parameterization, but this could be replaced by more universal or material dependent forms given to us by experiment or molecular dynamics simulations), with τ_{max} is the maximum stick stress at the interface before slip occurs.

Linear stability analysis of such equations for very thin films in the absence of noise (Tovstopyat-Nelip & Hentschel, 2000) suggests that at low driving velocities $V < V_1 = 2v_0$, where v_0 is the velocity at which the velocity-weakening surface stress has a minimum, there are three stationary solutions corresponding to films with constant density stuck to either the top or bottom plate or moving with the average velocity $V/2$. The first two solutions are stable and they are attractors of the system. The third solution is unstable. When the velocity of the upper plate reaches V_1 a pitchfork bifurcation in the dynamics occurs: the average velocity solution becomes stable and two new fixed-point solutions emerge with velocities $v_+ = V/2 + u$, and $v_- = V/2 - u$, which are both unstable. This situation exists for driving velocities $V_1 < V < V_2$. Finally, $V > V_2$ the two stick solutions disappear and we end up with only the average velocity which is stable. Thus we are dealing with a subcritical pitchfork bifurcation at $V = V_1$.

The presence of thermal noise changes the amplitude of such velocity and density fluctuations dramatically. When the driving velocity V is small enough, the stochastic force can excite the liquid sufficiently for the velocity distribution to cross the line $v(x) = V/2$, so the system does not evolve to any of the two stable stick attractors and the spatiotemporal fluctuating structure, determined by nonlinear properties of the fluid dynamics and induced by thermal noise, is developed. Scaling arguments assuming the compressibility of the film and the applied force are the only relevant parameters would suggest that this occurs when $\Gamma \sim c^3 V^2/a$. Thus, such noise-induced fluctuations will be most easily developed at low driving velocities. When the upper plate moves with significantly higher velocities, the thermal excitation becomes relatively small.

Numerical integration of such equations in the presence of noise show stick-slip dynamics, the development of strong density and velocity fluctuations, and a reduction in friction with increasing thermal noise. At low driving in the presence of noise the stick phase is unstable and the system tends to build the spatiotemporal structure with the velocity fluctuating between two attractors $v(x) = 0$ and $v(x) = V$. As the driving velocity is increased the coherent structure builds up showing pronounced structure in both the velocity and density fluctuations. The period of this structure is on the order of magnitude of 10 μm using reasonable thin film material parameters. At even higher driving velocities the amplitude of the resulting periodic structure in density fluctuations decreases again.

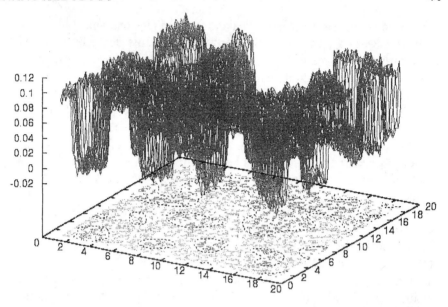

Figure 1. A snapshot of the x component of the average velocity field in the direction of driving in a two-dimensional thin film for driving velocities $V/c = 0.1$ and noise levels $a\Pi c^5 = 0.1$. Note the partial stick to the upper and lower plates.

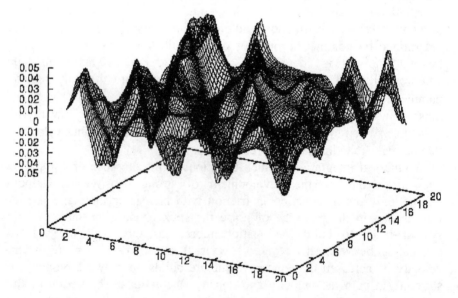

Figure 2. A snapshot of the relative density fluctuations $\log(\rho/\rho_{av})$ in the thin film for driving velocities $V/c = .1$ and noise levels $a\Pi c^5 = 0.1$. Note the mesoscopic density fluctuations due to compression and rarifaction of the thin film.

Specifically the case of one fluid layer coupled to both the top and bottom plates shows shear melting to a marked degree, while the velocity $v(x,t)$ and the density $\rho(x,t)$ in the direction of driving exhibit coherent spatiotemporal fluctuations at the micron lengthscale. In Figure 1 can be seen the velocity fluctuations in a thin film driven with a velocity $V/c = 0.1$. Note that portions of the thin film appear to be stuck either to the lower surface ($v = 0$) or to the upper driven surface ($v/c = 0.1$). The density fluctuations are also on a mesoscopic scale (see Fig. 2) as the velocity fluctuations alternatively compress and expand the thin film.

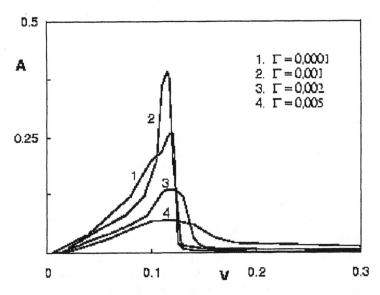

Figure 3. The dependence of the amplitude of the density fluctuations on a driving velocity V for different strengths of dimensionless thermal noise $a\Pi Hc^5$.

We observed in our simulations (see Fig. 3) that the induced density fluctuations are a strong function of the thermal noise and can change by several orders of magnitude. This is similar to the well-known phenomenon of stochastic resonance (Linder *et al.*, 1995) where the cooperation between the external drive and the thermal noise leads to the appearance of resonant drift in Langevin equations but here the cooperation between noise and drive creates well-pronounced spatiotemporal structures. Theoretically we predict (Tovstopyat-Nelip & Hentschel, 2000) that the density fluctuations will diverge as

$$<(\rho - \rho_{av})^2> \sim \Gamma/(V - V_1), \tag{6}$$

as the subcritical pitchfork bifurcation is approached and tend to a constant when V becomes large enough. All these facts are consistent with our numerical results.

4 RECOMMENDATIONS FOR FUTURE RESEARCH NEEDS

The recommendation for future research needs, especially over a five or ten year period, must be subjective, dependent as it is on the knowledge and interests of the individual asked. We believe that the single most important area of research lies in developing methods to deal with length scales of about 100 nm to 10 microns for long times (up to seconds rather than picoseconds) and in the uncovering of new physical phenomena that are characteristic of these length and time scales. This is a multiscale problem where the insights from molecular dynamics need to be incorporated into either continuum equations of motion or into kinetic Monte Carlo or Lattice Boltzmann techniques (Chen & Doolan, 1998) which solves a simplified form of the Boltzmann equation for the fluid flow in question on a lattice and which can reach longer time scales than those amenable to full molecular dynamics simulations. There are many interesting areas for such investigations we shall describe a few from the buffet table available.

For thin films flows, confinement and boundary effects may be of a longer range than is normally supposed. We know that for thin films the layers are quantized and a number of unusual phenomena have been observed both in experiment and molecular dynamics simulations. These include the fact that the exact number of layers n in the film and the material properties of the substrate will have a crucial influence on thin film dynamics. For strongly nonlinear phenomena such as shear melting, yield stresses, ductile flow, in plane ordering, slip dynamics between film layers, multi layer locking, and layer slippage no proper understanding of their influence on mesoscale flows exists. It would be interesting to study such anisotropic flows by setting up two-dimensional layered hydrodynamic flows in which mass and momentum can be exchanged between neighboring layers, and integrating numerically the resulting equations of motion as well as studying their stability.

Another important question is how breakdown of classical bulk hydrodynamic flows occurs in the presence of large surface stresses and thermal noise. The effects of noise can be very significant at the nanoscale. The center of mass velocity fluctuations of a thin fluid layer consisting of N atoms could scale as $v_{rms} \sim (kT/N\,m)^{1/2} \sim v_{shear}$ implying that the assumptions made in deriving the hydrodynamic equations of motion begin to loose their validity. For example at shear rates of 1m/sec, breakdown occurs for $N \sim 10^6$

atoms. Also in the case of stick-slip hydrodynamics the fluctuation-dissipation theorem yields a velocity-weakening effect noise strength. Clearly when the velocity and density fluctuate strongly, as described above, these fluctuations can exceed their corresponding average values and the hydrodynamic description breaks down. Another problem concerns nanoscale gas flows. In such flows the molecular length scales can be of the same order of magnitude as system size. Indeed, Knudsen numbers ranging between $.001 < K_n < 10$ often appear in MEMS flows (Ho & Tai, 1998). In this situation, again stick boundary conditions breakdown and slip is observed at the interface of microchannels and microcavities. In this situation the mass and heat transfer can be very different from macroscopic predictions. Again, the length and timescales for such flows may be too large for molecular dynamics, but Lattice Boltzmann methods may be able to simulate these regimes (Nie et al., 2000).

The question of the effect of system geometry on nanoscale flows needs to be investigated at both the atomic level and at the mesoscale. Using nanolithography, nanoscale channels, junctions, motors, and even complete turbine engines are now being manufactured. The flows in such structures are extremely sensitive to the exact dimensions of the device. What are more small changes in shear can lead to large changes in boundary conditions (Thompson and Troian, 1997). The interface itself is not a well-defined static boundary as in macroscopic hydrodynamics. It can fluctuate, or deform in response to the fluid flow leading to wear and device damage.

The question of how trapped surface charges or broken surface bonds will affect fluid flow in MEMS is also a very important question. All surfaces will have electrostatic charges on them and in the presence of ionic liquids will form electrical double layers as ions in the fluid screen the surface charges. These electrical double layers have a characteristic scale (the Debye length) that can vary from a few nanometers to microns, being inversely proportional to the ion concentration. Clearly if the Debye length is of the same scale as the system size, the characteristic flow will be significantly modified (Mohiuddin Mala et al., 1996). How such surface charge distributions affect flow in MEMS devices remains to be established.

Boundary lubrication is very closely related to the manner in which thin films wet surfaces. Wetting is rather complex phenomenon, for even the question of the existence of a dynamic contact angle θ_d between a spreading film and a substrate has to be treated with some care. A cursory definition of a macroscopic dynamic contact angle $\theta_{d,macro}$ as the angle between the nominal contact line and the substrate surface would suggest that this contact angle has a singularity at the spreading velocity U tends to zero. A microscopic examination of the region near the spreading droplet in the vicinity of the apparent interfacial triple point, however, would reveal that the nominal macroscopic contact line is actually preceded by a mesoscopically thin precursor film due to a disjoining pressure which reflects the existence of attractive long-range Van der Waals forces between the liquid and substrate phases (Joanny & de Gennes, 1984; de Gennes, 1985). Thus the apparent singularity is a macroscopic artifact and a separate microscopic dynamic contact angle $\theta_{d,macro}$ can be defined by the angle between the spreading surface and the asymptote between the macroscopic droplet and the tip of this precursor film. The thickness W of this precursor film, as opposed to its spreading length l is controlled by a balancing of the Van der Waals forces, which will tend to make this film microscopically thin, indeed of atomic dimensions, with the short-range capillary forces which will tend to discourage the existence of such a film at all.

Though the problem of a droplet spreading on a solid surface has been previously treated assuming the velocity field in the droplet has a parabolic profile as in Poiseuille flow (Joanny, 1990), it should be emphasized that this is a very idealized form of wetting dynamics. To study mathematically the dynamics of wetting it is not enough to assume Poiseuille flow for the droplet. Clearly stick boundary conditions on the substrate surface are incompatible with a moving contact line, but what are the correct boundary conditions? Possibilities would include stick-slip boundary conditions, a frictional force proportional to the surface velocity v introduced at the substrate surface resulting in a stress tensor $\tau_{x,z} = \gamma v$; another possibility would be that the main body of fluid in the droplet before the nominal contact line does indeed obey stick boundary conditions, but slip boundary condition are applied to the precursor film whose length diverges $l = vt$, and consequently the molecules in the precursor film are free to roll at the substrate surface.

Volatility and contamination will also strongly influence nanoscale wetting, and can lead to temperature and surface-tension gradients resulting in additional forces (the Marangoni effect). In fact the alteration in dynamics can be dramatic: for example if surfactant solutions are spread on moist surfaces large surface tension gradients will appear leading to a fingering of the contact line (Marmur and Lelah (1981)); this spreading is an order of magnitude faster than surface tension or gravity dominated wetting. Also if the substrate surface is not smooth then the wetting dynamics may resemble more closely flow in porous media than the creeping flow described here. Finally liquids have been assumed to be Newtonian; for non-Newtonian rheologies the shape of a spreading droplet may vary considerably from the cap like droplets seen in simple wetting. For example shear thinning liquids typically develop a protrusion which extends beyond the cap forming a macroscopic foot (Brochard and de Gennes, 1984)).

Finally we need to mention the very important problem of the effects of surface roughness on creeping flows at the nanoscale. Rough surfaces created by a variety of physical processes including wear, fracture, and corrosion are self-affine. Kinetic Monte Carlo methods can be used to generate such surfaces including the distribution and geometric properties of their asperities and their elastic and plastic development under loading and wear on long timescales. Creep, plastic flow, moving crystalline phases, and shear-induced melting transitions in such nonequilibrium rough interfaces in response to driving need to be studied in order to understand how mesoscale flows will wear MEMS devices. To study the flows themselves both Lattice Boltzmann techniques and numerical integration for creeping flows over such surfaces will need to be investigated. The morphology of a self-affine surface can be characterized by the roughness exponent (or the Hurst exponent) α (see Family & Vicsek, 1985), which is defined through the scaling of surface fluctuations

$$<|h(\vec{x}+\vec{\Delta})-h(\vec{x})|> \approx A\Delta^{\alpha}. \qquad (7)$$

For this important case the pressure field will no longer be Laplacian as is the usual case for creeping flows in Hele-Shaw cells but rather obeys

$$\vec{\nabla}.[((H-h(\vec{x}))^2/12\eta)\vec{\nabla}p] = 0. \qquad (8)$$

How does the interface develop as a function of the Hurst exponent α, and the size of the surface fluctuations A?

REFERENCES

J. V. Alsten and S. Granick, Molecular Tribometry of Ultrathin Liquid Films, Phys. Rev. Lett. 61, 2570-2573 (1988).

A.R.C. Baljon and M.O. Robbins, Adhesion and Friction in Thin Films, MRS Bulletin, Theory and Simulation of Polymers at Interfaces, 22,1, 22-26 (January, 1997).

A.-L. Barabasi and H. E. Stanley, Fractal Concepts in Surface Growth, (Cambridge Univ. Press, Cambridge, UK 1995).

A. A. Batista, J. M. Carlson, Bifurcations from steady sliding to stick-slip in boundary lubrication Phys. Rev. E57, 4986-4996 (1998).

T. Baumberger, F. Heslot, and B. Perrin, Crossover from creep to inertial motion in friction dynamics, Nature 367, 544-546 (1994).

T. Baumberger, C. Caroli, B. Perrin and O. Ronsin, Nonlinear analysis of the stick-slip bifurcation in the creep-controlled regime of dry friction, Phys. Rev. E51, 4005-4010 (1994).

B. Bhushan, J. N. Israelachvili, and U. Landman, Nanotribology: friction, wear and lubrication at the atomic scale, Nature 374, 607-616 (1995).

G. Binnig, H. Rohrer, Ch. Gerber and E. Weibel, Surface Studies by Scanning Tunneling Microscopy, Phys. Rev. Lett. 49, 57-61 (1982).

G. Binnig, H. Rohrer, Ch. Gerber and E. Weibel, 7x7 Reconstruction on Si(111) Resolved in Real Space, Phys. Rev. Lett. 50, 120-123 (1983).

Y. Braiman, F.Family, H. G. E. Hentschel, Array-enhanced friction in the periodic stick-slip motion of nonlinear oscillators Phys. Rev. E53, R3005-R3008 (1996).

Y. Braiman, F. Family, and H. G. E. Hentschel, Nonlinear friction in the periodic stick-slip motion of coupled oscillators Phys. Rev. B55, 5491-54 (1997).

F, Brochard, P.G. de Gennes, J. Phys. Lett. (Paris) **45**, L597 (1984).

J. M. Carlson and A. A. Batista, Constitutive relations for the friction between lubricated surfaces Phys. Rev. E53, 4153-4165 (1996).

S. Chen and G.D. Doolan, Lattice Boltzmann methods for fluid flows, Ann. Rev. Fluid Mech., 30, 329 (1991).

M. Cieplak, E. D. Smith, and M. O. Robbins, Molecular Origins of Friction: The Force on Adsorbed Layers Science 265, 1209-1212 (1994).

C.Daly and J.Krim, Sliding Friction of Solid Xenon Monolayers and Bilayers on Ag(111) Phys. Rev. Lett. 76, 803-806 (1996).

P.G. de Gennes, Rev. Mod. Phys. 57,827 (1985).

A. L. Demirel and S. Granick, Friction Fluctuations and Friction Memory in Stick-Slip Motion, Phys. Rev. Lett. 77, 4330-4333 (1996).

F. Family and T. Vicsek, Scaling of the Active Zone in the Eden Process on Percolation Networks and the Ballistic Deposition Model, J. Phys. A18, L75 (1985).

F. Family and T. Vicsek, Dynamics of Fractal Surfaces, (World-Scientific, Singapore, 1991).

J. Gao, W.D. Luedtke, and U. Landman, Layering transitions and dynamics of Confined liquid films, Phys. Rev. Lett. 79, 705-708 (1997).

M. L. Gee, P. M. McGuiggan, J. N. Israelachvili, and A. M. Homola J. Chem. Phys. 93, 1895 (1990).

S. Granick, Polymer Surface Dynamics, MRS Bulletin, Polymer Surfaces and Interfaces, 21,1, 33-36 (January, 1996).

F. Heslot, T. Baumberger, B. Perrin, B. Caroli and C. Caroli, Creep, stick-slip and dry friction dynamics:experiments and a heuristic model Phys. Rev E49, 4973-4988 (1994).

C-M Ho and Y-C Tai, Micro-Electro-Mechanical-Systems (MEMS) and Fluid Flow, Ann. Rev. Fluid Mech. 30, 579-612 (1995).

Y.Z. Hu and S. Granick, Microscopic study of thin film lubrication and its contribution to macroscopic tribology, Tribology Letters 5, 81-88 (1998).

J. N. Israelachvili, Intermolecular and Surface Forces (Academic Press, London, 1991).

J. N. Israelachvili, P.M. McGuiggan and A. M. Homola, Dynamic Properties of Molecularly Thin Liquid Films, Science 240, 189 (1988).

J.F. Joanny, P.G. de Gennes, C.R. Acad. Sci. (Paris) 299, 279. (1984).

J.F. Joanny, *Spreading Kinetics*, in Hydrodynamics of Dispersed Media, Eds. J.P. Hulin, A.M. Cazabat, E. Guyon, F. Carmona, North-Holland, Amsterdam, 17-27 (1990).

J. Klein and E. Kumacheva, Confinement-induced phase transitions in simple liquids, Science 269, 816-819 (1995).

J. Klein and E. Kumacheva, Simple liquids confined to molecularly thin layers. I. Confinement-induced liquid-to-solid phase transitions, J. Chem. Phys. 108, 6996-7009 (1998).

E. Kumacheva and J. Klein, Simple liquids confined to molecularly thin layers. II. Shear and frictional behaviour of solidified films, J. Chem. Phys. 108, 7010-7022 (1998).

J. Koplik, J.R. Banavar, and J.F. Willemsen, Molecular dynamics of fluid flow at solid surfaces, Phys. Fluids A1, 781-94 (1989).

J. Koplik and J.R. Banavar, Continuum deductions from molecular hydrodynamics, Ann. Rev. Fluid Mech. 27, 257-292 (1995).

J. Krim MRS Bulletin, Fundamentals of Friction 23, 6 (June, 1998).

J. Krim, and A. Widom, Damping of a crystal oscillator by an adsorbed monolayer and its relation to interfacial viscosity, Phys. Rev. B38, 12184-12189 (1988).

J. Krim, D. H. Solina, and R. Chiarello, Nanotribology of a Kr Monolayer: A Quatz-Crystal Microbalance study of Atomic-Scale Friction, Phys. Rev. Lett. 66, 181-184 (1991).

U. Landman, W.D. Luedtke, N.A. Burnham, and R.J. Colton, Atomistic Mechanisms and Dynamics of Adhesion, Nanoindentation, and Fracture Science 248, 454-461 (1991).

U. Landman, W. D. Luedtke, J. P. Gao, Atomic-Scale Issues in Tribology - Interfacial Junctions and Nano-elastohydrodynamics Langmuir 12, 4514-4528 (1996).

J.F. Lindner, B.K. Meadows, W.L. Ditto, M.E. Inchiosa, and A.R. Bulsara, Array Enhanced Stochastic Resonance and Spatiotemporal Synchronization Phys. Rev. Lett., 75, 3-6 (1995).

G. Luengo, J. Israelachvili, S. Granick, Generalized Effects in Confined Fluids - New Friction Map for Boundary Lubrication, Wear 200, 328-335 (1996).

C. Mak and J. Krim, Quartz-crystal microbalance studies of velocity dependence of interfacial friction Phys. Rev. B58, 5157-5159 (1998).

C. Mak and J. Krim, Quartz crystal microbalance studies of disorder-induced lubrication, Faraday Discussions 107, 389-397 (1998).

A. Marmur, M.D. Lelah, Chem. Eng. Comm. 13, 133 (1981).

G. Mohiuddin Mala, D. Li, J.D. Dale, Heat transfer and fluid flow in microchannels, ASME DSC 59, 127-36 (1996).

X. Nie, G.D. Doolan, and S. Chen, Lattice-Boltzmann Simulations of Fluid Flows in MEMS, (Preprint, 2000).

B. N. J. Persson, Theory of friction and boundary lubrication, Phys. Rev. B48, 18140-18158 (1993).

B. N. J. Persson, Theory of friction: friction dynamics for boundary lubricated surfaces, Phys. Rev. B55, 8004-8012 (1997).

G. Reiter, A. L. Demirel and S. Granick, From Static to Kinetic Friction in Confined Liquid Films Science 263, 1741-1744 (1994).

M. G. Rozman, M. Urbakh, and J. Klafter, Stick-Slip Motion and Force Fluctuations in a Driven Two-Wave Potential Phys. Rev. Lett. 77, 683-686 (1996).

M. G. Rozman, M. Urbakh, and J. Klafter, Origin of stick-slip motion in a driven two-wave potential, Phys. Rev. E54, 6485-6494 (1996).

E. D. Smith, M. O. Robbins, M. Cieplak, Friction on adsorbed monolayers Phys. Rev. B54, 8252-8260 (1996).

P.A. Thompson and M.O. Robbins, Origin of Stick-Slip Motion in Boundary Lubrication, Science 250, 792-794 (1990).

P.A. Thompson, and M.O. Robbins, Shear flow near solids: Epitaxial order and flow boundary conditions, Phys. Rev. A41, 6830-6837 (1990).

P.A. Thompson, G.S. Grest and M.O. Robbins, Phase Transitions and Universal Dynamics of Confined Films, Phys. Rev. Lett. 68, 3448-3451 (1992).

P.A. Thompson and S>M. Troian, A general boundary condition for liquid flow at solid surfaces, Nature 369, 360-362 (1997).

I. Tovstopyat-Nelip and H.G.E. Hentschel, Friction and Noise Induced Coherent Structures in Boundary Lubrication, Phys. Rev. E. 61, 3318-3324 (2000).

M. Van Dyke, M. , *An Album of Fluid Motion,* Parabolic Press, Stanford (1982).

Y. Xia and G.M. Whitesides, Soft Lithography, Ann. Rev. Mater. Sci. 28, 153-184 (1998).

H. Yoshizava, P. McGuiggan, amd J. N. Israelachvili, Identification of a Second Dynamic State During Stick-Slip Motion, Science 259, 1305-1307 (1993).

Chapter 7

MODELING ADHESIVE FORCES FOR ULTRA LOW FLYING HEAD DISK INTERFACES

Andreas A. Polycarpou
Department of Mechanical and Industrial Engineering
University of Illinois at Urbana-Champaign
Urbana, IL 61801

1 INTRODUCTION

This paper addresses a major issue in microtribology related to the head/disk interface (HDI) in magnetic storage. This is the issue of strong intermolecular (adhesive) forces that may be present at the interface with very smooth surfaces. More specifically, a model termed the sub-boundary lubrication (SBL) model is developed and applied to typical HDI's. It uses the Lennard-Jones attractive potential to characterize the intermolecular forces, and a statistical surface roughness model.

For extremely high-density recording, the head/disk spacing is expected to be reduced further, down to few nanometers as shown in Table 1 [1]. The success of conventional magnetic recording to areal densities of 100 Gbit/in^2 and beyond will depend on the successful design and implementation of low flying interfaces. The tribological issues that will be encountered for 100 Gbit/in^2 interfaces (flying height of approximately 6 nm) are enormous. In such ultra low flying height regimes, the mechanical integrity of the HDI is a main concern since head/disk contact is unavoidable. To minimize the risk of head disk contact super-smooth media and sliders that approach molecular dimensions will be used. From a tribological point of view, however, head/disk contact on super-smooth media presents a serious challenge, such as high adhesive/intermolecular forces.

In the present work, a quasi-dynamic adhesion model based on the SBL model is used to calculate the adhesion forces at the HDI. The normal separation investigated is of the order of 10 nm (nominal fly-height) down to contact recording. Three different levels of surface roughness are investigated. The first case is a rougher interface with a combined standard deviation of surface heights, $\sigma = 5.83$ nm. The second case is an intermediate rough case with $\sigma = 1.45$ nm, and the last case is a super smooth interface with a combined $\sigma = 0.83$ nm. The latter is expected to be used for extremely high-density recording. The model revealed the significance of the surface roughness on the adhesion forces as the fly-height decreases.

Table 1. Magnetic spacing requirements, Ref. [1]

Areal density, Gb.in^{-2}	Magnetic Spacing, nm	Disk Overcoat thickness, nm	Flying height, nm	PTR nominal value, nm	Slider Overcoat thickness, nm
4.5	45.72	6.80	28.45	4.32	6.20
6	40.64	5.90	25.40	3.81	5.40
7.4	36.32	5.30	22.20	3.81	4.90
9.5	32.00	4.70	19.20	3.81	4.30
12	28.19	4.10	16.40	3.81	3.80
15	24.89	3.60	14.50	3.81	3.00
20	21.84	3.00	13.40	3.35	2.20
30	20.83	2.80	13.00	3.20	1.90
40	19.56	2.60	12.20	3.05	1.80
60	18.1	2.4	11.7	2.4	1.6
100	10	2	6	1	1

2 CRITICAL ASSESMENT

The recording head or slider is flying over the disk with the use of an air bearing as shown schematically in Fig. 1 [2].

Air bearings are typically designed to fly at a nominal distance from the disk, called fly-height (FH), which on modern hard disk drives is around 20–30 nm. In general there are two types of air bearing designs; positive pressure air bearings which provide the positive pressure through the design of the rails, to be enough to counter balance the preload exerted by the suspension which is typically of the order of few grams. The second type of air bearing design is what is referred to as a negative pressure air bearing in which the design is such that in addition to the lift pressure it also provides a suction pressure (negative pressure). Nevertheless, the net lift off force is still equal to the pre-load.

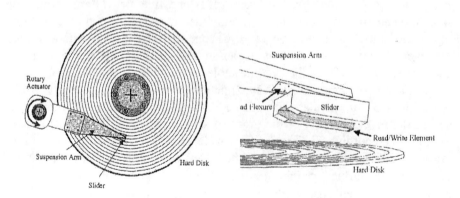

Figure 1. Schematic of the head/disk interface (from Ref. [2]).

The existing air bearing analyses usually do dot take into account the surface roughness or the intermolecular forces that may be present at the interface. Existing designs have served the magnetic industry fairly well with flying height down to around 15 nm, corresponding to areal densities of up to 10 Gbit/in^2. However as the fly height decreases further, additional surface forces, such as adhesion forces will probably become significant. These adhesion forces are insignificant for larger fly heights because their active range is very short and only become relevant at extremely small separations.

Typical forces at the head disk interface are schematically depicted in Fig. 2. In the case of negative pressure air bearings, the net air bearing force, F_a is the difference between the positive pressure and the suction pressure.

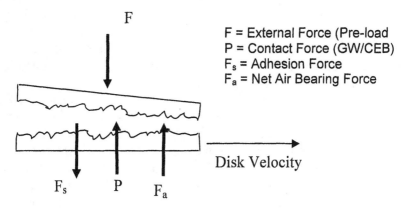

Figure 2. Schematic of the relevant forces at the head/disk interface.

During static contact, the air bearing force $F_a = 0$ and the external force (pre-load), $F = P - F_s$, where P is the contact load and is calculated from the CEB elastic/plastic contact model, and F_s is the adhesion force. These forces together with the tangential (friction) force at the interface were used by Polycarpou and Etsion to develop the SBL stiction model [3], [4]. They showed that the adhesion force becomes significant and may dominate the forces at the interface for smooth surfaces and thicker lubricants at the surfaces (including humidity).

During low flying and pseudo contact two different cases my arise:
(1) If adhesion force $F_s = 0$, then, $F = F_a$ which is currently assumed in models (assuming that the contact force is small, i.e., $P = 0$).
(2) If $F_s \neq 0$, then $F = F_a - F_s$ and for large F_s, interface will collapse (assuming $P = 0$).

Note that in real cases, the air bearing forces go from their actual values during nominal flying heights to zero during actual contact. The exact transition, however, is not well known.

3 MODEL DESCRIPTION

Concerning continuum modeling efforts of intermolecular attractive forces at an interface accounting for both the effects of surface roughness and liquid film thickness, a number of analytical meniscus models have been suggested, *e.g.*, [5]. Such models share the common assumption that stiction results from a significant increase in adhesion due to the capillary forces of the adsorbed film at the interface and are valid for static conditions only. Meniscus models are useful when the lubricant at the interface is abundant, relatively thick and mobile. However, when the lubricant thickness is extremely small, of the order of few monolayers that strongly adhered to the solids (immobile), the meniscus model breaks down [6]. For these cases, an alternative to the meniscus model has been suggested in [7]. This alternative model, termed sub-boundary lubrication (SBL), is more likely for strong lubricant-solid bond and extremely thin lubricant thickness, and was used in [3] and [4] to develop the SBL static friction model that accurately predicted experimental stiction measurements on magnetic thin film disks. The difference between the meniscus model, on the one extreme, and the SBL model, on the other extreme, is that the former contains liquid bridges with negative mean curvature and a pressure difference that will pull lubricant off the surrounding solid areas. On the other hand, the SBL model considers the case in which there is strong interaction between lubricant and substrate, *i.e.*, a large negative amount of energy is associated with the formation of a unit area of solid-lubricant interface. In such a case the energy cost for liquid bridge formation is too high. Meniscus formation is energetically unfavorable and lubricant layers remain essentially uniform and of quasi-solid nature.

Both the meniscus based models and the SBL stiction model were used for calculating the adhesive forces under static (no sliding) conditions. In this work we will use the SBL adhesion model for steady sliding conditions and extremely low fly heights. It is important to note that the Lennard-Jones potential is valid for both static and dynamic (sliding) conditions, whereas the capillary meniscus based models are strictly for static conditions-formation of meniscus bridges. Also, as the fly height decreases asperity contact may or may not occur. A brief description of the rough surface model, the contact force model, and the adhesive model follows.

4 ROUGH SURFACE MODEL

The surface topography of the interfaces is modeled with the Greenwood and Williamson (GW) [8] statistical model. Each rough surface is modeled by a collection of asperities having spherical summits with a uniform radius R and a standard deviation σ_s of their heights, u. For the basic assumptions of the GW model see Refs. [8] and [3]. Additional assumptions are made with respect to the presence of the thin lubricant layer between the contacting surfaces [3], [7]. These are:

1. Each surface is covered with a uniform layer of lubricant with small thickness compared to the asperity radius. Therefore, the topography of the resulting surfaces is the same as the original solid surfaces.

2. When surfaces come into contact, asperities penetrate and displace the lubricant. As a result, the asperity deformation, and hence the contact load at a given mean separation of the solids are the same as in the case of un-lubricated contact.

3. The strength of the bond at the lubricant-solid interface is greater than the cohesive strength of the lubricant. This implies that any adhesion force arising between the closely separated lubricant surfaces will be transmitted to the solid.

4. For geometrical calculations, the two uniform lubricant layers are equivalent to a single layer of thickness t, equal to the sum of the two thicknesses, $t_1 + t_2$, applied to the flat (see Fig. 3).

The above 4 assumptions along with the basic GW assumptions allow the two rough surfaces to be represented by an equivalent rough surface in contact with a smooth plane. The geometry of this model is shown in Fig. 3, where d the separation of the surfaces, measured from the mean of the asperity heights; h is the separation of the surfaces measured from the mean of the surface heights; y_s is the distance between the means of asperity and surface heights; $t_0 = t$ is the lubricant layer thickness.

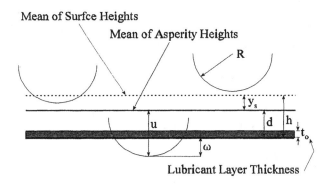

Figure 3. Contacting rough surface model.

5 CONTACT LOAD

The contacting surfaces are modeled using the CEB elastic-plastic model [9], which is an extension of the GW. The deformation that occurs in the contacting regions can be elastic, plastic or elastic-plastic, depending on the nominal pressure, surface roughness and material properties. The dimensionless contact load $P^* = P/(A_n E)$ is given by

$$P^*(h^*) = \beta \frac{4}{3}\sqrt{\frac{\sigma}{R}} \int_{h^*-y_s^*}^{h^*-y_s^*+\omega_c^*} \omega^{*\frac{3}{2}} \phi^*(u^*)\,du^* +$$

$$+ \beta \frac{\pi K H}{E^*} \int_{h^*-y_s^*+\omega_c^*}^{\infty} (2\omega^* - \omega_c^*)\phi^*(u^*)\,du^*. \qquad (1)$$

The first and second integrals are the contributions of the elastically and plastically deformed asperities, respectively. A_n is the nominal area of contact; $\beta = \eta R \sigma$; η is the areal density of asperities; K is the maximum contact pressure factor; $\phi^*(u^*)$ is the normalized distribution of the asperity heights which is assumed to be Gaussian. H is the hardness of the softer material; E^* is the equivalent elastic modulus; ω is the local interference of a contacting asperity normalized in the form

$$\omega^* = \omega/\sigma = u^* - d^* = u^* - h^* + y_s^*, \qquad (2)$$

ω_c is the critical interference of an asperity at the inception of plastic deformation and is another form of the plasticity index, ψ, defined by GW [8], i.e.,

$$\psi = \frac{2E^*}{\pi K H}\sqrt{\frac{\sigma_s}{R}}. \qquad (3)$$

6 ADHESION FORCE

The adhesion force originates from the Lennard-Jones interactive potential that forms the basis for the adhesion models in [3] and [7]. There are two major adhesion models of single asperity behavior. These are the JKR and the DMT models. Of these two models, the latter is more suitable for materials with relatively large ratio of elastic modulus to surface free energy, such as metals. On the other hand, the JKR model is more suited for component materials like rubber. In the present paper, an extension of the DMT model is used, and therefore the results are applicable to cases where the contacting solids are relatively stiff.

The interaction of the lubricant surface molecules is assumed to be governed by the Lennard-Jones potential, which corresponds to an attractive pressure relation,

$$p(Y) = \frac{8}{3}\frac{\Delta\gamma}{\varepsilon}\left[\left(\frac{\varepsilon}{Y(r)}\right)^3 - \left(\frac{\varepsilon}{Y(r)}\right)^9\right] \quad (4)$$

where $Y(r)$ is the separation of the lubricant surfaces at a radius r (see Fig. 4).

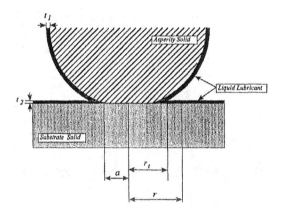

Figure 4. Geometry of a contacting asperity in the presence of sub-boundary lubrication.

There are two parameters in the Lennard-Jones potential. The first one is the inter-atomic spacing ε ($\varepsilon^* = \varepsilon/\sigma$), which is in the range of 0.3–0.5 nm. The second parameter is the energy of adhesion $\Delta\gamma_l = 2\gamma$, where γ is the surface tension of the liquid and is measure experimentally.

The total adhesion force between the contacting surfaces in the presence of sub-boundary lubrication can be separated into three components. These components are from the contributions of the completely non-contacting asperities, the solid non-contacting-lubricant contacting asperities, and the elastic-plastic solid contact. The total dimensionless adhesion force $F_s^* = F_s/(A_n E)$ is given below, where each integral corresponds to one of the three components described above,

$$F_s^*(h^*) = \frac{8\pi\eta R \Delta\gamma_l \varepsilon^{*2}}{3E^*}\left\{\int_{-\infty}^{h^*-y_s^*-t^*}\left[\frac{1}{(\varepsilon^*-\omega^*-t^*)^2}-\frac{0.25\,\varepsilon^{*6}}{(\varepsilon^*-\omega^*-t^*)^8}\right]\phi^*(u^*)\,du^* + \right.$$

$$\left. +\frac{3}{4\varepsilon^{*2}}\int_{h^*-y_s^*-t^*}^{h^*-y_s^*}\phi^*(u^*)\,du^* + 2\int_{h^*-y_s^*}^{\infty}\int_0^{\infty}\left[\frac{1}{(Z^*-t^*)^3}-\frac{\varepsilon^{*6}}{(Z^*-t^*)^9}\right]s^*\phi^*(u^*)\,ds^*\,du^*\right\}$$

(5)

Z is the profile outside the contact area for elastic contact between a sphere and a flat, obtained from the Hertz solution. For more details see Refs. [3] and [7]. A schematic of a rough surface in contact with a smooth surface with a thin lubricant is shown below in Fig. 5. From the asperities shown, only two asperities are in solid contact, one elastically (P_{EL}) and the other plastically (P_{PL}). For the adhesion force, 2 of the 6 asperities are non contacting the lube or the solid surface $(F_s)_n$, 2 other asperities are contacting the lube only $(F_s)_l$, and the last 2 are contacting the solid surface (one in elastic contact, $(F_s)_{cEL}$, and the other in plastic contact $(F_s)_{cPL}$.

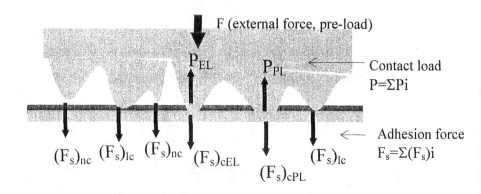

Figure 5. Schematic of a rough surface in contact with a smooth plane in the presence of sub-boundary lubrication, showing the relevant contact and adhesion forces.

7 RESULTS AND DISCUSSION

Simulations were carried out with 3 different surface roughness. The first case is relatively rough (σ=5.83 nm) for head/disk interfaces, especially for higher areal densities. The second case is an intermediate rough interface with a combined σ=1.45 nm, and the last case is an extremely smooth interface, with a combined σ=0.83 nm. Such smooth interfaces are expected to be used for areal densities of 100 Gbit/in^2 and beyond. Note that these roughness are for the combined interfaces. If for example the center line average, R_a, for the head is R_a = 0.57 nm, then the Ra for the disk has to be equal to R_a = 0.48 nm for a combined σ=0.83 nm, assuming that the surface heights follow a Gaussian distribution ($R_q = \sigma = (\pi/2)^{0.5} R_a$. The average radii of curvature of the asperities, R and the areal density of asperities were extracted from AFM roughness data, and are shown in Table 2. The combined Young's modulus, and the hardness of the softer material (DLC) are taken to be 85.29 GPa and 2.5 GPa, respectively

[4]. The thickness of the lubricant on the disk is also assumed to be $t = 2$ nm with a $\Delta\gamma = 4\ 0$ mN/m. The plasticity index ranges from $\psi = 1.9$ for the roughest interface (primarily plastic deformations) to $\psi = 0.4$ for the smoothest case indicating mostly elastic contact conditions.

Table 2. Surface roughness parameters.

Surface	Roughest	Intermediate	Smoothest
σ (nm):	5.83	1.45	0.83
R (μm^{-2})	5.82	6.22	8.04
η (μm^{-2})	5.2	10.4	11.1
ψ	1.90	0.61	0.40

Figure 6(a) shows the dimensionless contact load $P^* = P/A_n E$ and the dimensionless adhesion force, $F_a^* = F_a/A_n E$ versus the dimensionless separation (or fly-height divided by the standard deviation of surface heights σ) for the roughest case, $\psi = 1.9$. At all values of the dimensionless flying height, F_a^* is always less than P^*, that is the adhesion force is insignificant for this particular interface. Figure 6(b) shows the total dimensioness adhesion force and the contribution to the adhesive force from the non-contacting asperities, the lube contacting asperities and the elastic/plastic contacting asperities (see also Fig. 5). The largest contribution is from the lube contacting asperities. Figures 7(a) and 7(b) show the flying height in length units of nanometer (nm) versus the load and adhesion force in force units of grams in logarithmic and linear scales respectively for a nominal contact area of $A_n = 25{,}000$ μm^2. Also shown in these figures is a line indicating 3 times the value of the standard deviation of surface heights ($3\sigma = 3R_q$).

For $FH > 3R_q$, no significant contact occurs as expected from the Gaussian distribution of asperity heights, and also seen in Fig. 7. For $FH < 3R_q$ contact occurs and becomes more significant as the flying height decreases. From Fig. 7 it is clear that such a rough interface ($\sigma = 5.83$ nm) is not viable for ultrahigh recording densities since significant contact occurs at 17.5 nm (= $3R_q$). For example, at a $FH = 17.5$ nm, $F_s = 0.2$ g, and $P = 2$ g. Additional simulations with different lube thickness predict that adhesive forces are still insignificant even with thicker lubes of $t = 4$ nm.

Next, consider two smoother cases with $\psi = 0.61$ ($\sigma = 1.45$ nm) and $\psi = 0.4$ ($\sigma = 0.83$ nm) which are more realistic candidates for ultrahigh density recording. As it will be seen next, in both of these cases adhesion forces are very significant and may dominate the forces at the interface, and in some cases may even cause the collapse of the HDI. Figure 8 depicts the dimensionless adhesion and contact forces versus the dimensionless flying

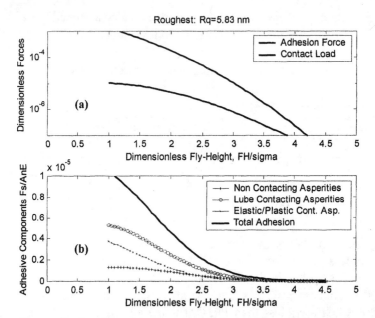

Figure 6. Roughest Interface, $\psi = 1.9$: (a) Dimensionless forces versus dimensionless separation (fly-height); (b) Dimensionless adhesion force components.

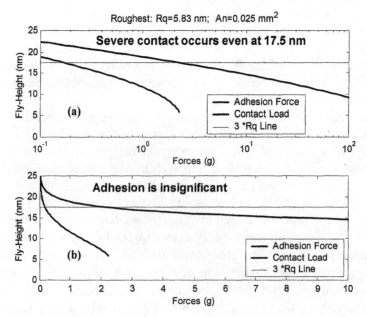

Figure 7. Roughest Interface, $\psi = 1.9$: Fly Height versus force. (a) Logarithmic scale; (b) Zoom linear scale.

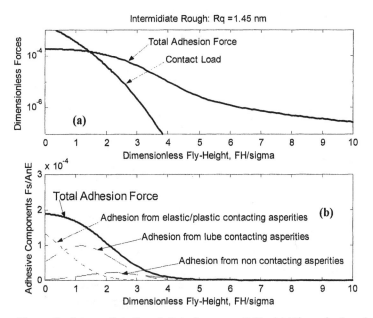

Figure 8. Intermediate Rough Interface, $\psi = 0.61$: (a) Dimensionless forces versus dimensionless separation (fly-height); (b) Dimensionless adhesion force components.

height for the intermediate rough case ($\sigma = 1.45$ nm). For the practical range of interest of the dimensionless flying height ($FH/\sigma > 3$) adhesion forces are always larger than the contact load. Only at very small dimensionless separations, $FH/\sigma = h/\sigma < 1.5$, $P^* > F_s^*$ which is under severe contact conditions and not of interest for ultra high recording densities.

Figure 9 shows the flying height versus forces for the same roughness as in Fig. 8 and an $A_n = 25,000$. The $3Rq$ line indicating when contact will start being significant is also shown at $3 \times 1.45 = 4.35$ nm. At a flying height of 6 nm the adhesion force is $F_s = 1.8$ g as also indicated in the figure. The implications of such a significant adhesion force at a fully flying interface are that the balance of forces will now be disturbed and the HDI may collapse. For $FH = 6$ nm, $F = 2.5$ g, $F_a = 2.5$ g, $P \approx 0$, and $F_s = 1.8$ g. Then referring to Fig. 2, $F \neq P + F_a - F_s$, and a net force of 1.8 g is pulling the slider to the disk.

Figures 10 and 11 are for the smoothest case considered ($\sigma = 0.83$ nm). The behavior of the adhesion force is very similar to the previous case. However, a closer examination of Fig. 11(b) which zooms into the area of interest, *i.e.*, $FH = 6$ nm reveals the somewhat counterintuitive result that $F_s = 1$ g which is less than the previous rougher HDI case ($\sigma = 1.45$ nm). The explanation to this result can be found by examining Fig. 12 that shows a schematic representation of the HDI at a flying height of 6 nm for both cases. As it can be seen, for the rougher case more asperities are in close proximity to the slider than the smoothest case that is about 2 nm below.

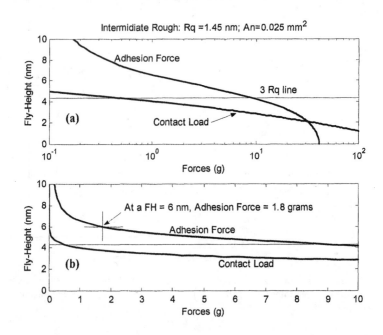

Figure 9. Intermediate Rough Interface, $\psi = 0.61$: Fly Height versus force. (a) Logarithmic scale; (b) Zoom linear scale.

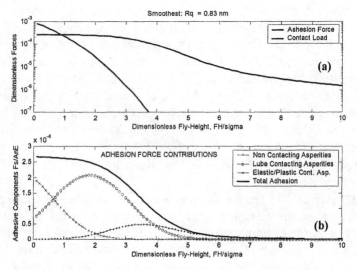

Figure 10. Smoothest Interface, $\psi = 0.40$: (a) Dimensionless forces versus dimensionless separation (flying height); (b) Dimensionless adhesion force components.

Adhesion Forces for Ultra Low Flying Head Disk Interfaces

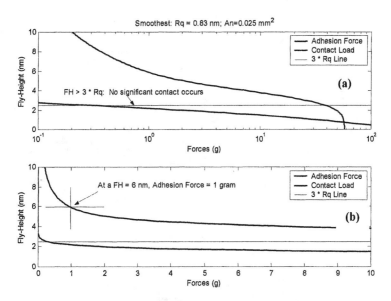

Figure 11. Smoothest Interface, $\psi = 0.4$: Fly Height versus force. (a) Logarithmic scale and (b) Zoom linear scale.

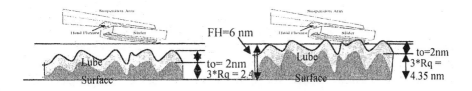

Figure 12. Schematic (to-scale) of the flying slider over (a) the smoothest disk (Rq=0.83 nm), and (b) over the intermediate rough disk.

The adhesion forces, for both cases are shown in Fig. 13. At a flying height of 8 nm both surfaces have about the same F_s. As flying height decreases to 4.5 nm the rougher case has higher adhesion force. One needs to go to flying height of about 3.2 nm before the smoothest surface will have larger F_s. Note however that at this flying height, the intermediate rough interface ($\sigma = 1.45$ nm) will be mostly a contacting interface.

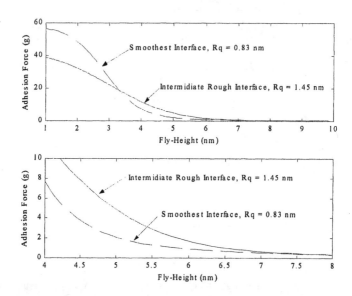

Figure 13. Adhesion forces for $\psi = 0.41$ ($R_q = 0.83$ nm) (- -) and for $\psi = 0.93$ ($R_q = 1.45$ nm): (a) Larger linear range and (b) Smaller linear range.

8 CONCLUSIONS

A quasi-dynamic adhesion model based on the SBL model has been presented and used to calculate the adhesion forces at the HDI. The normal separation investigated is of the order of 10 nm (nominal fly-height) down to contact recording. Three different levels of surface roughness were investigated. The first case is a rougher interface with a combined standard deviation of surface heights, $\sigma = 5.83$ nm. In this case, the contact force is always larger than the adhesive force and contact will first occur at a fly height of about 17 nm ($3R_q$), which clearly indicates that this is not a good choice for ultra high areal densities. The second case is an intermediate rough case with $\sigma = 1.45$ nm, and the last case is a super smooth interface with a combined $\sigma = 0.83$ nm. For both of these cases:

(a) At flying height of 10 nm or greater, there are no significant adhesive forces.
(b) For flying height between 10 to 5 nm, adhesive forces start to become significant, more so for the rougher case.
(c) At 5 nm flying height, no significant asperity contact occurs in either case.
(d) For flying height less than 5 nm, asperity contacts may occur and the contact force also becomes significant. Also, the air bearing

force diminishes, and adhesion may dominate, forcing the interface to collapse.

In summary, the quasi-steady adhesion model predicts that significant intermolecular forces exist at low fly-heights and super smooth surfaces.

9 RECOMMENDATIONS AND FUTURE RESEARCH NEEDS

Implications of this work is that super smooth surfaces with combined roughness of the order of less than 2 nm will probably have significant adhesion forces and need to be accounted for. This does not only apply to the Head/disk interface but many other applications in MEMS, bioengineering, and biomedical applications. If other surface forces are also dominant at the interface, *e.g.*, electrostatic, hydrodynamic forces, then they could also be added to the model.

REFERENCES

[1] Menon, A., "Critical Requirements for 100 Gbit/in^2 Head/media Interface," *Proceedings of the Symposium on Interface Tribology Towards 100 Gbit/in^2*, C.S. Bhatia A.A. Polycarpou, and A. Menon, eds. ASME TRIB-Vol 9, pp. 1-9 (1999).
[2] Talke, F. E. "On Tribological Problems in Magnetic Disk Recording Technology," *Wear*, Vol. 190, n 2, p 232-238 (1995).
[3] Polycarpou, A. A. and Etsion, I., "Static Friction of Contacting Real Surfaces in the Presence of Sub-Boundary Lubrication," *ASME Jour. of Trib.*, Vol. 120, pp. 296-303 (1997).
[4] Polycarpou, A.A., Etsion, I., "Comparison of the Static Friction Sub-Boundary Lubrication Model with Experimental Measurements on Thin Film Disks," *STLE Tribology Transactions*, Vol. 41 No. 2, pp. 217-224.
[5] Gui, J. and Marchon, B., "A Stiction Model for a Head-Disk Interface of a Rigid Disk Drive," *Jour. Appl. Phys.*, Vol. 78 (6), pp. 4206-4217 (1995).
[6] Israelachvili, J. N., *Intermolecular and Surface Forces*, Academic Press, second edition, San Diego, (1991).
[7] Stanley, H. M., Etsion, I. and Bogy, D. B., "Adhesion of Contacting Rough Surfaces in the Presence of Sub-Boundary Lubrication," *ASME Jour. of Trib.*, Vol. 112, pp 98-104, (1990).
[8] Greenwood, J. A. and Williamson, J. P. B., "Contact of Nominally Flat Surfaces," *Proc. of the Roy. Soc. of London*, Vol. A295, pp 300-319, (1966).
[9] Chang, W. R., Etsion, I. and Bogy, D. B., "An Elastic-Plastic Model for the Contact of Rough Surfaces," *ASME Jour. of Trib.*, Vol. 109, pp 257-263, (1987).

Chapter 8

SELF-LUBRICATING BUCKYBALLS AND BUCKYTUBES FOR NANOBEARINGS AND GEARS - SCIENCE OR SCIENCE F(R)ICTION?

M. N. Gardos
Raytheon Electronic Systems, Engineering Services Center
El Segundo, CA 90245

Abstract A review of the literature shows that the popular misconceptions of employing carbon nanotubes and carbon buckyballs as moving mechani-cal assembly components for Drexlerian nanomachinery are based on incomplete computer models missing the basic tenets of nanotribology. A large number of models fail to take into account (a) the effects of enhancing or retarding the $\pi - \pi^*$ orbital interaction of contacting flat or curved graphene planes by environmental influences, and (b) the war-ning signs of non-tribology-related data on high cohesive energy density and thus poor self-lubricating ability. A variety of tribometric data on carbon buckyballs, generated by test apparatus ranging from atomic force microscopy to continuum-mechanical testers, are also gleaned from the literature to compare the results of these experiments with the validity of the hypothesis brought forth by this paper. New materials approaches are reviewed for nanomachinery, using other fullerene-like chemistries besides carbon.

1 INTRODUCTION

In his 1986 book entitled "Engines of Creation," Eric Drexler has envisioned atomic-molecular level robotic engines ("nanobots") revolu-tionizing nearly all aspects of life. Since then, thanks mostly to the explosive development of microelectromechanical systems (MEMS) during the past decade, researchers have extrapolated from the now-doable microscopic (micrometers) to the yet-imaginary nano-electromechanical systems (nanometer-NEMS) regime. What was once considered science fiction is now optimistically viewed as manageable. The prospects seem so realistic to some that the Pentagon is now fund-ing nanotechnology research at the Mitre Corporation in McLean, VA, and the ambitious federal nanotechnology initiative is being kicked off this year.

The Drexlerian version of nanotechnology is the craft of const-ructing molecular engines smaller than a few hundred nanometers. A moving part may be any rigid molecular structure, a rotary bearing a σ bond of low steric hindrance and a rotary motor or a propeller a bacterial flagellum. Since any motor or engine (nano, micro or macro) must do useful work in realistic environments, we must examine the possible parasitic power losses that may prevent the "engines of creation" to create. This paper is an attempt to review the realities and myths surrounding some relevant aspects of this futuristic art form from the viewpoint of a practicing tribologist, as supported by known science fundamentals and relevant experimental data.

Since present Drexlerian nanotechnology is essentially limited to computer modeling of gears, shafts and other moving mechanical assemblies built up from individual atoms and molecules [1], the assumptions used to construct these models must include all fundamental tenets of atomic-molecular-level surface interactions. At the same time, atomic force microscopist-nanotribologists who are first to measure, and those numerical simulator counterparts who model such interactions must pay close attention to each other's efforts. Carbon-based bucky-balls and bucky(nano)tubes which have already been synthesized and which, by virtue of their engineeringly enticing shapes, are readily ima-gined as parts for NEMS engines [2,3] are especially ripe for scrutiny.

It is too bad that the notion of buckytube-buckyball bearings and gears is largely misconceived, lacking solid roots in molecular engineering. To inject a dose of reality into the proceedings, this writer is compelled to review the literature dealing with the modeling of such virtual NEMS assemblies. To a tribologist's dismay, hints of unfavorable results were found hidden in papers purporting viability. Problem areas generally mentioned as side comments can be, in fact, traced to the neglect or misunderstanding of the essential basics. The best example is the unfavorable chemical bonding between the rolled-up, graphene sheets of nested, uncapped buckytube pin/bushing combinations under load [1] or between loaded benzyne gear teeth (a) grafted radially on the outside diameter of buckytubes acting as interlocking, paddlewheel-like gear mechanisms or (b) attached axially to form rack-and-pinion combina-tions [2]. Just looking at the conceptualized figures presented in [3] where the benzene ring-type loaded gears mesh with each other in-register forcing the $\pi - \pi^*$ orbitals to overlap, satisfactory operation looks problematic. Moreover, NEMS parts might have to be working not only in ultrahigh vacuum (UHV, convenient for calculations), but also in various other atmospheric environments NEMS must operate. It is axiomatic that nanobots must do useful work in all of the envi-ronments of interest. What if the available power is insufficient to move anything? Thin layers of C_{60} buckyballs imagined as solid lub-ricant layers for nanobots or as self-lubricated balls in a NEMS rolling element bearing running with buckytube races are facing the same predicament.

2 CHEMICAL BONDING-BASED BARRIERS FOR BUCKYTUBE BEARINGS AND GEARS

Under the right environmental conditions a plain bearing combination of a single-walled, uncapped, zero-angle-chirality buckytube shaft turning inside of a slightly larger diameter buckytube bushing is super-rotary [2]. As reported, the theoretically useful energy of such ultrahigh-speed motion of the spinning shaft "is more restricted when the bearing is under load, suggesting that a very careful design of the nanobearing is needed to ensure optimum performance." Are there any *unloaded* bearings in power-generation or transfer devices, the size scale notwith-standing? In a related modeling exercise, the self-assembly of such plain bearings was examined [4], because these mechanisms cannot be realis-tically fabricated by manipulation with nanotweezers. The self-nesting trajectories of the buckytube shaft and bushing often resulted in "a false dock, because in the presence of small external perturbations (for example, non-bonded interactions with other nanomachine components or with solvent molecules) docking may not have occurred."

It is difficult to imagine such "non-bonded interactions" as negligibly small. How would it be possible that the presence of the graphitic edge sites of the tube ends, 10^{11}-times more reactive than a defect-free basal plane [5], would not affect rotational behavior during interaction with the surrounding atmosphere, liquid medium or with the neighboring mechanical parts? It is equally unlikely that the propensity of the sp^2-bonded (planar; 2-D) graphene planes to incorporate intercalates from a particular environment would not change the interfacial shear strength from those measured in UHV [6]. In the same vein, the sp^3-type (tetra-gonal; 3-D) bonding of diamond imparts strong, three-dimensional strength to diamond. As an imaginary MEMS or NEMS bearing made from some nanocrystalline or amorphous diamond shape, the most vul-nerable part of this material (or its silicon or silicon carbide counterpart) is the chemical state of the surface. The magnitude of adhesion and thus the coefficient of adhesive friction are essentially defined by the number of dangling (high-friction), reconstructed (reduced-friction) or adsorbate-passivated (low-friction) σ bonds. High-friction linking of the sliding counterfaces by unsaturated bonds on heating, some reduction in friction due to the reconstruction of these dangling bonds and their eventual passivation by benign adsorbates on cooling to provide the lowest possible friction are the main causes of radically increased and reduced interaction [7].

In the case of graphite, the sp^2-type (trigonal) bonding leads to high in-plane strength, but weaker bonds between the graphene sheets. Papers and books quoting only "weak" van der Waals bonding between the sheets seem to ignore the π - π^* (bonding-antibonding) orbital overlap in the hexagonal corners of the Brillouin zone of *in-register* graphitic planes. The strength of this interaction can be as high as 9 to 16 kcal/mol [6]. This value is significantly less than the ~100 kcal/mol of the strong C-C bonds that can form between sliding diamond interfaces, but is far from negligible. Albeit weaker, this electron-hole overlap is responsible for the high friction and wear of graphite in UHV. Graphite is an excellent solid lubricant only under terrestrial conditions, where sufficient amounts of water vapor are present in the atmosphere. Why? The water molecule acts as a weak donor intercalate, separating the π - π^* orbitals and thus reducing the intra- and intergranular friction of the lamellae. At higher temperatures in air or in vacuum where water readily desorbs, higher molecular weight donor/acceptor-type intercalates with better charge transfer characteristics must be incorporated in the lattice to achieve *and maintain* shear strength reduction [6].

Since ours is a polycrystalline world, layered lubricants like graphite do not exist as infinitely large, single-crystal sheets. The individual par-ticles of ground-up lubricant powders or the grains in highly consolidated blocks of graphite have a variety of broken bonds, vacancies and other defects even on their stable basal planes and, especially, at the highly reactive edge sites. How these active sites interact with each other, their substrate, the counterface and the thermal-atmospheric environment during intra- and interparticle slip determine the friction and wear life of a polycrystalline solid lubricant layer. That means that in addition to any π - π^* overlap, the bonds exposed on the edges or the few basal plane defects on the sliding surface of graphite will act similar to the interacting dangling σ bonds of silicon and diamond. They will impart high friction and cracking (incipient wear) in the wake of any high shear strength tribocontact.

It is, therefore, not difficult to imagine that the reactive edge sites of the modeled buckytube pin/bushing combinations might influence both the docking and the operation of the nested tubes. One would also expect some interaction with other NEMS assemblies or with the surrounding atmospheric medium. Indeed, the density of states of carbon nanotube ends was found to be high in UHV [8]. Even if the tubes were capped, the architecture, bonding arrangement thus the electron distribution at the caps and their resultant differences in reactivity (attraction) toward the surroundings can differ from one type of cap to another [9]. Once these tubes are mated with other nearby NEMS com-ponents, the various possible bonding interactions could impart excessive parasitic forces.

3 CHEMICAL BONDING-BASED BARRIERS FOR BUCKYBALL BEARING COMPONENTS

The soccer ball-like appearance of the C_{60} sphere, *i.e.*, the occasional pentagonal patches in the largely hexagonal surface arrangement, clearly show that the electron distribution around the pentagonal edges and corners (in analogy with the buckytube end-caps) would be different (higher) than at the common edges of neighboring hexagons. This should also translate into stronger interaction between the balls nested in their cubic lattice. Even without doing any tribometric work, a strong hint of poor solid lubricating capacity of an ordered C_{60} layers does come from the high heat of sublimation ($\Delta H°$) of C_{60} and C_{70} [10]. The 40.1 ± 1.3 kcal/mol value for the former and a 43.0 ± 2.2 kcal/mol for the latter are far too high for a van der Waals solid like the high molecular weight hydrocarbon coronene ($C_{24}H_{12}$) $\Delta H° = 14.1$ kcal/mol or ovaline ($C_{32}H_{14}$) $\Delta H° = 21.9$ kcal/mol. There is considerable attraction between the buckyballs, despite of the super-rotary phase of these balls spinning at ultrahigh speeds in the lattice.

Since C_{60} oligomer formation is theoretically possible [11], further shear strength increases may be anticipated on cross-linking. Environment-induced hardening of buckyball films may represent the footprint of even stronger attachment between balls. While it might not be sur-prising that implanting the layers with bismuth and nitrogen would lead to an increase in microhardness [12] via crosslinking of the broken carbon bonds, it is somewhat unexpected that simple photo-illumination can lead to the same result [13]. Superimposed is the effect of oxidation (C_{60} is vulnerable to oxygen) where exposure to air in the dark, followed by illumination for 2 hours, led to large hardness increases compared to the as-grown films and those stored in air and in darkness [14].

Is there any wonder that the published nano- and micro/macro-tribometric results on C_{60} (or for that matter, on C_{70}) are, on the whole, disappointing? A reasonably thorough literature survey is summarized in Table I, containing nano-tribometric data [15-22] generated with atomic force- or friction force microscopy (AFM-FFM) and a similar surface force apparatus (SFA). The corresponding micro- and macro-tribo-metric results are given in Table II [23-31]. The high friction values indicate a high shear strength material regardless of the test method-ology, test environment or the deposition technique employed. Even where the buckyballs were firmly grafted at the end of self-assembled monolayer chains to prevent their characteristically high bonding in the cubic lattice and allow some flexibility of movement during sliding, the AFM-type friction coefficients were still unexpectedly high [21,22]. The equivalent results in Table II [30] are just as discouraging. These data bode poorly for any promising use of buckyballs as NEMS com-ponents.

4 LUBRICATING NEMS ENGINES

As previously mentioned, the magnitude of adhesion and thus the adhesive friction of silicon and diamond are essentially defined by the number of dangling (high-friction), reconstructed (reduced-friction) or adsorbate-passivated (low-friction) surface bonds. Incipient linking of the sliding counterfaces by unsaturated bonds on heating, and their passi-vation by benign adsorbates on cooling, were suggested as the main causes of radically increased and reduced adhesion and friction, respectively [7]. In simpler terms used to explain the relevance of tribo-metric principles, one must find the right kind of lubricating gas or liquid medium within which the friction remains in acceptable ranges. The concept of friction is more relevant than wear, because NEMS wear does not exist in the conventional sense. Nanoscale "wear" means the des-truction of the molecular component such as breaking carbon bonds on the buckytube or ball surface, possibly fragmenting them into shell segments. Preventing any destruction by engineering the tribological surfaces from highly stable and low surface energy bonds appears to be the best solution.

A good static-case analogy of tribochemical dangling bonds generation and annihilation by the chemisorption of gases is the thermal disso-ciation and hydrogen-passivated deactivation of silicon defects (dangling Si σ bonds) at the Si/SiO_2 interface [32]. Surprisingly, the polarizable He atom can fulfill the same role [33]. The atomic-molecular-level funda-mentals of this microelectronics technique comprise the essence of the analogous concept of gas-lubricating MEMS-MMAs made from diamond and Si [7, 34-36]. It would be highly advantageous, if the atmospheric environment in which the NEMS operate could provide similar surface stability.

The secret of continuous and effective NEMS lubrication may indeed lie in employing gas or liquid molecules, which can tribocatalytically chemisorb on the NEMS component surfaces from the operating environment to form extremely thin, low surface energy layers. This has already been done for various allotropic forms of carbon-graphite turbo-pump seals on the macroscale, using hydrogen and helium as gas-phase "lubricants" [37]. The same general concept may be applied to carbon-graphite allotropes used in any other mechanical device, regardless of size.

Table I. Friction of C_{60} under nano-tribometric conditions.

Sample Prep.	Apparatus	Atm.	Coeff. of Friction	Reference
(?) C_{60}	SFM	RT air	0.8	[15]
sublimated C_{60}	FFM	RT air	0.82 ± 0.33	[16]
sublimated C_{60}	FFM	RT argon	0.67 ± 0.22	[16]
sublimated C_{60}	SFA	RT air	0.14 – 0.17	[17]
sublimated C_{60}	SFM	UHV	0.15	[18] [19]
sublimated C_{60}	FFM	UHV	1.0 - 2.0	[20]
SAM-grafted C_{60}	FFM	RT air	0.04-0.15	[21]
SAM-grafted C_{60}	AFM	(UHV?)	6–15× of graphite (at zero load)	[22]

Table II. Friction of C_{60} and C_{70} under unidirectional and oscillatory micro/macro-tribometric conditions.

Sample Prep.	Apparatus	Atm.	Coeff. of Friction	Reference
sublimated C_{60}	PIN/DISC (UNI.)	RT air	0.6	[23]
sublimated C_{60}	pin/disc (osc.)	RT air	0.12 – 0.5	[24] [25]
sublimated C_{60}	pin/disc (uni.)	RT air	0.4 – 0.5	[26]
benzene sol.	pin/disc (uni.)	RT air	0.6	[27]
sublimated C_{60}	pin/disc (uni.)	RT air	0.6 – 0.8	[28]
sublimated C_{70}	pin/disc (uni.)	RT air	0.5 – 0.9	[28] [29]
SAM-grafted C_{60}	pin/disc (uni.)	RT air	0.12 - 0.18	[30]
sublimated C_{60}	pin/disc (uni.)	UHV	0.3 – 0.6	[31]

As treated by a large number of publications, the mechanism of hydrogen abstraction with detrapping, recombination and retrapping from carbon-graphites has been a subject of great interest to a variety of researchers. Those in the nuclear industry are one example: trapping and releasing hydrogen from graphite or purposely deposited a-C:H (diamondlike carbon, a.k.a. DLC) film-type first-wall materials of fusion reactors is of great concern. The release of hydrogen from carbon-graphite occurs by local molecular recombination of the detrapped atoms from π-bonded and σ-bonded sites, followed by fast transport of the H_2 molecules to the surface. If a carbon-graphite material is made to slide at progressively higher temperatures and the hydrogen source is only from the bulk, first the weaker π-bonded atomic H and then the more strongly σ-bonded (C-H) hydrogen will detrap by heating under prog-ressively higher environmental temperatures. Effusing to the surface, they "lubricate" the surface bonds until desorbed [37].

All of the bulk-absorbed and surface-adsorbed hydrogen species can be eventually driven off, if the temperatures became high enough and/or the exposure sufficiently long. The effectiveness of hydrogen or helium to act as a gas-phase "lubricant" on the tribo- and heat-activated surface depends on the sp^2/sp^3 ratio of the carbon-graphite material. Low-ratio (more diamondlike) carbons more readily chemisorb these gases. Effective sorption also depends on what type of gas species (some can be in an excited state) will react with the dangling surface π and σ bonds. This concept of gas-phase lubrication has been previously discussed not only with carbon-graphite seal materials, but with DLC as well [39,40]. In addition to Si and diamond [7, 34-36, 40], similar passivation of surface bonds was applied to carborene polymers [41] and even to metal-based hard surface layers [42].

Experimental proof of this hypothetical NEMS lubrication method depends on the ingenuity of atomic force microscopists. Nevertheless, there is information in the literature already, hinting the applicability of this technique even on the nanoscale. Electron field emission from carbon nanotube caps was high, until the tip was heated to 900 K and the surface adsorbates were lost, resulting in an over 4-fold drop in the current. However, resorption occurred rapidly at room temperature and the adsorbate-enhanced current returned to its initial level in about one minute [43]. This phenomenon is highly analogous to the deep electron trap passivation by hydrogen at the Si/SiO_2 interface [32], as well as similar elimination of deep surface traps on diamond electron emitter surfaces by hydrogenation [44-46]. There is close connection between the fundamental electronic properties of matter and their tribological behavior [6], as manifested by the passivation of dangling surface bonds.

However, one must be extremely careful about the choice of adsorb-ates. Surface fluorination and an anticipated Teflon-like behavior would be extremely convenient, but the findings in [47] rule out the use of fluorinated C_{60}. Slow nucleophilic substitution of the fluorine by atmo-spheric moisture releases HF, which etches the surrounding environment. In vacuum, however, carbon nanostructures made from perfluorinated hydrocarbons [48] might be useful.

If a nanobot were designed for a medical purpose, it would have to demonstrate long-term survivability and efficiency in-vivo. In 1997, chemists at the Boston College synthesized the first NEMS Brownian motor in the shape of a molecular propeller (paddle wheel) driving a ratchet, which allows only unidirectional turning as the fluid molecules strike the paddles. The propeller has three benzene ring paddle blades, which also act as the gear teeth. A row of four benzene rings (fulfilling the role of the pawl) is positioned between two of the paddle blades, so the propeller cannot turn without pushing the pawl aside. Because of the twist in the pawl, it is easier to turn the wheel in the clockwise direction [49]. Unfortunately, the wheels spun equally in both directions, as the late Prof. Feynman's analysis predicted. Moreover, the motor output was essentially spent by external friction with the fluid medium.

Since there are even more complicated NEMS motors operating in biological entities (*e.g.*, for energy conversion [50] and as bacterial flagellum [51]), the hydrophilic-hydrophobic match of the rotor-stator interfaces and interaction with the solvating medium need careful molecular engineering to expend minimum energy toward life-sustaining locomotion. One example is the unavoidable hydrogen-bonding interaction between a benzene ring and a water molecule. It is almost 3 kcal/mol, because the water is a weak donor to the delocalized π electrons [6]. The same, relatively low (but not insignificant) bond strength limits the thermal-atmospheric stability of intercalated water within the graphene layers.

The electron distribution, ergo the atomic-level surface energy over a high molecular weight nanocomponent is usually uneven. Ergo, shape factor differences between components must also be taken into account. There are, for example, affinity differences between flat graphene planes of highly ordered pyrolytic graphite and the curved surfaces of the cylindrical nanotubes. Trying to explain the low contact resistance of carbon nanotubes, Tersoff [52] argued that the differences in electrical conductivity of nanotube versus gold (non-conductive) as opposed to nanotube versus aluminum (conductive) contact may be explained by the overlap (or the lack thereof) between the projected Fermi sphere of the free-electron metal (a circle) and the Fermi-level points at the hexa-gonal corners of the Brillouin zone of graphite. With gold there is no overlap, with aluminum there is. However, Delaney and Di Ventra [53] correctly pointed out that Tersoff's argument applies only to contacting flat sheets of graphene and gold (or aluminum). The nanotube wave functions are elongated-rectangular (*not* hexagonal), and there *is* overlap with the gold wave functions even in the absence of other scattering processes.

Therefore, the low contact resistance of nanotubes must have some other explanation.

5 OTHER NEMS BEARING MATERIALS AND SHAPES

Our engineering imagination is essentially limited to the geometric shapes of conventional moving mechanical assemblies with which we are familiar, even if the nanoscale required more maverick thinking. If we so confine ourselves, let us at least explore (a) likely fullerene-like parts made from materials other than carbon, and (b) the possibility of their own suitable environmental lubrication.

Nature has been making protozoans with fullerene-like siliceous skeletons for millions of years [54]. Similar structures, such as zeolites, can now be molecular-engineered [55]. Since (a) silicon- and boron-based amorphous materials such as glasses [56] and hexagonal-BN [57] can also be synthesized into similar structures, and (b) these materials lend themselves to manipulation into hydrophobic or hydrophilic surface chemistry states by grafting, the possibilities for NEMS structures can be expanded. One especially good candidate appears to be WS_2 (a well-known lamellar dichalcogenide solid lubricant) synthesized in nanotube form [58,59]. Used as an AFM tip, it was shown extremely inert, picking up no contaminants from the surroundings [60]. So far, this new inorganic fullerene was tried only as a lubricating oil additive [59,61], because during tribotesting in air it could not be confined between specimen interfaces as a dry powder: it appeared to "flow out of the contact zone" [62]. This kind of inertness and low interparticle bonding is expected from WS_2, considering the bonding electronic structure of Group VI dichalcogenide (MoS_2 and WS_2) packets versus that of the Group V equivalents (e.g., $NbSe_2$). The d_z^2 orbitals in the Group VI materials are doubly occupied (completely filled), resulting in no long-range bonding between the lamellae. Consequently, the metal atoms are staggered in the crystal structure and the shear strength is low. In contrast, the Group V material d_z^2 orbitals are only singly occupied (half-filled), resulting in long-range bonding between the aligned metal atoms and higher shear strength [6].

6 CONCLUSIONS

A review of the fundamentals and some related literature show that the notion of employing self-lubricating carbon nanotubes and buckyballs as NEMS moving mechanical assembly components leaves a lot to be desired. The main problem lies in the fact that Drexlerian nano-machinery are often based on incomplete computer models missing the key basic tenets of nanotribology. Certain models fail to take into account the effects of enhancing or retarding the $\pi - \pi^*$ orbital inter-action of contacting flat or curved graphene planes by environmental influences. The warning signs of non-tribology-related, reliable data on high cohesive energy density and thus poor self-lubricating ability, such as the high heat of sublimation of buckyballs, have also been ignored. Disappointing nano- and micro/macro-tribometric data on carbon bucky-balls, gleaned from the literature, are used to rationalize the validity of the thesis brought forth by this paper.

Backed by other related experimental data, it is suggested that continuous and effective NEMS lubrication may lie in employing gas or liquid molecules, which can tribo-catalytically chemisorb on the NEMS component surfaces from the operating environment to form extremely thin, low surface energy layers.

The time has also come to move from the realm of modeling to nanotribological experiments. Clearly, laboratory proof of this hypothetical NEMS lubrication method depends on (a) the ingenuity of atomic force microscopists, who are first in line to make the initial mea-surements, and (b) the materials scientist's finesse to molecular-engineer a variety of useful structures, which can be self-assembled and tested. This writer hopes that he will still be here when all this happens.

ACKNOWLEDGEMENTS

The author is deeply indebted to Dr. William L. Warren of the DARPA/DSO for his unflagging support for clarifying this misunderstood subject and helping to lay the groundwork for the National Nano-technology Initiative.

REFERENCES

1. Malsch, I. (1999) Nanotechnology in Europe: Scientific Trends and Organizational Dynamics, *Nanotechnology*, **10**, pp. 1-7.
2. Sohlberg, K., Tuzun, R.E., Sumpter, B.G., and Nold, D.W. (1997) Application of Rigid-body Dynamics and Semiclassical Mechanics to Molecular Bearings, *Nano-technology*, **8**, pp. 103-111.
3. Han, J., Globus, A., Jaffe, R., and Deardorff, G. (1997) Molecular Dynamics Simulations of Carbon Nanotube-based Gears, *Nanotechnology*, **8**, pp. 95-102.
4. Tuzun, R.E., Sohlberg, K., Noid, D.W., and Sumpter, B.G. (1998) Docking Envelopes for the Assembly of Molecular Bearings, *Nanotechnology*, **9**, pp. 37-48.
5. Gardos, M.N. (1987) Synergistic Effects of Graphite on the Friction and Wear of MoS_2 Films in Air, *Tribol. Trans.*, **31**, pp. 214-227.
6. Gardos, M.N. (1997) The Problem-solving Role of Basic Science in Solid Lubrication, in *New Directions in Tribology*, plenary and invited papers from the First World

Tribology Congress, 8-12 Sept. 1997, London, England, Mech. Eng. Publ. Ltd, Inst. Mech. Eng., London, pp. 229-250.
7. Gardos, M.N. (1999) Tribological Fundamentals of Polycrystalline Diamond Films, *Surf. & Coat. Technol.*, **113**, pp. 183-200.
8. Roth, S, Curran, S., Dusberg, G. (1998) Density of States and Tunneling Spectroscopy on Molecular Nanostructures, *Thin Solid Films*, **331**, pp. 45-50.
9. Iijima, S. (1994) Carbon Nanotubes, *MRS Bull.*, **XIX(9)**, pp. 43-49.
10. Pan, C., Sampson, M.P., Chai, Y., Hague, R.H., and Margrave, J.L. (1991) Heats of Sublimation from a Polycrystalline Mixture of C_{60} and C_{70}, *J. Phys. Chem.*, **95**, pp. 2944-2946.
11. Porezag, D., and Frauenheim, T. (1999) Structural and Vibrational Properties of C_{60} Oligomers, *Carbon*, **37**, pp. 463-470.
12. Foerster, C.E., Lepienski, C.M., Serbena, F.C., and Zawislak, F.C. (1999) Ion Irradiation Hardening of C_{60} Thin Films, *Thin Solid Films*, **340**, pp. 201-204.
13. Tachibana, M., Sakuma, H., and Kojima, K. (1997) Photo-illumination Hardening of C_{60} Crystals, *J. Appl. Phys.*, **82**, pp. 4253-4255.
14. Manika, I., Maniks, J., and Kalnacs, J. (1998) Atmosphere-induced Change in Microhardness and Plasticity of C_{60} Single Crystals and Polycrystalline Films, *Carbon*, **36**, pp. 641-644.
15. Mate, C.M. (1993) Nanotribology Studies of Carbon Surfaces by Force Microscopy, *Wear*, **168**, pp. 17-20.
16. Schwartz, U.D., Zwörnen, O., Köster, P., and Wiesendanger, R. (1997) Quantitative Analysis of the Frictional Properties at Low Loads, I. Carbon Compounds, *Phys. Rev. B*, **56**, pp. 6987-6996.
17. Luengo, G., Campbell, S.E., Srdanov, V.I., Wudl, F. and Israelachvili, J.N. (1997) Direct Measurement of the Adhesion and Friction of Smooth C_{60} Surfaces, *Chem. Mater.*, **9**, pp. 1166-1171.
18. Lüthi, R., Meyer, R., Haefke, H. and Güntherodt, H.J. (1994) Nanosled Experi-ments: Determination of Dissipation and Cohesive Energies of C_{60} with UHV-SFM, *Helv. Phys. Acta*, **67**, pp. 755-756.
19. Meyer, E., Lüthi, R., Howald, L., Gutmannsbauer, W., Haefke, H. and Güntherodt, H.J. (1996) Friction Force Microscopy on Well Defined Surfaces, *Nanotechnology*, **7**, pp. 340-344 (1996).
20. Schwartz, U.D., Allers, W, Gensterblum, G. and Wiesendanger, R. (1995) Low-load Friction Behavior of Epitaxial C_{60} Monolayers under Hertzian Contact, *Phys. Rev. B.*, **52**, pp. 14,976-14,984.
21. Tsukruk, V., Everson, M.P., Lander, L.M. and Brittain, W.J. (1996) Nanotribolo-gical Properties of Composite Molecular Films: C_{60} Anchored to a Self-Assembled Monolayer, *Langmuir*, **12**, pp. 3905-3911.
22. Lee, S., Shou, Y.S., Lee, R. and Perry, S.S. (2000) Structured Characterization and Frictional Properties of C_{60}-terminated Self-assembled Monolayers on Au (111), *Thin Solid Films*, **358**, pp. 152-158.
23. Blau, P.J. and Haberlin, C.H. (1992) An Investigation of the Microfrictional Behavior of C_{60} Particle Layers on Aluminum, *Thin Solid Films*, **219**, pp. 129-134.
24. Bhushan, B., Gupta, B.K., Van Cleef, G.W., Capp, C. and Coe, J.V. (1993) Sublimed C_{60} Films for Tribology, *Appl. Phys. Lett.*, **62**, pp. 3253-3255.
25. Bhushan, B., Gupta, B.K., Van Cleef, G.W., Capp, C. and Coe, J.V. (1993) Fullerene C_{60} Films for Solid Lubrication, *Tribol. Trans.*, **36**, pp. 573-580.
26. Bhattacharya, R.S., Rai, A.K., Zabinski, J.S. and McDevitt, N.T. (1994) Ion Beam Modification of Fullerene Films and their Frictional Behavior, *J. Mater. Res.*, **9**, pp. 1615-1618.

27. Zhao, W., Tang, J., Puri, A., Falster, A.U. and Simmons, Jr., W.B. (1995) Frictional Behavior of C_{60} Microparticle-coated Steel, *Mat. Res. Soc. Symp. Proc.*, **383**, pp. 313-318.
28. Zhao, W., Tang, J., Puri, A., Sweany, R.L., Li, Y. and Chen, L. (1996) Tribological Properties of Fullerenes C_{60} and C_{70} Microparticles, *J. Mater. Res.*, **11**, pp. 2749-2756.
29. Zhao, W., Tang, J., Li, Y. and Chen, L. (1996) High Friction Coefficient of Fullerene C_{70} Film, *Wear*, **198**, pp. 165-168.
30. Lander, L.M., Brittain, W.J., De Palma, V.A. and Girolmo, S.R. (1994) Friction and Wear of Surface-immobilized C_{60} Monolayers, *Chem. Mater.*, **7**, pp. 1437-1439.
31. Millard-Pinard, N., Martin, J.M., Belin, M. and Bernier, P. (1994) In-situ Raman Study of C_{60} Films under Shear Conditions, *MRS Fall Meeting*, Boston, MA,
32. Gadiyak, G.V. (1999) Physical Model and Numerical Results of Disssociation Kinetics of Hydrogen-passivated Si/SiO_2 Interface Defects, *Thin Solid Films*, **350**, pp. 147-152.
33. Stesmans, A., Afanas'ev, V.V. and Revesz, A.G. (1999) Suppression of Thermal Interface Degradation in (111) Si/SiO_2 by Noble Gases, *Appl. Phys. Lett.*, **74**, pp 1466-1468.
34. Gardos, M.N. and Gabelich, S.A. (1999) Atmospheric Effects of Friction, Friction Noise and Wear with Silicon and Diamond. Part I. Test Methodology, *Tribol. Lett.*, **6**, pp. 79-86.
35. Ibid, Part II. SEM Tribometry of Silicon in Vacuum and Hydrogen, pp. 87-102.
36. Ibid, Part III. SEM Tribometry of Polycrystalline Diamond in Vacuum and Hydrogen, pp. 103-112.
37. Gardos, M.N. et al, (1997) Crystal-structure-controlled Triobological Behavior of Carbon-Graphite Seal Materials in Partial Pressures of Helium and Hydrogen, Parts I, II and III, *Tribol. Lett.*, **3**, pp. 175-204.
38. Paulmier, D. et al, (1994) Influence of Active Gases on Tribological Behavior of Hard Carbon Coatings Containing Hydrogen, *Dia. Films and Technol.*, **4**, pp. 167-179.
39. Huu, T. Lee et al, (1997) Friction and Wear Properties of Hard Carbon Coatings at High Sliding Speed, *Wear*, **204**, pp. 442-446.
40. Gardos, M.N. (1998) Re(de)construction-induced Friction Signatures of Polished Polycrystalline Diamond Films in Vacuum and Hydrogen, *Tribol. Lett.*, **4**, pp. 175-188.
41. Korshak, V.V. et al, (1979) Pyropolycarboranes: The Base for Self-lubricating Materials with Gas Lubricant, *J. Appl. Polymer Sci.*, **23**, pp. 1915-1921.
42. Kula, P. (1996) The 'Self lubrication' by Hydrogen during Dry Friction of Hardened Surface Layers, *Wear*, **201**, pp. 155-162.
43. Dean, K.A. and Chalamala, B.R. (1999) Field Emission Microscopy of Carbon Nanotube Caps, *J. Appl. Phys.*, **85**, pp. 3832-3836.
44. Gavrilenko, V.I. (1993) Adsorption of Hydrogen on the 9001) Surface of Diamond, *Phys. Rev. B.*, **47**, pp. 9556-9560.
45. Bozeman, S.P., Baumann, P.K., Ward, V.L., Powers, M.J., Cuomo, J.J., Nemanich, R.J. and Dreifus, D.L. (1996) Electron Emission Measurements from CVD Diamond Surfaces, *Dia. & Related Mater.*, **5**, pp. 802-806.
46. Humphreys, T.P., Thomas, R.E., Malta, D.P., Posthill, J.B., Martini, M.J., Rudder, R.A., Hudson, G.C., Markunas, R.J. and Pettenkofer, C. (1997) The Role of Atomic Hydrogen and its Influence on the Enhancement of Secondary Electron Emission from C9001) Surfaces, *Appl. Phys. Lett.*, **70**, pp. 1257-1259.
47. Taylor, R., Avent, A.G., Dennis, T.J., Hare, J.P., Kroto, H.W., Walton, D.R.M., Holloway, J.H., Hope, E.G. and Langley, G.J. (1992) No Lubricants from Fluori-nated C_{60}, *Nature*, **355**, pp. 27-28.
48. Kavan, L. and Hlavaty, J. (1999) Carbon Nanostructures from Perfluorinated Hydrocarbons, *Carbon*, **37**, pp. 1863-1865.

49. Musser, G. (1999) Science and the Citizen. Physics. "Taming Maxwell's Demon", *Scientific American*, **280**, pp. 24.
50. Stock, D., Leslie, A.G.W. and Walker, J.E. (1999) Molecular Architecture of the Rotary Motor in ATP Synthase, *Science*, **286,** pp. 1700-1702.
51. Cuzel, P., Surette, M. and Leibler, S. (2000) An Ultrasensitive Bacterial Motor Revealed by Monitoring Signaling Proteins in Single Cells, *Science*, **287**, pp. 1652-1655.
52. Tersoff, J. (1999) Contact Resistance of Carbon Nanotubes, *Appl. Phys. Lett.*, **74**, pp. 2122-2124.
53. Delaney, P. and. Di Ventra, M. (1999) comment on Contact Resistance of Carbon Nanotubes, *Appl. Phys. Lett.*, **75**, pp. 4028-4029.
54. Curl, R.F. and Smalley, R.E. (1991) Fullerenes, *Sci. Amer.*, **265**, pp. 54-64.
55. Haberland, R. (1998) Transport Processes in Porous Media: Diffusion in Zeolites, *Thin Solid Films*, **330**, pp. 34-46.
56. Suenaga, K., Zhang, Y. and Iijima, S. (2000) Coiled Structure of Eccentric Coaxial Nanocable Made of Amorphous Boron and Silicon Oxide, *Appl. Phys. Lett.*, **76**, pp. 1564-1566.
57. Alexandre, S.S., Mazzoni, M.S.C. and. Chacham, H. (1999) Stability, Geometry, and Electronic Structure of the Boron Nitride $B_{36}N_{36}$ Fullerene, *Appl. Phys. Lett.*, **75**, pp. 61-63.
58. Remskar, M., Skraba, Z., Sanjinés, R. and Lévy, F. (1999) Syntactic Coalescence of WS_2 Nanotubes, *Appl. Phys. Lett.*, **74**, pp. 3633-3635.
59. Rapoport, L., Feldman, Y., Homyonfer, M., Cohen, H., Sloan, J., Hutchinson, J.L. and Tenne, R. (1999) Inorganic Fullerene-like Material as Additives to Lubricants: Structure-Function Relationship, *Wear*, pp. 225-229.
60. Rothschild, A., Cohen, S.R. and Tenne, R. (1999) WS_2 Nanotubes as Tips in Scanning Probe Microscopy, *Appl. Phys. Lett.*, **75**, pp. 4025-4027.
61. Rapoport, L., Feldman, Y., Homyonfer, M., Cohen, H., Cohen, S. and Tenne, R. (1999) Stability of Hollow Nanoparticles of WS_2 under Friction and Wear, paper presented at the 26th Leeds-Lyon Symp., 14-1 On Tribology, 14-17 Sept. 1999, U. of Leeds, England (in press).
62. Personal Communications with Prof. L. Rapoport, Center for Technological Education, Holon, Israel, Sept. 1999.

Chapter 9

CARBON NANOTUBES: OBJECTS OF WELL-DEFINED GEOMETRY FOR NEW STUDIES IN NANOTRIBOLOGY

Min-Feng Yu [1], Mark J. Dyer [2] and Rodney S. Ruoff [1]
[1] *Department of Physics, Washington University in St. Louis, St. Louis, MO 63130*
[2] *Zyvex LLC, 1321 North Plano Road, Suite 200, Richardson, TX 75081*

Carbon nanotubes, including both multi-walled carbon nanotubes (MWCNTs) and single-walled carbon nanotubes (SWCNTs), are fascinating low dimensional systems for studies in electronics and mechanics. Their applications in nano-electronic or nano-mechanical systems have been suggested. For example, 'on-nanotube' devices such as diode, bucky shuttle, or multiple terminal logic circuits have been treated by theory for electronic systems[1-4], and the use of nanotubes as nano-pistons, nano-syringes, and rotors for mechanical systems have also been modeled[5-7].

Experiment has shown that carbon nanotubes can be metallic or semiconducting one-dimensional wires and their quantum characteristics[8-10] have been explored. In contrast, there has been less experimental study of their mechanical and nanotribological properties[11,12], which is of interest because of the low dimensionality and the expected perfect or near-perfect crystalline structure of nanotubes. In order to use nanotubes in nano-mechanical systems, achieving an understanding of their nanotribological properties is very important. For example, how will nanotubes interact with each other, or with other surfaces, when in motion?

How either commensurate, or highly incommensurate, crystalline surfaces in motion interact is currently a topic of great interest in nanotribology research[13-15]. Such surface-to-surface experiments are difficult. They call for preparation of two atomically flat crystalline surfaces and the achievement of perfect alignment between the surfaces when separated by small distances. Studies on atomic-scale friction have mostly been carried out with lateral force microscopy where a sharp tip scans across an atom-resolved flat surface and the lateral force (or friction force between the tip and the surface) is recorded. Stick-slip motion has been observed in this point-to-surface sliding experiment[16]. It is possible to exploit the inherent structure of MWCNTs to achieve experiments on the sliding of structurally perfect or near-perfect nested shells. MWCNTs consist of coaxial cylindrical shells of rolled graphene sheets. The spacing between neighboring shells is ~ 0.34 nm, similar to the interlayer spacing in graphite. These shells thus interact with each other via van der Waals forces.

We previously reported tensile-loading experiments on individual MWCNTs using a nano-stressing stage onto which AFM tips are mounted to apply load and as force sensors[17]. In the experiment, a nano-stressing stage operated inside a scanning electron microscope (SEM) is developed for achieving the three dimensional manipulation and the measurement of the tensile stress-strain properties of individual nanotubes. A schematic showing the structure of the stage and a photo image are presented in Fig. 1a and Fig. 1b, respectively. Detailed information about the nano-stressing stage and the experimental procedure can be found elsewhere[17,18].

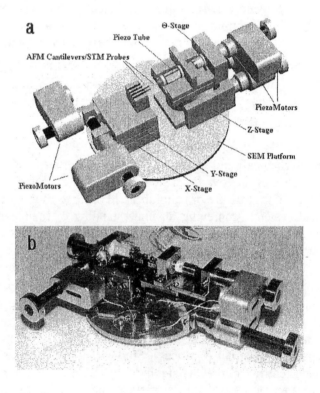

Figure 1. (a) A schematic showing the stage. (b) A photo image of the real stage that is about palm size.

The stage has four degrees of freedom, three translation and one rotation, which provide a coarse adjustment step size of 30nm and a less than 1nm step resolution for the fine adjustment by the incorporated piezo element. Atomic force microscope (AFM) probes are attached to the stages and the sharp probe tips are used as "fingers" to select, pick up, mount and stretch individual MWCNTs. The AFM probes are also used as force sensors to measure the applied force on MWCNTs by recording the deflection of the probe cantilever. In the tensile-loading measurement, the MWCNTs were found to fracture by a sword-in-sheath mechanism[17] as shown in Fig. 2, and thus the summed length of MWCNT fragments after fracture is seen to be significantly larger than the initial length of the MWCNT before fracture. The images thus indicate that the outermost shell of the MWCNT broke first (it carried the tensile load because of the attachment method[17]), and the internal shells then pulled out all the way from the end of the top fragment attached on the (top) AFM tip surface.

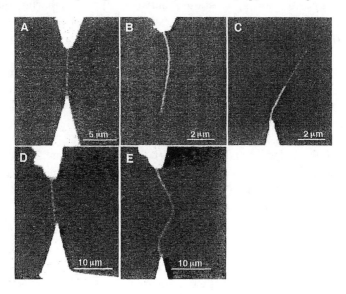

Figure 2. Fracture of two individually loaded MWCNTs. (A) a MWNT having a section length of ~6.9 μm under tensile load just prior to breaking. (B) After breaking, one fragment of the same MWCNT is attached to the upper AFM tip and has a length of ~6.6 μm. (C) The other fragment of the same MWCNT attached to the lower AFM tip has a length of ~5.9 μm. (D) Another MWCNT having a section length of ~11 μm under tensile load just prior to breaking. (E) The same MWCNT displaying an *S* shape immediately following fracture; notice that it now has a length of ~23 μm and the lower cantilever has deflected back to its initial (relaxed) position, indicating the partial pullout of the two nanotube fragments after breaking.

Figure 3. A schematic showing the realization of the sliding between nested shells in a MWCNT.

Thus one method of creating an experimental configuration for measuring the forces involved in the relative motion of nested cylinders in a MWCNT is to first tensile load the outermost shell until it breaks, and then controllably pull out the internal shells while continuously monitoring the force needed (Fig. 3). The full experiment can be recorded on video, and time-dependent force and distance data can then be collected for detailed analysis.

It is conceivable that different configurations for this kind of nanoscale tribology measurement can be realized with nanotubes by this nano-stressing stage, for example, sliding between two neighboring nanotubes, sliding between crossed nanotubes, and sliding between nanotube and surfaces of different materials. With the increasing availability of different kind of nanotubes, it is also possible to study the differences in tribology for different types. Since the final applications of nanotubes in nanotechnology will more or less depend on the understanding of the tribology properties of materials at the nanoscale, we project increasing activity in developing new tools for the nanoscale studies and in generating broad data sets for nanoscale materials.

We acknowledge partial support for this work by the Office of Naval Research and Defense Advanced Research Programs Agency under Navy grant N000014-99-1-0769, the NSF under the "New Tools and Methods for Nanotechnology grant NSF-DMR 9871874," and by Zyvex.

REFERENCES

1. Srivastava, D. & Menon, M. Carbon nanotubes: molecular electronic devices and interconnects. *Mater. Res. Soc. Symp. Proc.* **514**, 471 (1998).
2. Kwon, Y.-K., Tomanek, D. & Iijima, S. "Bucky Shuttle" Memory Device: Synthetic Approach and Molecular Dynamics Simulations. *Phys. Rev. Lett.* **82**, 1470-1473 (1999).
3. Menon, M. & Srivastava, D. Carbon nanotube based molecular electronic devices. *J. Mater. Res.* **13**, 2357-2362 (1998).
4. Treboux, G., Lapstun, P., Wu, Z. & Silverbrook, K. Interference-modulated conductance in a three-terminal nanotube system. *J. Phys. Chem. B* **103**, 8671-8674 (1999).
5. Tuzun, R. E., Noid, D. W., Sumpter, B. G. & Merkle, R. C. Dynamics of fluid flow inside carbon nanotubes. *Nanotechnology* **7**, 241-246 (1996).
6. Han, J., Globus, A., Jaffe, R. & Deardorff, G. Molecular dynamics simulations of carbon nanotube-based gears. *Nanotechnology* **8**, 103-111 (1997).

7. Tuzun, R. E., Sohlberg, K., Noid, D. W. & Sumpter, B. G. Docking envelopes for the assembly of molecular bearings. *Nanotechnology* **9**, 37-48 (1998).
8. Hamada, N., Sawada, S. & Oshiyama, A. New one-dimensional conductors: graphitic microtubules. *Phys. Rev. Lett.* **68**, 1579-81 (1992).
9. Tans, S. J. *et al.* Individual single-wall carbon nanotubes as quantum wires. *Nature* **386**, 474-477 (1997).
10. Tans, S. J., Verschueren, A. R. M. & Dekker, C. Room-temperature transistor based on a single carbon nanotube. *Nature* **393**, 49-51 (1998).
11. Falvo, M. R. *et al.* Nanometer-scale rolling and sliding of carbon nanotubes. *Nature* **397**, 236-238 (1999).
12. Buldum, A. & Lu, J. P. Atomic scale sliding and rolling of carbon nanotubes. *Phys. Rev. Lett.* **83**, 5050-5053 (1999).
13. Hirano, M., Shinjo, K., Kaneko, R. & Murata, Y. Anisotropy of friction forces in muscovite mica. *Phys. Rev. Lett.* **67**, 2642-2645 (1991).
14. Hirano, M., Shinjo, K., Kaneko, R. & Murata, Y. Observation of superlubricity by scanning tunneling microscopy. *Phys. Rev. Lett.* **78**, 1448-1451 (1997).
15. Muser, M. H. & Robbins, M. O. Conditions for static friction between flat crystalline surfaces. *Phys. Rev.* **61**, 2335-2342 (2000).
16. Bhushan, B. *Handbook of micro/nano tribology* (CRC Press, Inc., Boca Raton, 1995).
17. Yu, M.-F. *et al.* The strength and breaking mechanism of multiwalled carbon nanotubes under tensile load. *Science* **287**, 637-640 (2000).
18. Yu, M.-F. *et al.* Three-dimensional manipulation of carbon nanotubes under a scanning electron microscope. *Nanotechnology* **10**, 244 (1999).

Chapter 10

A NOVEL FRICTIONAL FORCE MICROSCOPE WITH 3-DIMENSIONAL FORCE DETECTION

M. Dienwiebel, J. A. Heimberg*, T. Zijlstra†, E. van der Drift†
D. J. Spaanderman‡, E. de Kuyper and J. W. M. Frenken
Kamerlingh Onnes Laboratory, Leiden University, 2300 RA Leiden, The Netherlands
* Naval Research Laboratory, Washington, DC 20375, USA
† DIMES, Delft University of Technology, 2600 GB Delft, The Netherlands
‡ FOM-Institute for Atomic and Molecular Physics, 1098 SJ Amsterdam, The Netherlands

1 INTRODUCTION

It has long been recognized that the tip-on-flat geometry of an atomic force microscope (AFM) closely resembles a realistic nanocontact, which can serve as a model in the investigation of friction at the atomic scale.[1] AFM's have been adapted to measure lateral forces down to the nanometer and nanonewton regimes by measuring the torsional response of the force probe. In the past, frictional force microscopes (FFM's) have produced predominantly qualitative results although recently some groups have produced quantitative nanotribology results.[2,3] Even as the techniques in traditional FFM become more refined, the basic problem remains that this method uses a force probe designed to be most sensitive to normal forces. One of the ramifications of this is that most cantilevers snap into contact as the tip-to-sample distance approaches the near contact regime so that hard contact is always made when measuring lateral forces. This near contact regime is of great technological importance as fly heights in disk drives decrease and as the length scales in MEMS are reduced. Cantilevers that do not suffer from the "snap-to-contact" problem have high torsional spring constants so that very small lateral forces are not detectable. In addition, significant coupling exists between torsional and normal motions of the cantilevers.[4] Finally, in most FFM's only one component of the lateral force is measured.

In this paper, we introduce the principles of a novel, quantitative FFM that measures both lateral forces with better sensitivity and at a near surface regime currently inaccessible to FFM's. We describe the relevant technical aspects of the instrument, and demonstrate its performance with frictional force maps for a W-tip on a highly oriented pyrolytic graphite surface

(HOPG) at ambient conditions. The force sensor, which forms the heart of the instrument has been described elsewhere.[5] A full account of the design characteristics and the performance specifications will be given in a separate publication.[6]

2 INSTRUMENTATION

2.1 General

The instrument consists of two stages: the fiber head assembly containing the detection system and a specialized lateral force cantilever, and the sample stage that allows for manipulation of the sample with respect to the tip and cantilever. The so-called "tribolever" is formed by the monocrystalline monolithic silicon structure shown in Figure 1. Four, high aspect ratio legs extend out from a central detection pyramid. A fiber optic interferometer reflects off of each pyramid face to track the motion of the pyramid. Combining the measured displacements, the instrument can follow the three-dimensional motion of the pyramid and thus obtain the three components of the forces on the tip. The scanning tip, which can be etched metal wire, for example tungsten, is threaded through the central hole of the pyramid and extends ~50 μm out from the base of the pyramid to interact with the surface.

2.2 The Tribolever

The entire tribolever structure is made in a Si wafer by a combination of dry and wet etching and oxidation steps. The central pyramid of the tribolever is formed via a KOH wet etch of the (100)-oriented Si wafer, which exposes the {111} planes of silicon. A special passivation technique necessary to protect the corners of the convex structure from underetching gives rise to the cross structure within the pyramid (Figure 1), in which an etched metal tip was glued[4]. The faces are highly reflective, oriented at well-defined angles and ideal for fiber optic interferometry. It is important to note that interferometry does not suffer from the problems associated with reflected beam detection techniques used in AFM: diffraction and dependence of the measured displacement on both beam spot location and the exact placement of tip on cantilever. The shape and dimensions of the four legs have been chosen and tested using finite element analysis[7] so that both lateral spring constants were equivalent and significantly lower than the torsional spring constant of single board, AFM cantilevers[5]. A typical torsional, AFM spring constant is k_X(AFM) = 71.6 N/m,[8] whereas the calculated tribolever lateral spring constants were k_X(tribolever) =

k_y(tribolever) = 1.48 N/m. In the vertical direction, k_Z(AFM) = 0.2 N/m while, the calculated value for the tribolever was k_Z(tribolever) = 25.8 N/m. Additionally, the calculated coupling between the three orthogonal directions in the tribolever is as low as ~10^{-4}% and the coupling between torsional (out of scanning plane) and normal forces was essentially zero.

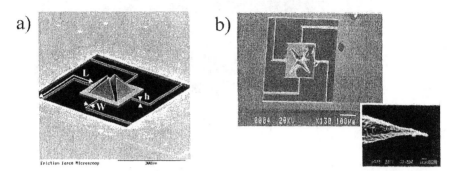

Figure 1. Two SEM micrographs of the tribolever. a) Front side of the sensor pointing away from the sample. The dimensions of the legs are L = 350 μm, w = 1.4 μm and h = 10.6 μm. b) shows the rear side of the tribolever with a W-tip extending approx. 50 μm out from the pyramid base (see inset).

2.3 Detection

The four optical fibers and the pyramid form two pairs of opposing interferometers, each pair driven by a single laser diode. Each fiber pair is sensitive to the displacement in the Z direction and to one component, either X or Y, of the lateral displacement. By adding or subtracting the two recorded displacements in one such pair, *e.g.*, the X-pair, we obtain the displacement components of the pyramid in the Z direction and the X direction respectively.

The distance of the end face of each fiber with respect to the pyramid face was adjusted with the use of an inertial piezo Nanomotor®.[9] The position of the fiber axis can be adjusted in a plane parallel to the pyramid face by the use of flexure hinges in the fiber head, in order to make each fiber aim at the center of its pyramid face (see figure 2). The silicon chip with the integrated force sensor also contains a kinematic mount (hole-groove-flat combination), which defines the sensor position with respect to three small ruby spheres, glued to the bottom plate of the fiber head. The reproducibility of the sensor position is such that adjustments of the flexure hinges to reposition the fibers are not necessary between tribolevers. A second kinematic mount is used between the detection part (fiber head with tribolever chip) and the sample stage.

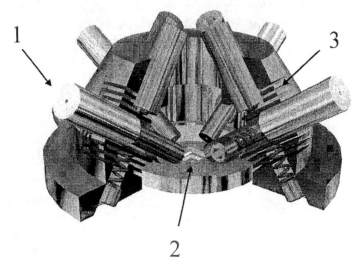

Figure 2. The fiber head assembly. An invar housing contains 1: the Nanomotors® (positioning of the four glass fibers), 2: The silicon chip with the tribolever, 3: flexure hinges for coarse adjustment of the Nanomotors with the glass fibers.

2.4 The Sample Stage

Because of the complexity and size of fiber positioning head/cantilever stage, it is kept stationary during operation. The sample sits on a traditional scan tube which rests inside a set of nested inertial motors that allow for four-dimensional motion of the sample with respect to the tip: X, Y, Z and rotational. The scanner is directly coupled to the Z coarse approach motor. The Z motor then sits in the center of an X-Y-Φ motor that allow for long-range manipulation of sample with respect to the tip. The Z and XY motors are similar to those discussed in the literature[10] however, the nested design is new, as is the Φ-motor which allows for rotational adjustment in the plane parallel to the scan. It is ultimately to be used to rotate tip and sample lattice planes with respect to each other to measure forces that are introduced when tip and sample lattices are slightly mismatched, or when the sample surface is asymmetric.[11]

3 PERFORMANCE

3.1 Calibration

For calibration of the spring constants in the X-, Y- and Z-direction of the tribolever, the sensor is placed on a special support that can be easily excited acoustically. By this means we can measure the resonance spectrum of the cantilever (and the support). We measure the resonance spectrum for a series of different small masses, placed on the backside of the pyramid. The spring constants can then be determined by measuring the frequency shifts of the X-, Y- and Z-resonances as a function of added mass.[12] A typical calibration result is shown in figure 3. The measured spring constants were k_x =1.689±0.033 N/m, k_y = 1.685±0.035 N/m. The measured spring constant in the Z-direction is k_z =10.318±0.123 N/m, this is about a factor 2 lower than calculated. The reason for the large deviation between the calculated and measured normal spring constant is due to a thin diaphragm that supports pyramid on the silicon chip. This diaphragm is the result of a wet etch step that forms a wide, recessed window to allow room for the detection fibers access to the pyramid. Due to minor differences in the leg dimensions from sensor to sensor, the spring constants can differ by as much as a factor two. Therefore each sensor is calibrated separately prior to experiments.

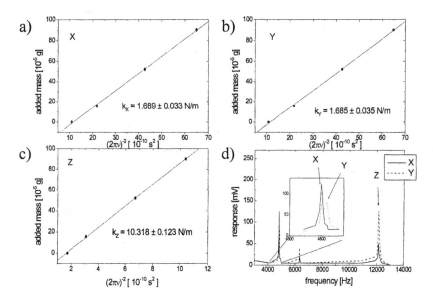

Figure 3. Inverse squared resonance frequencies as function of the added mass for the X- (a), Y- (b) and Z- (c) direction of the tribolever. The slope of the linear fit to the inverse squared frequencies gives the spring constant. From the inverse-squared frequency at zero added mass we obtain the effective mass. Graph (d) shows a typical resonance spectrum of the tribolever/support system without added masses. The unlabeled peaks do not depend on the added mass and are associated with resonances in the support.

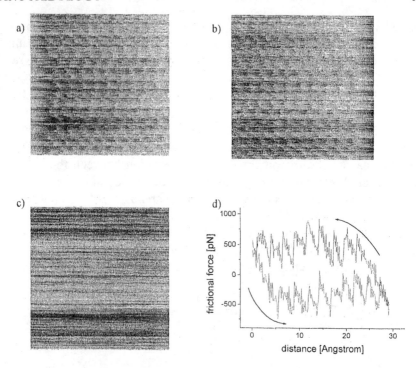

Figure 4. Test measurements of HOPG. Panel (a) shows a lateral force map in "forward" scan direction, panel (b) the opposite scan direction. Panel (c) shows the topography images measured simultaneously. Graph (d) shows a line scan through images (a) and (b). Parameters of the images (a)-(d): Scan size 2.9×2.9 nm^2, Normal force approx. −35 nN, The gray scale in panel (a) and (b) corresponds to a force range of 1.08 nN

3.2 Test Results

Figure 4 shows results of a measurement of a tungsten tip sliding on a HOPG sample at a constant normal force of about −35 nN (attractive force).[13] Figure 4 (a) shows a map of the X-component of the lateral force parallel to the fast scan direction, when the tip moves forward (from left to right). It shows variations in the frictional force with the unit cell periodicity of HOPG. Figure 4(b) shows the friction when the tip sliding in the opposite direction. A single forward and backward line is plotted in Fig. 4(d), showing a frictional force loop with sawtooth-like force variations, typical for atomic-scale stick/slip motion of the scanning tip1.[14] The total amplitude in the friction force loop is in the order of 1 nN. The present noise level on the lateral force measurement is below 50 pN. Finally Fig. 4(c) shows a micrograph of the topography (feedback signal in the Z-direction), measured simultaneously with the forward lateral force image. We cannot see atomic-scale variations in the topography. Note, however, that the force resolution in the normal direction is about 240 pN. The absence of Z-modulation demonstrates decoupling of lateral and normal force components.

4 SUMMARY AND OUTLOOK

We have built a novel special-purpose FFM. First tests show a high sensitivity of 50 pN in both lateral directions together with high stiffness along the Z-direction. In the future we plan to investigate a number of important questions that are presently inaccessible to standard FFM's: At which tip-sample separation distance does friction start? What is the dependence of the friction force on the relative orientation of the contacting surfaces? What is the relation between friction force, loading force and contact area, especially in the regime between a single-atom contacts and large-area contacts and multi-asperity contacts?

ACKNOWLEDGMENTS

We acknowledge the important contributions of Dr. H. J. Hug, University of Basel to the construction of the sample stage and the interferometer This project is part of the research program of the "Stichting voor Fundamenteel Onderzoek der Materie (FOM)," which is financially supported by the "Nederlandse Organisatie voor Wetenschappelijk Onderzoek (NWO)."

REFERENCES

1. B. Bhushan, J. Israelachvili, and U. Landman, Nature **374**, 607 (1995).
2. R.W. Carpick, N. Agraït, D.F. Ogletree, M. Salmeron, J. Vac. Sci. Technol. B, **14(2)**, 1289 (1996)
3. U. D. Schwartz, O. Zwörner, P. Köster, and R. Wiesendanger, in *Micro-Nanotribology and Its Applications* (B. Bhushan ed., Kluwer Academic Publ., 1997), Vol. NATO ASI Series E, Vol. 330, p.233
4. D. F. Ogletree, R. W. Carpick, and M. Salmeron, Rev. Sci. Instrum. **67**, 3298 (1996).
5. T. Zijlstra, J. A. Heimberg, E. van der Drift, D. Glastra van Loon, M. Dienwiebel, L.E.M. de Groot, and J. W. M. Frenken, Sensors & Actuators A, in press.
6. J. A. Heimberg, M. Dienwiebel, , T. Zijstra, E. van der Drift, D. J. Spaanderman ,E. de Kuyper, and J. W. M. Frenken, in preparation.
7. ANSYS©, a finite element analysis program, distributed by Swanson Analysis Systems, Inc., Houston, TX, USA.
8. U. Rabe, K. Janser, and W. Arnold, Rev. Sci. Instrum. **67**, 3281 (1996).
9. Klocke Nanotechnik GmbH, Horbacher Str. 128, D-52072 Aachen, Germany
10. H. J. Hug, B. Stiefel, P. J. A. van Schendel, A. Moser, S. Martin, and H.-J. Güntherodt, Rev. Sci. Instrum. **70**, 3625, (1999)

11. M. Liley, D. Gourdon, D. Stamou, U. Meseth, T. M. Fischer, C. Lautz, H. Stahlberg, H. Vogel, N. A. Burnham, and C. Duschl, Science, **280**, 273 (1998).
12. J. P. Cleveland, S. Manne, D. Bocek, and P. K. Hansma, Rev. Sci. Instrum. **64**, 403, (1993).
13. The spring constant in Z-direction for the tribolever used during the test measurements could not been determined experimentally because the resonance frequency was too high to be excited acoustically. Instead we estimated the normal spring constant here by calculating them from the leg dimensions. The spring constants for the used tribolever are k_x=5.31 N/m, k_Y=5.5 N/m and k_Z≈25 N/m.
14. C. M. Mate, G. M. McClelland, R. Erlandsson, and S. Chiang, Phys. Rev. Lett. **59**, 1942, (1987)

Chapter 11

QUANTIFICATION OF FRICTION IN MICROSYSTEM CONTACTS

Michael T. Dugger
Sandia National Laboratories
Albuquerque, NM 87185-0889
mtdugge@sandia.gov

Abstract Contact between moving structures in micromachines occurs at interfaces that are rough on the nanometer scale. While surface height distributions differ from other engineering applications, these interactions are far from single asperity contacts. The pressure-velocity regime in which microsystems operate is also different than in other engineering applications. Experimental methods for nanometer displacement and nanonewton force measurement are critical to the fundamental understanding of energy dissipation mechanisms at the molecular scale. However, experimental investigation of surface interactions in microsystems requires tools beyond scanning probe microscopy and the surface forces apparatus. New tools are needed to define the technology design space, to investigate aging and failure mechanisms in real microsystems, and to validate models of friction and wear in microsystems derived from molecular-scale simulations. To this end, devices for quantifying friction in simplified microsystem contacts have been developed. Since the experimental apparatus are also microsystems, the surface morphology and chemistry duplicate exactly those found in more complicated systems having contacting surfaces. A surface micromachined device has been used to investigate the friction and wear of adsorbed silane monolayers and selective tungsten treatment in micromachine contacts. Silane films decrease friction and permit longer sliding contact than possible with untreated surfaces, but eventually degrade to permit debris generation in environments containing water vapor. Selective tungsten treatment results in reduced wear in environments containing water vapor. A methodology for developing a predictive capability for friction between microsystem surfaces will be presented, and areas requiring further developments in metrology, experimentation, and modeling will be discussed.

1 INTRODUCTION

Maturation of microsystem technology requires the ability to develop designs with predictable performance and reliability, both of which rely on a scientifically based understanding of friction and wear. Investigation of tribological phenomena in microsystems presents significant new challenges for experimentalists. Even so, the ability to fabricate complex machines with micrometer-sized features offers great potential for new electromechanical systems. With such micro-electromechanical systems (MEMS), it is possible to duplicate the functionality of conventional mechanical assemblies in packages having greatly reduced volume and mass, as well as devise new systems with the ability to sense, process information, and act on their surroundings without human intervention.

Surfaces left behind by the processing methods used to create microsystems are frequently very different from those created by other fabrication methods, both in terms of chemistry and morphology. For example, the polycrystalline silicon surfaces created during surface micromachining of silicon create planar and sidewall surfaces with peak-to-valley roughness of 0.3 and 30 nm, respectively [1]. Electroplating into lithographically defined molds creates metallic surfaces in LIGA with asperities that are 10-30 nm in both the lateral and height dimensions [2]. In both cases, the structural members are elementally pure or simple alloys, unlike the multiphase polycrystalline materials used in macromachines. The microstructure and asperity geometry on microsystem surfaces is therefore unlike that found on engineering materials used in conventional mechanical systems.

Since the surface area to volume ratio of a mechanical element varies as the reciprocal of its characteristic length, the surface area to volume ratio in microsystems is several orders of magnitude larger than that in macroscale mechanical systems. Combined with the fact that available motive forces via electrostatics, differential thermal expansion, fluid pressure, etc. are small, this leads to the result that *surface interactions which are typically uncontrolled and ignored in macrosystems play a major role in the behavior of microsystems.*

During investigation of both fundamental mechanisms of friction and wear, as well as relationships between processing, performance, and reliability, questions arise about the validity of the experimental or numerical technique in simulating the behavior of the actual system under study. This question will be explored here in the context of fundamental mechanisms of energy dissipation, and in the context of establishing physics-based relationships between processing, performance and reliability of real microsystems. This chapter describes a philosophy for investigation of tribological phenomena in microsystems. Central to the ability to conduct experiments and validate simulations of asperity-level surface interactions is the development of an experimental device that can duplicate MEMS surfaces and operate at relevant interfacial pressures and velocities. This paper will describe such an experimental device, which allows quantitative measurement of friction under realistic operating conditions and at shear rates accessible by atomistic simulation.

2 SIZE SCALE OF SURFACE INTERACTIONS

Contact between moving structures in micromachines occurs at interfaces that are rough on the nanometer scale. The magnitude of the roughness is a function of the processing steps that created the surface. For example, in surface micromachined devices the planar surface of polycrystalline silicon parallel to the wafer surface is defined by the nucleation and growth of silicon from the vapor phase during chemical vapor deposition, or by chemical mechanical polishing procedures used to planarize the surface. Forces are frequently transmitted from structure to structure in a MEMS device through interactions between sidewall surfaces, perpendicular to the wafer surface. In this case, the sidewall roughness can be governed by the etch used to remove silicon from unwanted areas, or from the etch used to pattern the sacrificial oxide into which silicon is later deposited.

A simple estimate of the size and density of real areas of contact between microsystem surfaces can be obtained using the measured roughness parameters from these surfaces. Based on a simple assumption of Hertzian contact of asperities having the same shape and Gaussian height distribution [3], and ignoring additional refinements due to adhesive interaction between surfaces [4], contact between surfaces having rms roughness of 10 nm and modulus of 165 GPa (typical of polycrystalline silicon) will yield an interface with a few points of contact. This contrasts with macroscale mechanical systems, which may have millions of points of contact between structures. Therefore, while surface height distributions are different than in macroscopic devices, these interactions are far from single asperity contacts as explored using scanning probe microscopy.

3 MICROSYSTEM P-V REGIME

In addition to the mechanics and size scale of interactions between microsystem materials, the contact pressure and time scale of surface interactions must also be considered. Available experimental techniques can provide valuable information about some fundamental aspects of energy dissipation and wear at microsystem contacts. For example, scanning probe microscopy has proven to be a useful tool for exploring molecular scale contacts at interfaces containing monolayers of lubricants [5-10]. While these studies have provided critical insight into chain length, terminal group chemistry, and orientation on energy dissipation processes in single asperity contacts, the interfacial sliding speeds are low (typically < 1 mm/sec) and the contact pressures are high, in the range of 1 to several GPa. Significant understanding of tribology in constrained fluid systems has also been achieved using the surface forces apparatus, developed by Israelachvili [11]. This technique permits measurement of the normal and lateral forces between atomically flat surfaces separated by thin layers of lubricants, while the separation of the surfaces is precisely measured using optical interferometry. Due to the large areas of contact, the pressure is lower in these measurements than in scanning probe methods, but the maximum interfacial velocities remain in the mm/sec range.

Numerical methods offer a means of examining energy dissipation mechanisms at much higher shear rates. Molecular dynamics simulations have been performed on systems relevant to silicon micromachines treated with alkylsilane monolayers [12]. While molecular dynamics simulation offers a means to examine very short relaxation time events, interfacial velocities below 1 m/sec are intractable for systems of interest in microsystem tribology with present computational capabilities, due to the number of atoms required and long run times involved. The relationship between the above experimental techniques, numerical simulation, and the operating conditions of surface micromachines in P-V space is shown in Figure 1.

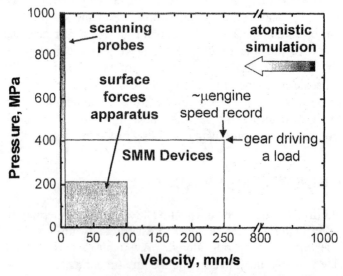

Figure 1. Experimental techniques, atomistic simulation, and operating regime of surface micromachines in pressure-velocity space.

In contrast with low shear rate experimental techniques involving single asperity contact, a quartz crystal microbalance (QCM) has been used to examine energy dissipation in molecules adsorbed on surfaces at very high shear rates [13]. Initial investigations examined energy dissipation by shear as the atoms moved on the surface in the absence of contact with another solid body. The QCM technique has recently been combined with a scanning tunneling microscope tip to enable investigation of interactions between the tip and the surface over a wide range of interfacial velocities and contact pressures [14]. This may offer the best means of performing experiments and simulations under similar conditions of contact, albeit with some constraints on materials and surface morphology. Since the QCM frequency exceeds the rate at which the tunneling tip feedback loop can respond, the force between the tip and surface varies with the local topography.

Investigation of surface interactions in microsystems under relevant contact conditions requires new experimental tools beyond scanning probe microscopy and the surface forces apparatus. To this end, techniques have been developed to quantify friction in simplified micromachine contacts. Prasad [15] developed a micromachined device for performing sliding experiments between micromachined structures, based on the clamping force required to permit motion of a beam. Lim [16] developed a micromachined device in which static and dynamic friction values could be extracted between planar surfaces of polycrystalline silicon by observing the displacement amplitude in an optical microscope. This technique is adequate for measuring friction during discrete sliding events, but is impractical for investigating the evolution in friction behavior with time during a long-term sliding experiment. Other important phenomena, such as reaction film formation on sliding surfaces, and wear debris formation, evolution and trapping, are best investigated with a dynamic test device.

4 SIDEWALL MICROTRIBOMETER

There has been considerable development of micromachines that sense or move out of the plane of deposition used during fabrication of the devices. Still, actuation in microsystems frequently occurs in a direction parallel to the plane of deposition and forces must be transmitted between sidewall surfaces on structures in this plane. This fact, and the desire to obtain dynamic friction data between actual device surfaces to investigate the evolution of friction and wear in these systems, has prompted the development of the surface micromachined (SMM) sidewall tribometer. The device is shown in Fig. 2. It is driven using two electrostatic comb drives. One is used to pull a suspended beam into contact with a fixed semi-cylindrical post, and the other to oscillate the beam against the post under a load. The contact geometry is equivalent to a cylinder against a flat surface, with the pressure distribution skewed slightly toward the top of the post by a 3-degree tilt of the semi-cylindrical surface away from the surface normal due to processing. Since the experimental apparatus is itself a

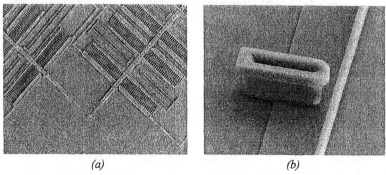

Figure 2. Overall view of the surface micromachined sidewall tribometer (a), and a detail at the beam and post where contact is made (b).

micromachine, the surface morphology and chemistry duplicate exactly those found in more complicated systems having contacting surfaces. In this case, sliding occurs at an isolated point of contact having an open geometry so that surfaces can be examined after testing for signs of wear. The device operates under pressure-velocity conditions representative of other SMM devices.

Dynamic friction data is obtained from the device by examining the motion of the device while the post and beam are in contact, and comparing this to the amplitude of motion of the device in the absence of contact. In order to quantify friction, the displacement amplitude of the device must first be determined both with and without contact. This is accomplished as shown schematically in Fig. 3. A waveform is applied to the oscillating comb to cause the beam to move side to side with respect to the post. For our initial experiments, a simple square wave was used to move the beam rapidly between two equilibrium positions, and leave it at rest at those positions until the next voltage change. While the beam is at each limit of motion, a strobe is triggered and an image of the beam and post is acquired

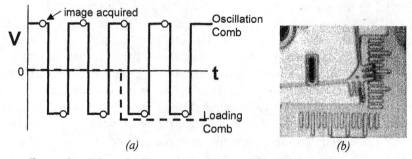

Figure 3. Schematic illustration of the voltage signals used to drive the sidewall tribometer (a), and an example of the image data acquired while the device is running (b).

with a CCD camera and frame grabber. Several samples of the beam position without contact are acquired to determine the amplitude of the beam motion without friction at the beam/post interface. Then a DC voltage is applied on the loading comb to bring the beam into contact with the post. The image sampling is continued at each phase after loading as the device is run for many cycles. At the conclusion of the test, a data directory containing hundreds to thousands of images is present. These images are processed by a macro, which determines the position of the "dot" at the end of the beam (Fig. 3b) with respect to the post for each image. Subtracting coordinates from images representing the limits of motion of the device enable the displacement amplitude to be determined as a function of number of oscillatory cycles of the beam.

The image processing macro creates a text file containing the displacement of the beam as a function of time. This data can be expressed in terms of forces through a mechanical analysis of the test structure. A detailed description of how forces are extracted from the displacement data is provided elsewhere [17-18]. In summary, an expression for friction coefficient can be derived which contains terms related to the spring constants of the comb drives and beams in the structure, voltages applied to the combs to actuate the device, lengths from the design, voltage versus displacement determined during calibration, and displacements measured by image processing described above. The main source of error in converting measured displacement into force is the determination of spring constants for the comb drives and beams in the structure. These can be determined by calculation based on the dimensions of spring elements, or by measurement of resonant frequency of the comb drives. In the former case, the stiffness $k = 3EI/L^3$, where E is the elastic modulus of the material, L is the length of the beam, and the moment of inertia $I = bh^3/12$. In this relationship, b is the thickness of the polysilicon film that makes up the beam element, and h is the width of the beam element parallel to the direction in which it is deflected. Since h is cubed in the expression for the moment of the beam, uncertainty in this measurement has a large impact on the calculated stiffness. However, in the expression for friction coefficient $\mu = F/N$, the expressions for friction force F and normal force N both contain these stiffness terms. If the beams are designed with the same dimensions, errors associated with uncertainty in stiffness will tend to cancel in the expression for friction coefficient. We estimate that the value of friction coefficient determined using the sidewall tribometer is within 10% of the actual value.

A small chamber was constructed in order to permit quantitative friction measurements in controlled environments at atmospheric pressure. The device houses a single 24-pin dual inline package (DIP), and uses high vacuum materials and fittings to provide a clean, leak-free environment. A wide variety of gas compositions can be flowed through the chamber, and sensors allow oxygen and water vapor to be measured with resolution of a few parts per million. The body and lid of the cell are constructed from 6061 aluminum. A 10-pin electrical feed-through on a 3.38 cm diameter metal-sealed vacuum flange provides signals to a 24-pin socket at the center of the cell. The socket is positioned so that a sidewall tribometer mounted inside the package will sit near the center of the cell, and close to the lid. The number of devices that may be run depends upon the number of signal lines required per device. For example, each friction test structure requires 5 signal lines to operate, so two devices may be run simultaneously with different drive signals within this de-vice. Input and output gas connections are made using either stainless steel or perfluoroalkoxy tubing. A sapphire window 1.78 cm in diameter in the lid allows the structures to be viewed with a microscope while they are running in the controlled environment. The free volume of the environmental cell is 250 cm^3. For controlled environment experiments, gas was provided to the input port of the cell from cylinders of ultra pure compressed gas at a supply pressure of 10 psig, which ran first through a drying/humidifying manifold. This manifold contained separate columns of desiccant and deionized water. Adjusting the relative flow rate of supply gas to the desiccant and deionized water columns permitted control of the water vapor content in the supply gas. The exhaust line of the cell was connected to a chilled mirror dew point monitor (General Eastern Hygro M4/D-2 sensor) and an electrochemical oxygen analyzer (Delta-F, model FA31111XA) in series. Under ambient conditions of 12.5 psia and 25°C, water vapor sensitivity was 222±5 ppmv (0.6±0.1 %RH). Under the same conditions, oxygen concentration sensitivity was 2±0.5 ppmv.

5 INVESTIGATION OF SURFACE TREATMENTS

Adhesion between polycrystalline silicon members or the substrate (referred to as "stiction") can lead to failure of MEMS. This adhesion can occur during processing, when the hydrophilic surfaces present after wet chemical release etching are dried in air and capillary forces bring the surfaces into contact. Adhesion may also occur in use, between surfaces left in contact for long periods of time or brought together by mechanical shock. Mastrangelo [19] discussed the mechanisms of adhesion-related failure, and suggested that low energy monolayer coatings, such as silane films, are the most effective and reliable adhesion prevention methods. Maboudian and Howe [20-21] reviewed several methods for adhesion reduction including chemisorbed molecular monolayer coatings, concluding that back-end processes such as packaging are likely to limit the applicability of adhesion reducing treatments, and more work is necessary to understand the impact of these surface treatments on friction and wear.

Owing to the relatively poor tribological properties of pure silicon, surface treatments are also of interest to improve their friction and wear behavior in dynamic contacts. Chemisorbed monolayers have allowed successful fabrication and release of compliant structures, but it is not evident that this treatment is adequate to assure reliable performance of sliding surfaces. Evolution in SMM technology has been toward increased stiffness in all directions, particularly out of plane, and toward actuators capable of delivering more force. This trend tends to minimize the importance of surface treatments solely as fabrication aids, and emphasize the need for durability of surface treatments at high contact pressure. In-plane actuators result in most forces being transmitted between sidewall surfaces where processing tends to result in greater roughness than on planar surfaces. We have used the sidewall tribometer in a preliminary investigation of the dynamic friction and wear behavior of silane monolayers and selective tungsten coatings for SMM devices.

After fabrication, sidewall tribometers were released using an oxide etch (H_2O:HF:HCl at 100:10:1, room temperature for 30 minutes). All chemicals were analytical reagent and low particle count electronic grades. Filtered high purity deionized water was used in solutions as well as for rinsing between the treatment steps.

For perfluorodecyltrichlorosilane ($CF_3(CF_2)_7(CH_2)_2SiCl_3$, abbreviated PFTS) deposition, devices were reoxidized (H_2O:H_2O_2 at 1:1, room temperature for 10 minutes) to provide surface -OH sites on the silicon for bonding of the silane molecules. Devices were kept wet and transferred to the coating solution by solvent exchange in the following sequence: H_2O to isopropyl alcohol to 1,1,2-trimethylpentane to the coating solution. Coating solutions consisted of 0.001 M silane in 1,1,2-trimethylpentane. Anhydrous solvents were used to prepare coating solutions within a nitrogen-purged glovebox. Solution volume was typically 200 ml in a glass beaker, and samples were transferred to the beaker in the same crystal carriers as used for release etching. After 30 minutes in the coating tank, devices were moved to a rinse tank that contained only the solvent. After all devices had been moved, the rinse tank was removed from the glovebox and the carrier transferred to an alcohol bath and then to pure water. Devices were removed from water with all structures free due to the hydrophobic nature of the resulting silane film (water contact angle 105-110°).

PFTS-coated parts were packaged in 24-pin DIPs and tested in the environmental cell described in section 4 above. Ambient atmospheric pressure was 12.5 psia, and ambient temperature was 25°C. Dry air was supplied to a manifold at 10 psig, and flow meters controlled the flow of this supply gas to desiccant or deionized water columns. Gas exiting from these columns was mixed to generate a controlled water vapor concentration. Experiments described here were performed in dry air (370 ppmv water vapor or less) and air with 40 % relative humidity (15,000 ppmv).

For selective tungsten deposition, the release etch was performed as described above for an entire wafer full of devices, the wafer was rinsed in water, and then allowed to dry in air. The wafer was transferred to a chemical vapor deposition (CVD) tool, where any surface oxide was removed in situ and tungsten was deposited using WF_6 plasma at high temperature. Deposition conditions are described in detail elsewhere [22]. The reaction between WF_6 and silicon consumes silicon and deposits a layer of tungsten about 20 nm thick. The reaction is self-limiting, and a conformal layer of tungsten is formed on all silicon surfaces to this thickness. Wafers were sectioned and electrical contact to sidewall tribometers was made using probes on a microelectronics probe station. Devices were run in laboratory air at 20% relative humidity.

Friction and wear tests were performed using a contact force of 20 µN. Devices were run at 100 Hz for approximately 10^6 total accumulated cycles. The sliding distance was typically 30-40 µm per cycle (twice the track length of 15-20 µm).

Representative behavior of the friction coefficient as a function of oscillatory cycles for the PFTS-coated structures is shown in Fig. 4. The device run in dry air shows a low and stable friction coefficient throughout

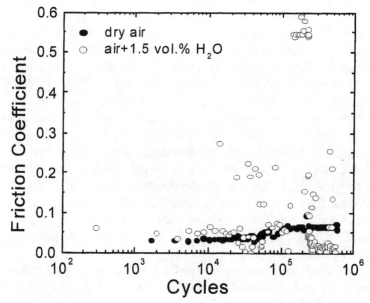

Figure 4. Friction coefficient versus cycles for PFTS-coated sidewall tribometers in dry and humid air.

the test, with a value of 0.06 at the end of the test. This value is in good agreement with Srinivasan et al. [23], where a friction coefficient of 0.08 was measured between PFTS-treated planar surfaces in air. Figure 4 shows that with water vapor present, the friction behavior of the devices was erratic, with values between zero and 0.3 observed during sliding. After about 10^5 cycles, devices run in humid air would stick when the friction force exceeded the actuation force provided by the comb drives (points near µ=0.55 in the data). The device could usually be made to continue operating by briefly removing and reapplying the normal load, but would begin to stick again shortly after reapplying the load. Accumulation of material could be observed by optical microscopy in the contact region of the beam in humid environments, at the ends of the contact area. No such accumulation was observed for tests run in dry conditions.

The contact regions of the beam and post for tests in dry and humid air are shown in Fig. 5. Scanning electron microscope micrographs of the worn surfaces from devices tested in dry air revealed very little damage or debris accumulation. A small amount of debris accumulation can be seen on the top of the beam, but very little damage has occurred to the beam surface. In contrast, the worn surface of a beam from a test in humid air shows a large amount of wear debris generation, so much so that thinning of the beam can be seen in the image. Other researchers examining environmental effects on micromachined device performance have observed a decrease in wear rate with increase in humidity [24]. The differences in wear behavior can be

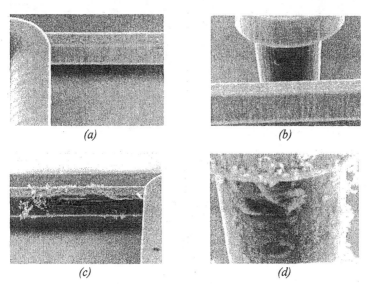

Figure 5. Scanning electron microscope images showing the wear surfaces on the beam (a) and post (b) of the sidewall tribometer run in dry air and humid air (c and d).

explained in terms of the known effect of water on the tribology of oxide ceramics. Adsorbed water can react with the surface of silicon ceramics to form a hydroxide layer. This forms a low friction surface film on the oxide and tends to decrease wear rate. Water may also promote stress-corrosion cracking in oxide ceramics. In this case, debris is formed and wear rates higher than in dry air are produced [25]. The active wear mode will therefore be determined by the pressure at contact points and the debris generation and trapping characteristics of the interface. It is not yet known how the PFTS film influences the above processes, and what the degradation mechanisms of the coating are.

Friction coefficient as a function of oscillatory cycles for the selective tungsten coated structures is shown in Fig. 6. In this case, The friction coefficient remained low and consistent for the entire duration of the experiment in humid air. Contacting surfaces of the beam and post show no evidence of wear (inset, Fig. 6). X-ray photoelectron spectroscopy measurements of the composition of the tungsten coated surfaces indicate that after 14 days exposure to air (the time after which these experiments were run), the surface contains about 33 atomic percent oxygen and 17 percent W (as WO_3), 28 percent carbon, and the balance nitrogen and fluorine. While the sidewall tribometer results suggest that selective tungsten is a more wear resistant surface than the PFTS-coated silicon,

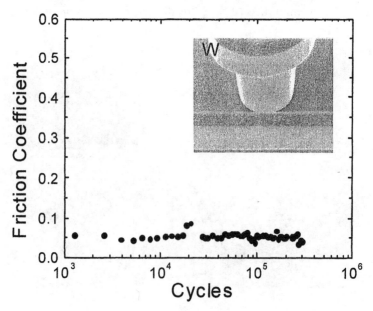

Figure 6. Friction behavior and scanning electron microscope image showing the wear surfaces on the post (inset) of a tungsten-coated sidewall tribometer.

tribological behavior is governed by the oxide and the large amount of adsorbed carbon. Additional work is under way to investigate the role of adsorbed hydrocarbons in the friction and wear of this surface treatment.

6 NEEDS IN MICROSYSTEM TRIBOLOGY

Quantitative microsystem tribology is needed to develop design data and a fundamental understanding of the performance and reliability of surface treatments for MEMS. In the area of metrology, on-chip methods of electrically sensing displacement (at nanometer levels) and force (at nanonewton levels) are needed to permit testing of many devices in parallel with real-time determination of friction forces. Surface composition diagnostics with high surface sensitivity and submicron spatial resolution are critical to understanding the tribochemical processes that occur during wear.

In the area of materials research, an understanding of environmental effects on friction and wear is required for microsystems. Effects of long term aging on surface chemistry and tribological behavior must be understood to insure the reliability of devices. A bridging of length and time scales is needed, to integrate the energy dissipation and wear phenomena at single points of contact into a description of an interface with multiple asperity contacts.

A combination of modeling tools and experimental techniques that permit simulation and experimentation at similar event time and spatial scales is required to develop fully validated models of energy dissipation and wear in monolayer lubricants

7 CONCLUSIONS

Microsystem contacts have molecular dimensions, but still represent contact of rough surfaces, such that contact occurs at a few locations. Scanning probe and surface forces apparatus experiments do not emulate the operating space of micromachines, and some constraints remain on the materials in contact with these techniques.

Micromachined structures are ideal for examining friction and wear in these systems, since they operate at relevant contact parameters and are constructed from materials used in device applications. A friction tester has been developed which permits quantification of friction in SMM sidewall contacts. This permits examination of environmental effects and coating performance at isolated contacts, can define microsystem design space, and provide data for comparison with numerical simulations.

ACKNOWLEDGEMENTS

The author is grateful for the experimental support provided by Tony Ohlhausen and Greg Poulter at Sandia. Discussions with Mark Stevens, Marc Polosky and Sam Miller at Sandia were enlightening and greatly appreciated. This work was supported by the United States Department of Energy under contract DE-AC04-94AL85000. Sandia is a multiprogram laboratory operated by Sandia Corporation, a Lockheed Martin Company, for the United States Department of Energy.

REFERENCES

1. Srinivasan, U., Foster, J.D., Habib, U., Howe, R.T., Maboudian, R. Senft, D.C., and Dugger, M.T., Transducer Research Foundation, Inc., June 7-12 1998, Hilton Head, SC.
2. Prasad, S.V., Hall, A.C., and Dugger, M.T., *Report SAND2000-1702*, Albuquerque, NM, Sandia National Laboratories, 2000.
3. Greenwood, J.A. and Williamson, J.B., *Proc. R. Soc. London, Ser. A* 1966, **295**, pp. 300-319.
4. Johnson, K.L., Kendall, K. and Roberts, A.D., *Proc. R. Soc. London, Ser. A*, 1971, **324**, pp. 301-313.
5. Salmeron, M., Xu, L., Hu, J. and Dai, Q., *MRS Bulletin*, 1997, **22**, pp. 36-41.
6. Terada, Y., Harada, M., Ikehara, T. and Nishi, T., *J. Appl. Phys.* 2000, **87**, pp. 2803-2807.
7. Perry, S.S., Mate, C.M., White, R.L. and Somorjai, G.A., *IEEE Trans. Magnetics*, 1996, **32**, pp. 115-121.
8. Wei, Z.Q., Wang, C.F., Zhu, C.Q., Xu, B. and Bai, C.L., *Surface Science*, 2000, **459**, pp. 401-412.
9. Kim, Y., Kim, K.-S., Park, M. and Jeong, J., *Thin Solid Films*, 1999, **341**, pp. 91-93.
10. Meyer, E., Overney, R., Brodbeck, D., Howald, L., Luthi, R., Frommer, J. and Guntherodt, H.-J., *Phys. Rev. Letters*, 1992, **69**, pp. 1777-1780.
11. Israelachvili, J.N and McGuiggan, P.M., *J. Mater. Res.*, 1990, **5**, pp. 2223.
12. Tutein, A.B., Stuart, S.J. and Harrison, J.A., *Langmuir* 2000, **16**, pp. 291-296 and references therein.
13. Krim, J. and Widom, A., Phys. Rev. 1983, 38B, pp. 12184-12189.
14. Borovsky, B., Abdelmaksoud, M., and Krim, J., this issue.
15. Prasad, R., MacDonald, N. and Taylor, D., *Transducers '95*, Stockholm, July 1995.
16. Lim, M.G., Chang, J.C., Schultz, D.P., Howe, R.T and White, R.M., *IEEE MicroElectro Mechanical Systems Workshop*, Napa Valley, CA, USA, pp. 82-88, 1990.
17. Dugger, M.T., Senft, D.C. and Nelson, G.C., in *Microstructure and Tribology of Polymer Surfaces*, V.V. Tsukruk and K.J. Wahl, eds., American Chemical Society, Washington, DC, 1999, pp. 455-473.
18. Dugger M.T., Review of Scientific Instruments, in preparation.
19. Mastrangelo, C.H. *Tribology Letters* 1997, **3**, pp.223-238.
20. Maboudian, R.; Howe, R.T. *J. Vac. Sci. Technol.* 1997, **B15**, pp.1-20.
21. Maboudian, R.; Howe, R.T. *Tribology Letters* 1997, **3**, pp.215-221.
22. Mani, S.S., Fleming, J.G., Walraven, J.A., Sniegowski, J.J., DeBoer, M.P., Irwin, L.W., Tanner, D.M., LaVan, D.A., Dugger, M.T., Jakubczak, J. and Miller, W. M., *Proceedings of the International Reliability Physics Symposium*, San Jose, CA, April 10-13, 2000.

23. Srinivasan, U.; Houston, M.R.; Howe, R.T.; Maboudian, R. *Transducers '97*, Chicago, IL, 1997.
24. Tanner, D.M.; Miller, W.M.; Eaton, W.P.; Irwin, L.W.; Peterson, K.A.; Dugger, M.T.; Senft, D.C.; Smith, N.F.; Tangyunyong P.; Miller, S.L. *Proceedings of the International Reliability Physics Symposium*, Reno, NV, 1998.
25. Lancaster, J.K., Mashal, Y.A.-H. and Atkins, A., *J. Phys. D: Appl. Phys.* 1992, **25**, pp. A205.

Chapter 12

NANOENGINEERING AND TRIBOPHYSICS FOR MICROELECTROMECHANICAL SYSTEMS

K. Komvopoulos
Department of Mechanical Engineering, University of California
Berkeley, California 94720

Abstract The rapidly emerging field of microelectromechanical systems (MEMS) is expected to lead to new technologies with impacting effects on science and engineering. The interdisciplinary nature of MEMS devices has generated a high demand for integrating basic knowledge of mechanical, electrical, chemical, and thermal phenomena encountered at the microscale and the identification of novel methods for the development of versatile micromachines. As the growth of MEMS continues to increase at a high rate, micromachine reliability and long-term durability are expected to assume even greater importance. Due to the low stiffness of most micromachine devices, the development of high attractive forces at MEMS interfaces often leads to permanent surface adhesion, a phenomenon known as stiction. The efficacy of various surface modification methods to reduce the magnitude of adhesion forces at MEMS interfaces is addressed, and simulation results for the micromachine stiffness required to overcome stiction are presented for different material systems with varying surface roughness. Surface texturing, fabrication of microbumps, surface passivation with hydrophobic thin solid films, and alteration of the surface chemical state by the adsorption of low surface energy substances are shown to be effective surface modification methods for controlling adhesion at MEMS interfaces. In addition, an appraisal of important research issues and nanotechnology developments in MEMS is presented in the context of fundamental understanding of nanoscale material behavior and instrumentation for testing at micromachine levels.

1 INTRODUCTION

The field of miniaturized devices, known as microelectromechanical systems (MEMS), sprung out from the integrated circuit (IC) industry about two decades ago. The rapid growth of this field, illustrated by the trend in MEMS device revenue shown in Fig. 1, has generated significant interest in the implementation of microdevices in various commercial applications. During the early stage of this novel technology the main emphasis was on the manufacturing of microsensors that can accurately sense the operating environment of macrodevices.

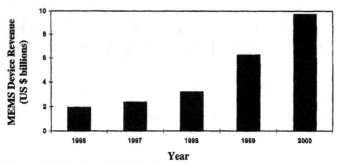

Figure 1. Evolution of MEMS device revenues.

As the field evolved, increasing demands for both real-time sensing and actuation led in the late '80s to the development of microactuators. Using IC-compatible processes, a wide range of planar polysilicon micromechanisms, such as gears, joints, cranks, and springs, were fabricated and assembled without the need of handling such small elements individually [1,2]. Micromotors with rotors of diameters in the range of 60-120 μm were electrostatically driven to continuous rotation [3]. However, the produced rotational speeds were only a small fraction of what should be achievable if only natural frequency were to limit the response of the rotor, a result attributed to frictional and viscous drag effects. Using a laser-based measurement system to monitor the motion of high-speed rotating micromechanisms, lifetimes of a few minutes were observed at steady-state speeds [4], presumably due to the rapid formation of wear debris at interfacial regions. Intermittent contact between polysilicon stators and silicon nitride rotors of micromotors has been reported to yield a steady-state motion characterized by distinct starts and pauses, believed to be due to airborne dust particles, electrostatically attracted from the air, or wear debris generated during sliding at the rotor-hub contact interface [5]. A significant effect of bearing wear on the gear ratio has been reported for polysilicon wobble micromotors [6]. High adhesion forces at micromachine interfaces may overcome the available restoring forces rendering MEMS devices inoperable, a phenomenon referred to as stiction [7].

As micromachine devices have begun to transition from the laboratory environment to commercial products, several challenging issues have to be resolved before commercialization [8]. In particular, understanding of surface phenomena is of paramount importance to the design of reliable and robust MEMS devices. Representative examples where surface interactions are of particular significance to microdevice performance are the digital micromirror device (DMD), scratch drive actuator (SDA), and wobble micromotor.

The operation of the DMD is based on a single-degree-of-freedom aluminum micromirror on a set of torsion bars (representing a single pixel) that can be switched between two positions corresponding to rotations of ±10° in order to reflect the light from an illuminator to a screen [9,10]. A schematic of the DMD is shown in Fig. 2. Each hermetically sealed chip has 0.5–1.2 million moving parts that are required to successfully undergo approximately 450 million contact cycles over its lifetime [10]. Aluminum electrodes actuate the DMD. In each half-cycle, a corner of the mirror makes contact with an electrode over a 20×20 nm area. In early designs the mirror tip was allowed to strike the electrode and, as a consequence, the two surfaces often remained stuck, hence seizing the operation of the particular pixel. Current designs limit the contact to spring tips (referred to as landing tips in Fig. 2) on the yoke that can dissipate some of the impact energy ($\sim 9 \times 10^{15}$ Joules over the contact area) and allow the tips to slide along the landing area as they bend. This design avoids the application of a bias voltage to reset the micromirrors whenever they are stuck. To reduce the required restoration force, it is desirable to use smaller contact areas and surface treatments providing a low surface energy substance at micromachine interfaces.

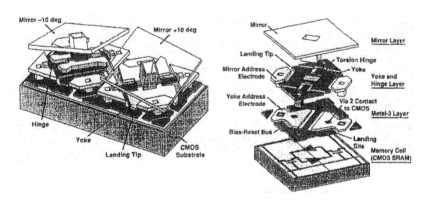

Figure 2. Schematic of a pair of actuated DMD mirrors (left) and an exploded view of a single pixel (right) [10].

The SDA is a stepper actuator that provides high forces (of the order of 100 μN) in small displacement increments of about 0.1 μm [11]. A schematic view of a SDA and the steps involved in a single operating cycle are shown in Fig. 3. Tribological interactions control both the operation and failure of this device. For efficient operation, the SDA must pivot on the edge of its bushing [step (iii) in Fig. 3]. Any slipping will limit the transfer of force and the occurrence of a lateral displacement. While no-slip contact between the bushing and insulator requires high friction, the plate pulled down by electrostatic forces possesses a large surface area and may remain stuck to the insulator after the voltage is removed. Knowledge of the friction behavior is therefore critical to ensuring no-slip pivoting of the bushing and full release of the plate after each cycle.

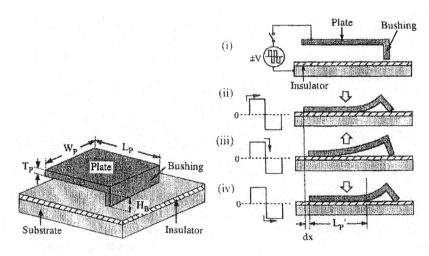

Figure 3. Schematic view of a SDA (left) and the steps involved in a single operating cycle (right) [11].

Although near-term commercial applications for micromotors are not apparent, significant effort has been devoted to the design and testing of different motor designs, with a prime goal to generate sufficient torque to overcome friction forces. Of particular interest is the wobble motor due to the different tribological considerations affecting its performance. For example, a high friction coefficient at the rotor-stator interface is required to avoid slip and high friction-induced torque in the opposite direction of motion. Another concern is the effect of wear on the ratio of the angular speeds of the stator poles and the rotor (gear ratio). The evolution of wear at the bearing and rotor surfaces has been found to greatly affect the gear ratio of wobble micromotors [12]. Figure 4(a) shows a schematic of the motor demonstrating its operating principle, Fig. 4(b) shows a cross section of the motor after the release process, and Fig. 4(c) shows the rotor geometry in various configurations. It has been argued that this axially-driven motor design exhibits two main advantages over the traditional side-driven wobble motors: (i) the generated torque is of the order of 400–600 pN·m, depending on the number of poles, compared to tens of pN·m for side-driven motors, and (ii) exhibits lower friction forces due to the occurrence of rolling instead of sliding motion [12]. It was also reported that the greatest variation in the gear ratio was encountered during the first 0.5 million cycles. The gear ratio initially increased, then decreased and finally stabilized at the steady state. The increase of the gear ratio in the early stage is most likely due to bearing wear and the subsequent decrease is due to rotor wear. When the surfaces were smoothed out ("running in" stage), the gear ratio reached a steady-state value and the motors operated for additional tens of thousand to a few million of cycles until the commencement of failure, most likely due to stiction caused by the increased real contact areas and high surface energy of the polished surfaces. It has been observed that intermittent stiction followed by terminal stiction is the principal failure mechanism in micromotors [13]. Thus, in order to prolong the life of wobble motors, as well as other micromotors, it is necessary to control the work of adhesion at interfacial regions.

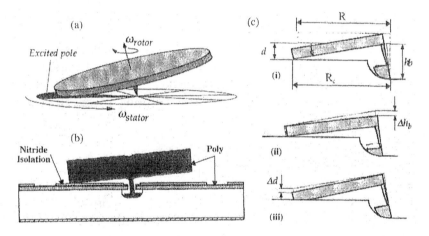

Figure 4. (a) Schematic of a wobble motor showing the mode of operation, (b) cross sectional view of the motor after the release etch step, and (c) wear effects on the geometry of the rotor element: (i) undeformed, (ii) worn bearing, and (iii) worn motor [12].

From the above studies it can be recognized that friction and wear phenomena may greatly affect the performance of miniaturized devices. In view of the high compliances and large surface areas of MEMS, the resulting frictional forces present a major challenge in the design of such devices because these forces can degrade both the performance and the life of the microdevice. Consequently, tribological issues must be addressed to ensure reliable long-term operation at different environments. Therefore, the main objective of this article is to provide a critical appraisal of the present knowledge of the different stiction and friction mechanisms encountered at MEMS interfaces and to discuss tribological solutions within the context of MEMS microfabrication constraints.

2 STICTION MECHANISMS

High adhesive forces between micromachine surfaces often occur immediately after the release-etch drying process and during operation. The relatively low kinetic energies of MEMS (typically in the range of 10^{-18}–10^{-6} J) may not be sufficient to overcome release-related and in-use stiction. Thus, the nature and magnitudes of attractive forces arising at micromachine interfaces is of particular importance to the design of anti-stiction microdevices.

The standard fabrication procedure of MEMS devices comprises the successive deposition and patterning of thin films. The last step in the fabrication process involves the removal of the sacrificial layer(s) surrounding the micromachine. Since this is normally accomplished by an isotropic aqueous chemistry etch, rinsing and drying are required in order to complete the process. It has been reported that a chemical reaction at silicon surfaces immersed in water results in the formation of precipitates, which can dissolve in water and redeposit at narrow gaps during the drying process to form a solid bridge [14]. Since the strength of the produced solid deposit is significantly greater than the available restoring force, the suspended microstructure remains stuck to the substrate. The formation of a chemical oxide layer to protect the silicon surfaces presents an effective alternative for overcoming adhesion due to this mechanism.

The large surface-to-volume ratio and high surface energies of the released microstructures often render them prone to in-use stiction due to attractive forces resulting from capillary, van der Waals, electrostatic, and hydrogen bridging effects [7,14-16]. The results of a simple analysis for perfectly smooth silicon surfaces, water liquid, and applied voltage of 1 V, shown in Fig. 5(a), demonstrate that the capillary, L_{cp}, electrostatic force, L_{el}, and van der Waals, L_{vdw}, attractive forces exceed the typical range of micromachine restoring forces at surface gaps less than 10 nm [7,14]. This is because capillary, electrostatic, and van der Waals forces are inversely proportional to the first, second, and third power, respectively, of the surface separation distance. Fabrication of dimples on the under cut surface and/or a slight upward curvature induced by the residual stress of a Si_3N_4 thin layer deposited on suspended microstructures are effective means for reducing the apparent contact area at micromachine interfaces [14]. Recent analytical studies accounting for the topographies of the approaching surfaces have shown a remarkable deviation from the results for the ideal case of atomically smooth surfaces at surface separation distances about twice the root-mean-square (rms) roughness of the composite surface [17]. Figure 5(b) shows analytical results for the critical cantilever beam stiffness versus the rms surface roughness for three different material systems comprising silicon, diamond-like carbon (DLC), and platinum. The simulation was performed for 1 µm² apparent contact area, 1 µm initial surface separation distance (i.e., free-standing distance), 20°C temperature, 50% relative humidity, and 1 V applied voltage. The significant increase of the various attractive forces for rms roughness values less than about 3 nm is reflected by the dramatic increase of the critical stiffness. It is noted that the most hydrophobic material system DLC/DLC yields the lowest critical stiffness. The results of the above analysis are supported by experimental measurements of the critical stiffness of polysilicon microcantilevers [18]. Increasing the rms roughness (texturing) by plasma etching in the range of 1–50 nm was found to reduce the critical stiffness from about 1.0 N/m down to 0.17 N/m. These studies illustrate the profound effect of the surface topography on the magnitude of the stiction force and reveal that very smooth micromachine surfaces yield a higher stiction probability. A detailed analysis of the various stiction mechanisms can be found in recent publications [7,19].

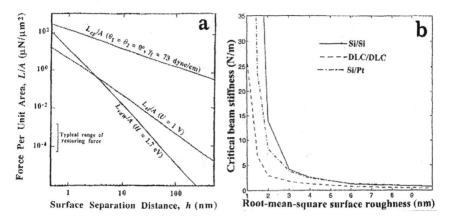

Figure 5. (a) Variation of interfacial attractive forces with surface separation distance for smooth surfaces [7,14], and (b) Critical stiffness versus surface roughness for different material systems [17].

The previous studies suggest that in order to avoid stiction it is essential to roughen the interfacial regions and reduce the work of adhesion at MEMS interfaces. Understanding the nature of micromachine stiction is often impeded by lack of information about the work of adhesion of MEMS interfaces. A method to quantify stiction at the microdevice level is to electrostatically pull microcantilever beams in contact with the underlying substrate by applying an appropriate voltage to a narrow actuation pad (placed near the anchor region in order to avoid the occurrence of an electrostatic attractive force over the contact region) and determine the work of adhesion based on the length of the shorter beam permanently attached to the substrate. The applied voltage may be insufficient to pull down the shorter and stiffer beams. Once the beams have been brought into contact with the underlying substrate, the actuation force is removed and some of the beams spring back, detaching themselves from the underlying surface. For beams shorter than a characteristic length, the available restoring force is sufficient to free them from the adhesion force keeping them attached to the ground layer. However, beams longer than this characteristic length will remain attached to the substrate surface. Thus, the value of this characteristic length, referred to as the detachment length, is defined as the beam length above which, all the beams adhere to the substrate. Based on this definition and energy balance considerations for the elastic energy stored in the deflected beam and the work of adhesion at the beam/substrate interface, the interface work of adhesion can be determined from the following relationship

$$\gamma_{ab} = \frac{3Eh^2t^3}{2s^4}, \qquad (1)$$

where E is the elastic modulus of polysilicon, h is the spacing between the free-standing beam and the substrate, and t is the beam thickness. The beam is adhered to the substrate ground plane over a distance $d = l - s$ from its tip, where l is the beam length. Eq. (1) is based on the assumption that the two contacting surfaces are perfectly smooth. The above calculation of the work of adhesion is based on the observation that the beam is attached to the substrate over a certain length d. Previous studies [20,21], based on the assumption that only the tip of the beam contacts the substrate, give for the work of adhesion,

$$\gamma_{ab} = \frac{3Eh^2t^3}{8l^4}. \qquad (2)$$

Comparison of Eqs. (1) and (2) shows that the work of adhesion is underestimated by using Eq. (2). The tip attachment is not likely to be observed with conventional static stiction test structures. SEM observations indicated that the attachment length d of the shortest adhering beam was usually around 0.1 times the beam length of the polysilicon microstructures. This implies that the remaining unattached length s of the shortest attached beam is close the length l of the longest detached beam. However, because of the additional factor 4 and the longer length l (instead of the shorter length s) in the denominator of Eq. (2), this relationship may underestimate the work of adhesion, by as much as an order of magnitude compared to Eq. (1). Results obtained using Eqs. (1) and (2) are discussed below.

Furthermore, polysilicon micromachine surfaces are rough and comprise asperities of various sizes. Thus, the topography of the microstructure surface affects the effective contact area and is responsible for the occurrence of capillary (meniscus), electrostatic, and van der Waals attractive forces and repulsive deformation forces between opposing asperities [7,19], as described previously. Therefore, the method described above yields estimates of the apparent work of adhesion. Because of the rough nature of the surface topography, a rough surface characterization was developed using fractal geometry to facilitate the more accurate analysis and prediction of surface stiction phenomena [17].

The SEM micrographs in Fig. 6 show an array of microcantilever beams, some of which are permanently attached to the substrate surface. The attachment lengths of polysilicon beams released with different treatments were determined based on SEM observations. Using Eq. (1), the work of adhesion was calculated for each of the corresponding lengths. In this calculation, the beam length attached to the substrate was measured to be approximately 10 µm, which is about 0.1 times the beam length of the conventionally dried microstructures, as evidenced from in Figs. 6(b) and 6(c). The attached length is assumed to be uniform for beams of similar lengths. This length is one of the key parameters used to determine the apparent work of adhesion.

Results for the apparent work of adhesion obtained from the attachment length measurements are compared in Tables 1 and 2. Table 1 shows the attachment lengths and corresponding apparent work of adhesion after drying the beam structures with the CO_2 supercritical-point drying method, and Table 2 shows similar results corresponding to conventional drying processes. The attachment lengths of the former samples are equal to approximately 260 µm, whereas the attachment lengths for the latter samples are about 70 µm. The calculated work of adhesion is about two orders of magnitude lower for the structures released by the supercritical drying process. This is indicative of the advantage afforded by using this release-drying method to reduce the interfacial adhesion energy. Earlier studies [21,23] suggested that the value of the work of adhesion for polysilicon beams under conventional drying is in the range of 100-300 mJ/m^2, which agrees well with the results shown in Table 2. As mentioned previously, the earlier results were obtained using Eq. (2). However, for shorter beams and/or hydrophilic surfaces the previous results underestimate the work of adhesion by a factor greater than 4. For hydrophobic surfaces, where the attachment length is longer and the discrepancy between l and d is larger, the work of adhesion approximated by Eq. (2) tends to underestimate even more (by as much as an order of magnitude) the work of adhesion compared to that obtained with Eq. (1).

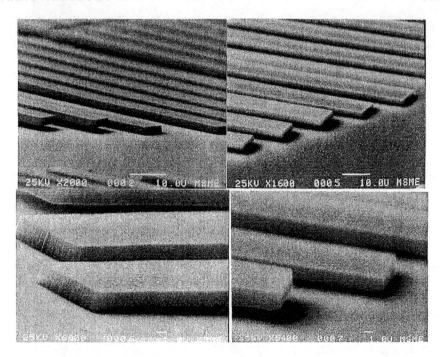

Figure 6. SEM micrographs of microcantilever beams: (a) and (b) beams free-standing above and adhering to the underlying substrate; (c) and (d) closer view of the shortest length attached beams.

The test microstructures were etch-released using three different kinds of wet etching. The maximum width of the beam structures was 15 µm. The etch time was decided considering the 100% overetch time and the etch rates of each solution. Tables 1 and 2 show that the release-etch effect on the attachment length is not discernible from the standard deviation values of the attachment length. Concentrated (49%) HF easily peels off a thin layer of polysilicon, reducing the thickness and the stiffness correspondingly, whereas this does not occur with 10:1 HF and 5:1 BHF etch solutions. Increasing the etch time in the concentrated HF process tends to decrease the required elastic restoring force. However, the change in the attachment length is within the experimental scatter. The standard deviation tends to increase with increasing etch time (Table 1). Although the selectivity between oxide and polysilicon is excellent, concentrated HF does attack polysilicon at a faster rate. The different roughness produced from each etch treatment is considered to be the prime reason for the differences in the attachment lengths shown in Tables 1 and 2, and, in turn, the work of adhesion. These results point out the importance of the surface topography on the determination of the work of adhesion and the magnitude of the stiction force, demonstrated in recent analytical studies [17,19].

Table 1. Attachment length and apparent work of adhesion of beam test microstructures exposed to different etch processes and released by the CO_2 critical-point drying method.

Etchant	Etch time (min)	Attachment length (μm)	Apparent work of adhesion (mJ/m^2)
Concentrated HF (49%)	3	273.8 ± 9.9	1.45
	7	263.8 ± 13.2	1.68
	11	241.3 ± 26.7	2.41
10:1 HF	70	262.5 ± 15.2	1.72
5:1 BHF	75	263.8 ± 12.2	1.68

Table 2. Attachment length and apparent work of adhesion of beam test microstructures exposed to different etch processes and released by the conventional evaporation method.

Etchant	Etch time (min)	Attachment length (μm)	Apparent work of adhesion (mJ/m^2)
Concentrated HF (49%)	3	78.8 ± 9.3	211.6
	7	75.0 ± 7.1	257.9
10:1 HF	70	67.5 ± 10.9	393.1
5:1 BHF	75	72.5 ± 4.3	295.4

To further quantify in-use stiction, an appropriate voltage can be applied in order to electrostatically pull the beams into contact with the underlying surface. The minimum voltage required to achieve this can be determined by balancing the beam restoring force, F_{res} [Eq. (3)] and the electrostatic force, F_{el} [Eq. (4)], which depends on the beam length and width,

$$F_{res} = \frac{Ehwt^3}{4l^3}, \qquad (3)$$

$$F_{el} = -\frac{1}{2}\alpha\varepsilon_0 V^2 \frac{wb}{h^2}. \qquad (4)$$

From Eqs. (3) and (4),

$$V = \sqrt{\frac{Eh^3 t^3}{2\alpha\varepsilon_0 bl^3}}, \qquad (5)$$

where w is the beam width, α is the fringing field coefficient, ε_0 is the dielectric constant (8.85×10^{-12} F/m), V is the applied voltage, and b is the width of the electrostatic actuation pad. For the adjacent beams whose spacing is comparable to their width, $\alpha \approx 2$. From Eq. (5), the minimum electrostatic voltages for pulling down the beams with lengths 100, 200, and 300 μm are calculated to be 32, 11.3, and 6.1 V, respectively. Repetitive actuation of the beams in the vertical direction provides a means of monitoring changes in the surface topography (*e.g.*, flattening of asperities due to repetitive contact) and the chemical state of the contacting surfaces (*e.g.*, the removal of a previously deposited low-surface energy monolayer) as a function of number of contact cycles, reflected through the changes in the magnitude of the attachment beam length, which is used to determine the apparent work of adhesion [Eq. (1)].

3 DYNAMIC FRICTION CHARACTERISTICS

The static friction coefficients of silicon-based materials have been found to depend on various environmental factors, such as the type of gas and pressure and the relative humidity [23]. Although no effect was found for samples exposed to argon, the significant variations of the static friction coefficient (between 0.2 and 0.75) encountered in nitrogen and oxygen atmospheres were associated with the lubrication effect of nitrogen and the increase of surface adhesion due to the adsorption of oxygen. Static friction coefficients in the range of 0.2–0.7 have been reported for SiN_x and SiO_2 surfaces slid against identical surfaces and very low friction (0.04–0.06) for surfaces coated with diamond-like carbon (DLC) [24]. Studies of the dynamic performance of polysilicon micromechanisms have revealed friction coefficient values for polysilicon on silicon between 0.25 and 0.35 [4]. Friction studies of IC-processed micromotors showed that the dynamic friction coefficient between polysilicon stators and Si_3N_4 rotors is in the range of 0.21–0.38 [5]. Much higher friction coefficients (0.7–0.9) were measured for polysilicon sliding against SiN in a pin-on-disk apparatus, presumably due to the different contact conditions, and low friction was observed in the presence of DLC and fluorocarbon films [25]. These controversies in the friction properties of silicon-based materials may be largely due to inappropriate testing conditions (scale effects).

Recent work on the dynamic friction characteristics of Si(100) surfaces performed with a friction force microscope (FFM) and a surface force microscope (SFM) revealed a remarkable scale effect on the friction properties [26]. To examine the effects of contact load and contact area on the friction behavior of silicon, silicon tips of nominal radius 150–200 nm and contact loads in the range of 5–700 nN were used in the FFM tests, and diamond tips of radius 100 nm and 16 μm and contact loads between 2 and 1000 μN were used in the SFM tests. These experiments simulated the interaction of an asperity with the silicon surface under different contact loads (or pressures) for the cases of similar (i.e., silicon on silicon) and dissimilar (i.e., diamond on silicon) material pairs. Representative results from these studies demonstrating the effects of normal load and contact pressure on the dynamic friction force of silicon are shown in Fig. 7. The slope of the line fit through the data is the dynamic friction coefficient μ. The results reveal a significant increase in friction with increasing contact pressure.

Three basically different friction regimes with significantly different coefficients of friction can be distinguished. The FFM measurements show a very low coefficient of friction ($\mu \approx 0.03$) throughout the entire load range [Fig. 7(a)]. A significantly higher coefficient of friction ($\mu \approx 0.25$) was produced in the SFM experiments with a 100-nm tip radius probe [Fig. 7(b)], whereas in the SFM experiments involving a relatively blunt indenter [Fig. 7(c)], the coefficient of friction was much lower ($\mu \approx 0.11$).

Despite the appreciably high contact pressures generated in the FFM experiments, a transition from elastic to plastic deformation did not occur. Surface imaging of silicon surfaces tested with the FFM did not reveal any discernible changes in the surface topography due to sliding, confirming the absence of plowing scratches. The very low coefficient of friction obtained throughout this load range (Fig. 7(a)) is attributed to the presence of a low surface energy, thin oxide film and the smoothness of the silicon substrate. However, Fig. 7(b) shows that a significantly higher friction coefficient was produced with the sharp SFM probe tip for loads ranging from 10 to 70 μN. Surface scanning with the same probe tip at very light loads (of the order of 0.5 μN) yielded conclusive evidence about the dominant role of plowing under these contact conditions. Figure 7(c) shows that an appreciably lower friction coefficient was obtained with the blunt SFM probe tip for loads in the range of 2–1000 μN, apparently due to the less severe plowing effect at lower contact pressures. However, the fact that the friction coefficient is not

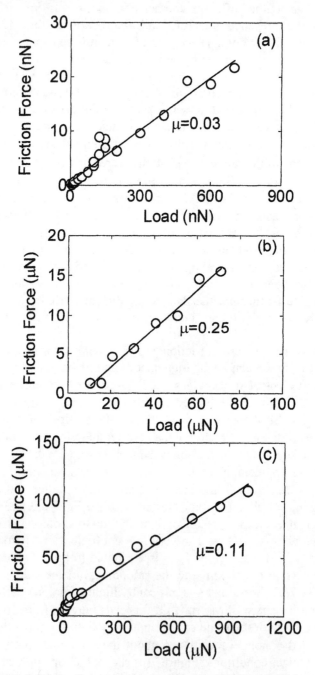

Figure 7. Friction force versus contact load for Si(100): (a) FFM tests (150 nm tip radius), (b) SFM tests (100 nm tip radius), and (c) SFM tests (16 μm tip radius) [26].

as low as in Fig. 7(a) suggests that the work dissipated due to plowing through the oxide film is greater than that due to surface adhesion.

AFM studies demonstrated that the lower friction values corresponded to mild plastic shearing of the native oxide layer, whereas the higher friction was due to penetration and rupture of the oxide film and plowing through the bulk silicon. These studies revealed that a transition from nanoscale wear process dominated by adhesion to a microscale wear process characterized by plowing occurred as the contact load increased between 2 nN and 1000 μN.

From the above studies it may be concluded that friction (and wear) testing must be performed at scales and contact conditions representative of MEMS devices.

4 MICROMACHINES FOR STICTION AND FRICTION TESTING

Micromachines for stiction and friction testing are essentially surface microstructures requiring six to eight masking levels for fabrication, depending on the optional steps needed to obtain different material systems. The general process flow for developing microstructures for stiction/friction testing is shown schematically in Fig. 8. A substrate typically consisting of n-type Si coated with ~300-nm-thick layers of thermal SiO_2 and low-pressure chemical vapor deposition (LPCVD) SiN (or low-stress Si_3N_4) is used to support the structures build on its surface [Fig. 8(a)]. The first processing step involves blanket deposition or growth of different materials on the substrate surface, such as a 400-nm-thick n- or p-type polysilicon layer [Fig. 8(b)]. A low-temperature-oxide (LTO) mask is used to deposit ~150 nm of SiN to form a pad [Fig. 8(c)]. Subsequently, a PSG sacrificial layer ~2 mm thick is deposited, annealed, and patterned to form dimple molds [Fig. 8(d)], in which a 150-nm-thick SiN layer may be formed using a LTO mask to develop another pad [Fig. 8(e)]. The next step involves etching locally the PSG layer to generate openings for mechanical anchoring and/or electrical contacts for the structural material [Fig. 8(f)]. This is followed by the deposition of a ~2.5-mm undoped polysilicon layer encased in a 500-nm-thick PSG layer [Fig. 8(g)]. Nitrogen annealing at ~650 °C for 1 h promotes diffusion of phosphorous from PSG into polysilicon leading to the formation of n+ polysilicon. Repetition of the previous steps with different materials results in a multi-layered polysilicon structure anchored to the substrate and encased in PSG. Release etching is usually the last step in machining a surface microstructure. In the release etch, aqueous or buffered HF etches the PSG leaving polysilicon and SiN virtually untouched. Removal of all the surrounding PSG and drying of the surfaces to remove the rinse solvent uncovers a free-standing polysilicon microstructure with the desired degrees of mechanical freedom [Fig. 8(h)].

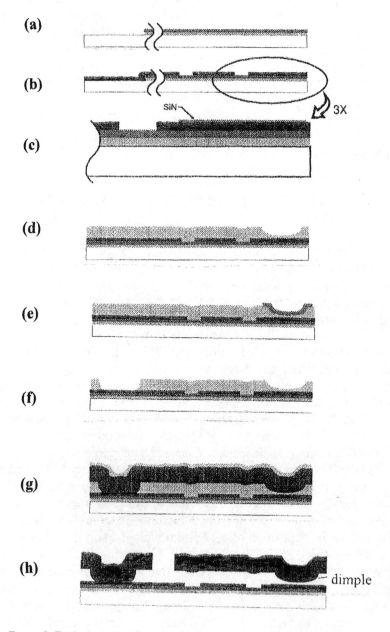

Figure 8. Typical process flow for fabricating micromachines for stiction and friction testing.

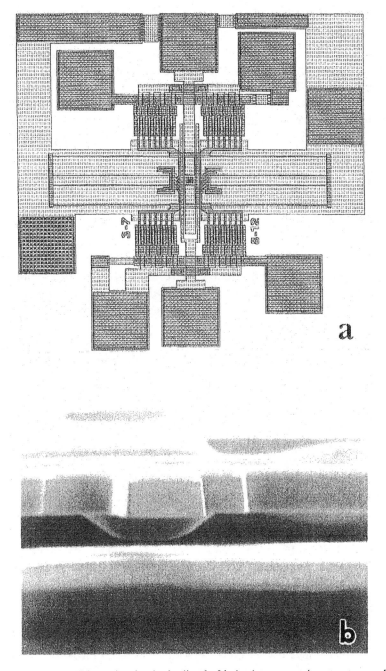

Figure 9. (a) Schematic of a single-dimple friction/wear test microstructure and (b) cross-section SEM view showing the dimple at the bottom of the suspended microstructure (shuttle) used to restrict contact over a small area.

Different material systems for friction and wear testing can be obtained using the above surface micromachining process scheme. Figure 9(a) shows a schematic of a generic polysilicon microstructure (shuttle) suspended over a substrate (typically of silicon nitride, polysilicon, or crystalline silicon) by a folded-beam truss suspension developed with the fabrication process shown in Fig. 8. The substrate is either electrically shorted to the shuttle to ensure the absence of any voltage across the friction couple or the polysilicon links of this circuit are broken to apply a voltage. The shuttle can be displaced in the lateral and/or vertical directions by applying a voltage between the shuttle and the electrodes beneath the shuttle. The lateral displacement of the shuttle is accomplished by applying a voltage between two electrostatic inter-digitated comb drives, one stationary and the other an integral part of the shuttle. Contact between the shuttle and the underlying substrate is restricted to a small spherical dimple at the bottom of the shuttle with a typical radius of 2–4 μm [Fig. 9(b)]. The length of the folded-cantilever truss suspension can be varied to produce dimpled microstructures with a wide range of restoring forces for stiction studies. Figure 10(a) shows a schematic of a typical test microstructure with a well-defined contact area, *i.e.*, an approximately 2-μm-radius hemispherical dimple at a distance from the underlying substrate of ~1.5–2 μm. When the dimple is brought into contact with the substrate, the microstructure may remain stuck to the substrate if the adhesive force, F_{ad}, at the contact interface is greater than the restoring force, F_{res}, generated by the flexed folded suspension [Fig. 10(b)]. Multiple microstructures can be fabricated to cover a wide range of suspension compliances. This provides a convenient means of determining the magnitude of adhesive forces. Figure 11 shows the restoring force of *n*-type polysilicon test microstructures versus the suspension length for an interfacial clearance of ~2 μm. Since the restoring force is inversely proportional to the third power of the cantilever beam length, its magnitude can be easily varied by several orders of magnitude to facilitate the measurement of adhesive forces at the contact interface of different material microsystems. For example, the typical restoring force of a lateral resonator is in the range of 0.05–2 μN. Based on the aforementioned fabrication process, surface microstructures with restoring forces in the range of 3 nN to 15 μN and apparent dimple contact areas between 10 μm^2 and 10^4 μm^2 can be fabricated to allow the measurement of attractive (adhesive) apparent pressures in the range of 0.00–1 μN/μm^2.

Static and dynamic coefficient of friction testing at the micromachine level can be achieved with micromachines similar to that shown in Fig. 9(a) [7,19,25]. By applying a dc bias voltage to one of the comb drives, the suspended shuttle is first displaced laterally by a distance D_x, thus generating a restoring force in the truss suspension $F_{res} = k_x \cdot D_x$, where k_x is

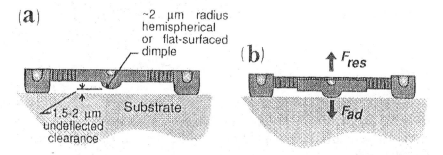

Figure 10. Schematics of a test microstructure with a single-dimple shuttle: (a) undeflected and (b) deflected and stuck conditions of shuttle.

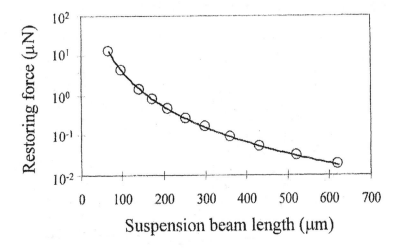

Figure 11. Restoring force versus length of a polysilicon cantilever deflected from its free-standing position by 2 µm.

the spring constant of the folded-beam suspension in the lateral direction determined empirically by resonating the microstructure (Fig. 12(a)). For polysilicon cantilever-beams of length 200 µm and 2 × 2 µm cross-section area, the spring constant is linear up to displacements of ~10 µm. The microstructure (shuttle) is then lowered toward the substrate by applying a bias (clamping) voltage to the underlying electrode until the dimple touches the test surface. Further increase of the voltage gives rise to a normal force, N. Thus, the shuttle is subjected to the suspension force, F_{res}, the normal force, N, and the friction force, F_f, at the dimple/test surface interface (Fig. 12(b)). Slip at the dimple/test surface interface will occur when $F_f = F_{res}$. Since the magnitude of F_{res} is fixed and F_f depends on the coefficient of friction f and the normal load N ($F_f = fN$), the slip condition can be observed by ramping down the clamping voltage to reduce the magnitude of the normal force. The static coefficient of friction is given by [25]

$$f = \frac{4\pi^2 \omega_0^2 M \Delta x}{\dfrac{a_z \varepsilon_0 A_c V_c^2}{2 z_d^2} - k_z (h - z_d)}, \quad (6)$$

where ω_0 is the fundamental resonant frequency, M is the total effective mass, a_z is a constant assuming values in the range of 1–1.5, ε_0 is the electrical permittivity of the intervening medium, A_c is the overlap area of the clamping electrode and the shuttle, V_c is the critical value of the clamping voltage at the onset of slip, k_z is the suspension spring constant in the vertical direction, and h and z_d are, respectively, the distances of the free-standing shuttle and the dimple from the substrate surface.

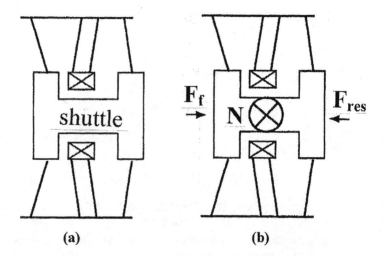

Figure 12. Schematic of a single-dimple friction/wear test microstructure: (a) displaced and (b) displaced and clamped conditions [25].

For dynamic friction (and wear) testing, the shuttle can be set to oscillatory motion (while the dimple is in contact with the substrate surface) if one of the comb drives is driven with a voltage. In the absence of viscous damping, the forced-vibration equation of the shuttle is given by

$$M \frac{d^2 x}{dt^2} + k_x x + fN \, \text{sgn}\left(\frac{dx}{dt}\right) = F_c(t), \quad (7)$$

where $F_c(t)$ is the forcing function corresponding to the comb attractive force. The amplitude and phase shift of the motion relative to $F_c(t)$ can be obtained in terms of the dynamic coefficient of friction f. The transfer function of the undamped lateral motion of the shuttle $H(\omega)$ can be obtained with a resonator after operation for a few seconds and is given by [7]

$$H(\omega) = \left(\frac{|F_c|}{k_x}\right)\omega_0 \sqrt{\frac{1}{\omega_0^2 - \omega^2} - \left(\frac{fN}{\omega F_c}\right)^2 \frac{\sin^2(\pi\omega_0/\omega)}{[1+\cos(\pi\omega_0/\omega)]^2}},$$

where ω and $|F_c|$ represent the frequency and amplitude of the comb electrostatic force, respectively. Both, the amplitude and the resonant frequency decrease with increasing coefficient of friction.

5 LUBRICATION OF MEMS INTERFACES

In conventional lubrication, low friction substances (solid or liquid) are applied to a system by external means, which cannot be adopted in MEMS devices due to the very small dimensions. For example, liquid lubricants may introduce capillary and viscous shear mechanisms, which may lead to energy dissipation and, eventually, to seizure due to excessive viscous drag forces. Thus, ultrathin liquid films deposited at micromachine interfaces prior to the release-drying process are the ideal lubricants for this technology. In particular, deposition of hydrophobic (low surface energy) densely packed organic molecules on micromachine surfaces is desirable because the typical gaps between suspended microstructures and the underlying substrate are on the order of 1–2 µm. Because such well-organized monolayers tend to rapidly adsorb on appropriate surfaces, they are referred to as self-assembled monolayers (SAM). Among the various types of SAM coatings, long-chain hydrocarbon and fluorocarbon molecules are especially suitable for MEMS because they form chemically bound and highly ordered monolayers on hydroxyl-terminated surfaces, such as those of oxidized polysilicon, silicon nitride, and aluminum, *i.e.*, the main structural materials in MEMS. A complete coverage of smooth surfaces of n-type silicon by densely packed monolayers of n-octadecyltrichlorosilane (OTS), n-$C_{18}H_{37}SiCl_3$, has been reported to occur by chemisorption of silane compounds incorporated in solutions [27]. It has been suggested that the driving force for the formation of an OTS monolayer results from the attraction of $SiCl_3$ head groups by the water surface layer, and that stabilization is a consequence of hydrolysis and cross-linking between head groups [28]. Adsorption of OTS molecules on silica surfaces produces a dense monolayer with alkane chains normal to the surface [29]. The amount of water adsorbed on fumed silica surfaces controls the adsorption of OTS molecules [30]. Experiments with glass spheres slid against p-type smooth silicon wafers coated with n-$C_{18}H_{37}SiCl_3$, n-$C_{11}H_{23}SiCl_3$ (UTS), and n-

$C_6F_{13}CH_2CH_2SiCl_3$ (FTS) revealed the occurrence of insignificant wear, relatively low friction coefficients for OTS and UTS monolayers ($\mu = 0.07$ and 0.09, respectively), and significantly higher friction ($\mu = 0.16$) and wear for FTS monolayers, presumably due to incomplete coverage of the silicon surface by the thinner (1 nm) FTS monolayers [31].

The presence of close-packed trichlorosilane monolayers with approximately vertical alkyl chain axis possessing low surface energies and large wetting contact angles [27-29] at micromachine interfaces should reduce stiction by suppressing the effects of capillary and van der Waals attractive forces. SAM films were introduced into the MEMS fabrication process fairly recently for the purpose of reducing microstructure stiction [14,31-33]. It was found that self-assembly from a non-aqueous solution leads to complete surface coverage by a chemically bonded monolayer. The process flow for depositing the OTS monolayer prior to the release-drying process is given in Table 3 [14,33].

Table 3. Self-assembled monolayer treatment [14,33].

Processing step	Treatment
Removal of sacrificial PSG layer	Aqueous HF release etch
Surface oxidation	DI water rinse RCA SC-1 soak DI water rinse
Water removal	Two-step methanol rinse Two-step CCl_4 rinse
SAM formation	OTS or FTS precursor solution treatment
Post-rinsing	Two-step CCl_4 rinse Two-step methanol rinse
Drying	IR lamp dry or oven bake in nitrogen

It has been postulated that when the precursor molecules contact the OH-terminated substrate surface, which is covered by a physisorbed water monolayer, the $SiCl_3$ bonds hydrolyze [Fig. 13(a)] and the OH groups of the SAM precursor and water monolayer condense to form Si-O-Si siloxane linkages [Fig. 13(b)]. In addition, the remaining Si-OH bonds on the silane molecules condense to form more siloxane bonds. Subsequently, the OTS monolayer is covalently bound both to itself (cross linkage) and the substrate surface (grafting) [Fig. 13(c)]. With this treatment, the chemical state of exposed silicon and silicon oxide surfaces can be modified by the adsorption of SAM films possessing different functional head groups, such as alkyl chains (CH_3), ester, and vinyl groups.

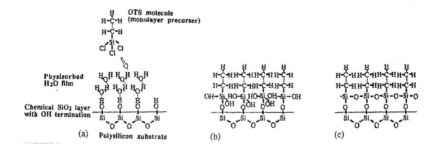

Figure 13. A plausible mechanism of the formation of the OTS monolayer on oxidized silicon surfaces: (a) liquid phase application and hydrolysis, (b) adsorption and formation of siloxane bonds, and (c) cross-linkage of molecules through covalent bonding [7].

The use of ultrathin, hydrophobic films to reduce adhesion and stiction in MEMS has been demonstrated in several recent studies. The integration of the OTS monolayer treatment into post-release rinse processing was found to reduce significantly microstructure stiction during the dry-release process [14], most likely due to the weak van der Waals and capillary forces produced in the presence of the low surface energy (20.6 mN/m) and large water contact angle (~111°) of OTS monolayers [31]. The relatively low friction coefficient of OTS-coated silicon surfaces slid against glass [63] indicates that this monolayer should also enhance the reliability of microdynamic devices. However, relatively little is known about the durability of Sam monolayers under micromachine operation conditions.

The atomic force microscope has been used by several researchers to study the effect of various surface chemical treatments on the magnitude of the adhesion force at silicon contacts. Alley et al. [32] treated smooth Si(100) substrates with concentrated HF and 5:1 H_2SO_4:H_2O_2 solutions to obtain hydrophobic (Si-H terminated) and hydrophilic (Si-OH terminated) surfaces and OTS-coated and uncoated sharp Si tips. A process similar to that shown in Table 3 was used to treat the AFM tip. The adhesion force between hydrophilic silicon surfaces was found to be equal to 29 nN, and was considered to be primarily due to the effect of the capillary force resulting from water adsorption at the tip/substrate contact interface due to the presence of the highly hydrophilic SiO_2 layer. However, reducing the surface energy by coating the AFM tip with an OTS monolayer produced an adhesion force of 5 nN, i.e., a decrease by a factor approximately equal to six. This small force is believed to be mainly due to the van der Waals force. Similar trends and behaviors were observed by Houston et al. [34-36] and Srinivasan et al. [37] for HF- and NH_4F-treated Si(100) substrates and silicon nitride AFM tips. It was reported that the NH_4F treatment of Si(100) samples cleaned with an RCA solution produced hydrophobic, H-terminated Si surfaces with contact angles of ~84° and adhesion forces equal to ~4 nN.

The above studies have demonstrated that profound improvements in the reliability of MEMS devices, as it pertains to seizure due to excessive interfacial adhesive forces, can be accomplished with microfabrication-compatible surface treatments producing anti-stiction SAM coatings. However, further basic research is necessary in order to fully understand the behavior of such monolayers under various micromachine operation conditions and environments.

6 CONCLUSIONS

From the previous discussion and analysis of the various surface phenomena encountered at MEMS interfaces, it is apparent that remarkable progress has been made toward reducing the magnitudes of high adhesion forces at microdevice interfaces. This has been accomplished by various means, such as the use of microdimples to reduce the apparent contact area, surface texturing to reduce the real contact area, various chemical treatments (*e.g.*, surface passivation by NH_4F) to produce hydrophobic surface behaviors, and deposition of low surface energy solid coatings (*e.g.*, DLC) and liquid films (*e.g.*, OTS) to reduce the work of adhesion. A characteristic recent example of the progress and innovation in the MEMS field is the first bionic chip [38]. This innovation uses a biological cell in a circuit as an electrical diode (or switch) to allow current to flow through the device at certain voltages. At a certain voltage (depending on the type of cell), the pores on the cell membrane open up and current starts to flow through the cell. The cell is placed in a hole in the center of a chip and is kept alive with the infusion of nutrients. With the proper computer control, the bionic chip can open and close the cell membrane in milliseconds, thus allowing very precise control never before achieved. It is speculated that with this chip it might be possible to introduce DNA, extract proteins, and administer medicines without affecting other kinds of neighboring cells.

Despite numerous research developments, many challenging obstacles must be overcome before the full potential of MEMS devices can be realized. One of the major challenges in this filed is to produce stand-alone microsystems including the actuator, drive, control electronics, and a sensing mechanism to provide feedback of the system's performance. There are serious issues related to tribology, mechanics, surface chemistry, and materials science in the operation and manufacturing of many MEMS devices and these issues are preventing an even faster commercialization. Very little is understood about surface interaction and mechanical behaviors in the micro- to nano-scales of materials used in the fabrication of MEMS devices. The MEMS community needs to be exposed to state-of-the-art tribology and vice versa. Better tribological understanding of MEMS will advance the state of the art in micromachining and IC industry in general. Specifically, basic knowledge of surface behaviors at different environments will contribute to a more accurate prediction of the performance of micromachined pressure sensors, accelerometers, and gyros, better understanding of stiction and in-use wear performance of micromirrors and micromotors, and insight into the significance of surface topography in microfluidics.

REFERENCES

1. M. Mehregany, K. J. Gabriel, and W. S. N. Trimmer, IEEE Trans. Electron Devices **35**, 719 (1988).
2. L.-S. Fan, Y.-C. Tai, and R. S. Muller, IEEE Trans. Electron Devices **35**, 724 (1988).
3. W. C. Tang, T.-C. Nguyen, and R. T. Howe, Sensors and Actuators **20**, 25 (1989).
4. K. J. Gabriel, F. Behi, R. Mahadevan, and M. Mehregany, Sensors and Actuators A **21-13**, 184 (1990).
5. Y.-C. Tai and R. Muller, Sensors and Actuators A **21-13**, 180 (1990).
6. M. Mehregany, S. D. Senturia, and J. H. Lang, IEEE Solid-State and Actuator Workshop, Hilton Head Island, SC, June 4-7, 1990, pp. 17-22.
7. K. Komvopoulos, Wear **200**, 305 (1996).
8. K. Komvopoulos, Proc. Micro-Electro-Mechanical Systems (MEMS), ASME **DSC-66**, New York, NY, 1998, pp. 261-264.
9. J. B. Sampsell, Transducers '93, 7th Int. Conf. Solid-State Sensors and Actuators, June 7-10, 1993, Yokohama, Japan, pp. 24-27.
10. L. J. Hornbeck, Micro-Optical Technologies for Measurement, Sensors, and Microsystems, SPIE Proc. **2783**, 1996, pp. 2-13.
11. T. Akiyama, D. Collard, and H. Fujita, J. Microelectromechanical Systems **6**, 10 (1997).
12. R. Legtenberg, E. Berenschot, J. van Baar, and M. Elwenspoek, J. Microelectromechanical Systems **7**, 79 (1998); **7**, 87 (1998).
13. S. L. Miller, G. LaVigne, M. S. Rodgers, J. J. Sniegowski, J. P. Walters, and P. J. McWhorter, SPIE Proc. **3224**, 24 (1997).
14. R. L. Alley, G. J. Cuan, R. T. Howe, and K. Komvopoulos, IEEE Solid-State and Actuator Workshop, Hilton Head Island, SC, June 21-25, 1992, pp. 202-207.
15. P. R. Scheeper, J. A. Voorthuyzen, W. Olthuis, and P. Bergveld, Sensors and Actuators A **30**, 231 (1992).
16. R. Legtenberg, J. Elders, and M. Elwenspoek, Transducers '93, 7th Int. Conf. Solid-State Sensors and Actuators, June 7-10, 1993, Yokohama, Japan, pp. 198-201.
17. K. Komvopoulos and W. Yan, J. Tribol. **119**, 391 (1997); **120**, 808 (1998).
18. R. L. Alley, P. Mai, K. Komvopoulos, and R. T. Howe, Transducers '93, 7th Int. Conf. Solid-State Sensors and Actuators, June 7-10, 1993, Yokohama, Japan, pp. 288-291.
19. K. Komvopoulos, Materials and Device Characterization in Micromachining, SPIE Proc. **3512**, 106 (1998).

20. C. H. Mastrangelo and C. H. Hsu, IEEE Solid-State and Actuator Workshop, Hilton Head Island, SC, June 22-25, 1992, pp. 208-212.
21. R. Maboudian and R. T. Howe, J. Vac. Sci. Technol. A **15**, 1 (1997).
22. K. Deng, W. H. Ko, and G. M. Michal, Transducers '91, 6th Int. Conf. Solid-State Sensors and Actuators, 24-27 June, 1991, San Francisco, CA, pp.213-216.
23. K. Deng and W. H. Ko, IEEE Solid-State and Actuator Workshop, Hilton Head Island, SC, June 22-25, 1992, pp. 98-101.
24. S. Suzuki, T. Matsuura, M. Uchizawa, S. Yura, and H. Shibata, IEEE Micro Electro Mechanical Systems Proc., 30 Jan.-2 Feb., 1991, Nara, Japan, pp. 143-147.
25. M. G. Lim, J. C. Chang, D. P. Schultz, R. T. Howe, and R. M. White, IEEE Micro Electro Mechanical Systems Proc., 11-14 Feb., 1990, Napa Valley, CA, pp. 82-88.
26. S. Niederberger, D. H. Gracias, K. Komvopoulos, and G. Somorjai, J. Appl. Phys. **87**, 3143 (2000).
27. M. Pomerantz, A. Segmüller, L. Netzer, and J. Sagiv, Thin Solid Films **132**, 153 (1985).
28. H. O. Finklea, L. R. Robinson, A. Blackburn, B. Richter, D. Allara, and T. Bright, Langmuir **2**, 239 (1986).
29. P. Guyot-Sionnest, R. Superfine, J. H. Hunt, and J. R. Shen, Chem. Phys. Lett. **144**, 1 (1988).
30. C.P. Tripp and M. L. Hair, Langmuir **8**, 1120 (1992).
31. V. DePalma and N. Tillman, Langmuir **5**, 868 (1989).
32. R. L. Alley, K. Komvopoulos, and R. T. Howe, J. Appl. Phys. **76**, 5731 (1994).
33. R. L. Alley, R. T. Howe, and K. Komvopoulos, U.S. Patent No. 5,403,665 (1995).
34. M. R. Houston, R. T. Howe, K. Komvopoulos, and R. Maboudian, Mater. Res. Soc. Symp. Proc. **383**, 391 (1995).
35. M. R. Houston, R. Maboudian, and R. T. Howe, 8th Int. Conf. Solid-State Sensors and Actuators, June 25-29, 1995, Stockholm, Sweden, pp. 210-213.
36. M. R. Houston, R. Maboudian, and R. T. Howe, Solid-State Sensor and Actuator Workshop, Hilton Head, SC, June 2-6, 1996, pp. 42-47.
37. U. Srinivasan, M. R. Houston, R. T. Howe, and R. Maboudian, Int. Conf. Solid-State Sensors and Actuators, Chicago, IL, June 16-19, 1997, pp. 1399-1402.
38. B. Rubinski and Y. Huang, Biomedical Microdevices, Feb. 2000.

Chapter 13

MATERIALS AND RELIABILITY ISSUES IN MEMS AND MICROSYSTEMS

Aris Christou
Materials Science and Engineering, University of Maryland
College Park, MD 20742

1 INTRODUCTION TO MICROELECTRO-MECHANICAL SYSTEMS

The recent evolution in microelectronics of combining electrical and mechanical functions has brought about an exciting new field – microelectromechanical system (MEMS) [1, 2]. Miniature structures developed by new fabrication techniques on semiconductor wafers make possible new devices that have the potential to revolutionize instrumentation and control systems. At the University of Wisconsin (Madison), an air driven microelectromechanical generator has been developed by Henry Guckel. Using parts that are a fraction of the thickness of human hair, a generator was designed giving an output of five volts and weighing less than five grams. In separate developments, a team at the University of Michigan (Ann Arbor) has built atomic-force microscopes that enable insights into surface science and produce miniature probes for use in advanced prostheses. At the Berkeley Sensor & Actuator Center at the University of California, researchers have built a microgripper capable of handling micron-sized structures. The heart of the gripper is a novel microstructure that consists of interdigitated fingers, or cantilevers, which are activated electrostatically to move the gripper arms. Such a device has potential applications in biomedicine and micro-telerobotics. In the sections that follow, the research issues related to materials and microsystems are discussed. The research problems that must be solved are presented as well as recommendations for research support by industry and federal agencies.

Despite the vast variety of MEMS devices being proposed and fabricated, most of the devices fall into the categories of sensors and actuators, the most important constituents of instrumentation and control systems. The research on MEMS sensors has dated from the late 1960s, and extensive efforts have been made for the fabrication of MEMS sensors to become a mature technology application, as accelerometers which are used in automobile air-bag systems, temperature and pressure flowmeters, and neural microprobes for biomedical study. The promise of the MEMS sensors is that batch fabricated silicon wafers are very low in cost and very sophisticated in feature size, incorporating the use of on-chip circuitry. The micromachined sensors can be produced today with high yield and merged with integrated electronics both in monolithic chips and hybrid multichip assemblies. For some types of sensors, accuracy can be as high as 16 bits and VLSI interface circuits are defined to allow features such as self-testing and digital compensation. One the other hand, consistent progress has also been made in the area of actuators. Microgrippers, piezoelectric micromotors, micropumps, magnetic microactuators, *etc.* have been successfully fabricated in laboratories. Though some problems remain to be solved in the area of microactuators all of the essential elements of simple MEMS are in place and complete microelectromechanical systems have been proposed with applications ranging from microrobots for security and medical applications to sophisticated positioning systems for assembly tasks at the submillimeter level.

The fabrication techniques used in MEMS are similar to those used by the electronics industry. The microminiatures are made using three microfabrication processes – surface micromachining, bulk micromachining and LIGA processing. These processes employ methods such as photolithography, material deposition, chemical etching, electroplating, and X-ray radiation to shape the mechanical and electronic structures. Although micromachining has been used for over 20 years as a processing technique in the fabrication of sensors, only recently has it been extended to the commercial production of mechanical features from silicon. In the past five years there has been a heightened interest in the use of micromachining technology. Improved etching technologies, deep UV lithography, X-ray lithography for the LIGA process, and ion projection lithography make it possible to attain accuracy for features with high aspect ratios.

While progress has been made in MEMS there are still many technical challenges which must be solved. In the area of microactuators better driver mechanisms are required. Nonplanar technologies, with assembly techniques at the submillimeter and micron levels are also required. Workstation based design systems along with a database of material, structural, and performance information must be developed. These problems are being addressed and the progress achieved will be discussed in this chapter.

In the sections that follow, applications of MEMS are presented (MEMS Devices section), and several MEMS technologies are discussed (MEMS Technology section). The architecture and current status of CAD tools for MEMS processing simulation and design is also included.

2 MEMS DEVICES

The MEMS devices developed to date are sensors and actuators. The principles of these MEMS devices are presented including a description of the structures for sensors and actuators.

2.1 Pressure Sensors

The important category of MEMS devices is that of the silicon pressure sensor which has been identified to have applications in many areas including transportation, health care, and industrial process control. Substantial progress has been made on silicon-based pressure sensors in recent years as a result of advances in etching technologies to form thin silicon membranes. Most of the micromachined silicon pressure sensors contain silicon membranes. Depending on the pressure sensing mechanism they can be classified into two types of devices – piezoresistive sensors and capacitive coupled sensors.

Figure 1 (a) shows the schematic of a piezoresistive sensor which contains a full bridge of diffused resistors to measure stress at four points on the diaphragm and to convert it to an electrical output signal. For a diffused resistor subjected to parallel and perpendicular stress components, $\sigma_\|$, $\sigma_\perp$ and the resistance change is [3] given in terms of the B-type coefficients, where $B_\|$ and $B_\perp$ are the piezoresistive coefficients parallel and perpendicular to the resistor alignment, depending on the crystal orientation. Thus by arranging

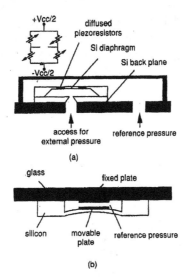

Figure 1. Schematics of (a) a piezoresistive and (b) a capacitive pressure sensor.

the alignment of the diffused resistors in the bridge, two of them increase in value and the other two decrease in value upon application of the stress. A voltage signal is generated due to the resistance variation and the voltage signal is calibrated to obtain the pressure applied. In the capacitive type sensor shown in Figure 1 (b), the diaphragm is used as a movable plate of a parallel plate capacitor. The pressure is determined from variation of the capacitance of the parallel capacitor.

By varying the diameter and thickness of the silicon diaphragms of 0–200 MPa have been fabricated. The bridge voltages are usually 5 to 10 volts and the sensitivity of silicon pressure sensors vary from 10 mV/kPa for low pressure to 0.001 mV/kPa for high pressure devices. Temperature compensation circuits are also added to the pressure sensor which can result in less than 1% sensitivity for sensors working in the range between 0°C and 50°C.

2.2 Accelerometers

Several silicon accelerometers based on the diaphragm or cantilever structure from micromachining have been fabricated [4]. One of the most interesting structures is a cantilever piezoresistive accelerometer, which is illustrated schematically in Figure 2. The center layer of the glass-silicon-glass sandwich structure, which consists of a mass block and a cantilever, is the heart of the device. Fabrication of the accelerometer is a batch process utilizing standard IC photolithographic and diffusion techniques in addition to the special techniques required to shape the silicon and glass. The silicon element and the top and bottom glass covers are fabricated separately in wafer form and then bonded together. The micro-machining technology will be discussed in more detail in section 3, while the details of the fabrication technology have been reviewed in previous sections and authors.

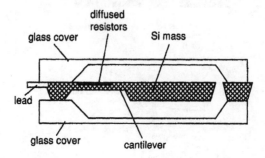

Figure 2. Structure of a cantilever piezoresistive accelerometer.

The cantilever has a *p*-diffused path which forms a resistor of about 100 Σ/cm. Under acceleration, the mass block exerts stresses on the cantilever. The stresses will induce a change in the resistance of the diffused resistor. The magnitude of the change depends on the crystallographic orientations of the silicon and stress. The variation in the resistance is then detected by an on-wafer Wheatstone bridge circuit. By using of silicon micro-machining technology the entire accelerometer can be made with weight under 0.02 gram and as small as 2×3×0.6 mm. The sensitivity of such accelerometers can be as low as 0.01 g with accuracy under 1 percent over a range of 100g, where *g* is the acceleration of gravity. The accelerometer also has good frequency response with a bandwidth of 100 MHz. The performance characteristics of such accelerometers can be varied over a wide range to meet the needs of different applications. One of the most interesting applications, for example, is in the automobile air-bag systems, which at the present time represents the largest application product.

2.3 Shear Stress Sensor

A second example of a MEMS sensor is a floating-element liquid shear stress sensor which has been recently developed at MIT by the group of Martin A. Schmidt [5], and has many potential applications such as flowmeters. The floating-element sensor (120×140×5 μm) has been designed for high shear stress (1–100 kPas) and high pressure environments (up to 6600 PSi) and utilizes a piezoresistive transduction scheme. Figure 3 shows the schematic diagram of the sensor, which consists of a plate (120×140 μm) and four tetheres (30×10 μm). The tethers function both as mechanical supports for the plates and a s resistors in the transduction scheme. The plate and the tethers are constructed from a 5 micron thick lightly doped silicon layer and are suspended 1.4 μm above another surface. A flow over the floating-element structure and parallel to the length of the tethers generates a shear stress on top of the suspended plat. Assuming the plat moves as a rigid body, the shear stress forces the suspended plat in the direction of the flow. Two of the tethers experience compressive stresses and the other two tensile stresses. These stresses generate axial strain fields throughout the tether structure, which introduce a change of resistance of the tether due to the piezoresistive properties of single crystalline silicon.

2.4 Biosensors and Neural Probes

Micro-machined structures are emerging as useful instruments in medicine, where precision sensing is critical. An electromechanical sensor array that is small enough to fit inside a blood vessel has been recently developed. The device is inserted into an artery within a catheter that has an

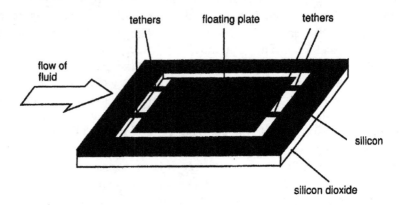

Figure 3. Structure of a shear stress sensor.

inside diameter of 650 µm. It measures the levels of oxygen, carbon dioxide, and PH in the blood. The firm also developed a sensor for monitoring the carbon dioxide and oxygen that a patient inhales during surgery.

Even more significant to the medical profession may be the development of micro-machined neural probes at the University of Michigan. In the past, sensors have been used to record impulses from one site in the brain at a time. The new neural probe developed is able to record single brain cell activity from 30 sites the brain at the circuit level, which

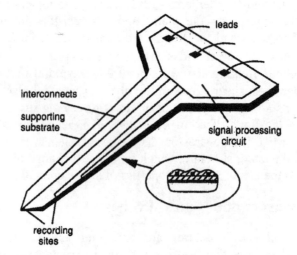

Figure 4. Structure of a neural probe.

would help scientists learn how to treat neural disorders. Such a probe will also help to develop a neural-electronic interface which will control auditory, visual, and neuromuscular prostheses. The schematic of the neural probe is illustrated in Fig. 4.

3 MICROACTUATORS

3.1 Microgripper

Recent progress has been made in the MEMS area in the construction of micro- grippers, which are capable of handling micron-sized objects and have applications in biomedicals as well as in micro-robotics. A typical microgripper consists of a fixed closure driver and two movable jaws. The jaws are closed by an electrostatic force from an electrostatic voltage applied across them and the closure driver.

Figure 5 shows the schematic of a microgripper developed at the University of California [6], by Richard S. Muller. The microgripper consists of a silicon die (7×5 mm), a 1.5 mm long support cantilever, made from boron-doped silicon substrate material, and a 400-μm long polysilicon overhanging gripper extending from the end of the support cantilever.

The silicon die is snapped free along a backside V-groove with a portable vacuum pen. The microgripper and its foundation die is mounted on a positioner and electrically connected using two large contacts that are provided for wire bonding. The microgripper is reported to exert 40 nN of force on the specimen with an applied voltage of 40 V.

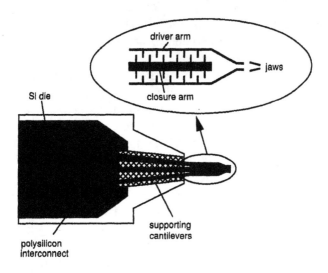

Figure 5. A silicon-fabricated microgripper.

3.2 Piezoelectric Micromotors

Another application of MEMS as a microactuator is micromotors for microrobots. Interest in microbots has been driven by recent success in developing intelligence architectures for mobile robots, which can be compiled efficiently into parallel networks on silicon. Various types of micromotors have been fabricated. A silicon center rotor is free to move around a center pole that is bonded to the glass substrate. A stopper is located between the glass substrate and the silicon center pole hub to prevent the rotor from separating from the center pole. During operation the oppositely placed stators are sequentially stepped with applied voltages, an electrostatic force is exerted.

4 MEMS TECHNOLOGY

4.1 Bulk Micromachining

Early silicon products were fabricated using bulk micromachining technology with isotropic wet chemical etchants such as hydrofluoric, nitric, and acetic acid mixture. The isotropic etchants have no preferential etch rate to crystal orientation and attack different planes at the same rates. Isotropic etching has problems with etch control, selectability, and precision. Most of the chemicals used today are wet etching anisotropic etchants which selectively etch the <100> and <110> directions of silicon. Etching in a <111> direction is much slower than the <100 > and <110> directions (typically 50 times slower). The use of anisotropic etchants results in vertical sidewalls for the <110> silicon substrate and at an angle of 54.7' for the <100> orientation. Lateral geometries and etch profile can be precisely controlled through anisotropic etching.

In a typical surface micromachining process (see Fig. 6), a silicon wafer is first oxidized in a low-pressure chemical vapor deposition (LPCVD) system, which grows silicon dioxide on the surface. Chemicals are then used to etch away parts of the silicon dioxide and to define a required shape. The structure is put through a second LPCVD process, in which a silicon layer grows on top of the silicon dioxide and adheres to the wafer in the areas where silicon dioxide has been etched away. In subsequent etching steps, the shape of the layer is defined, and the film of silicon dioxide is dissolved.

In the above technique, the end component is a suspended mass, or a miniature bridge, a geometric feature common to many surface micromachined components. The silicon dioxide is known as the sacrificial layer for the sake of fabricating a suspended structure. In addition, chemical vapor deposition, sputtering and direct evaporation are used to deposit thin

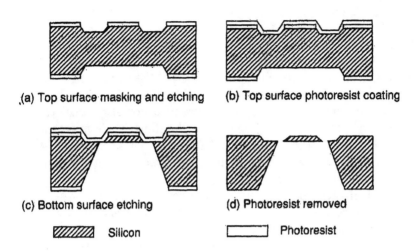

Figure 6. Typical bulk micromachining process.

films on a chip. Polysilicon, silicon nitride, and tungsten, among other materials, are used for the structures of surface micromachined parts while phosphorus-doped silicon functions as the sacrificial layer. There has been strong interest in polysilicon micromachined parts recently, however some problems still exist. Polysilicon structures of thickness larger than several microns are difficult to fabricate due to the larger intrinsic stress which results from the thicker polysilicon layers.

The advantage of surface micromachining is that the elements are fabricated in pre-assembled form. This avoids assembling individual mechanical parts by hand. The bulk micromachining technique, in contrast, cannot provide such integrity of micromachined parts. However, the surface micromachining process tends to introduce intrinsic stress as the structure layer is grown on the sacrificial layer. Such stress causes structure warp and limits the thickness of structure to be fabricated with in several microns. Recently an alternative process has been developed for single crystal silicon micromotors [8] based on the combination of both bulk micro-machining and surface micromachining techniques. The new approach utilizes deep etch stop (boron) diffusion and deep trench etching to define the vertical structure of the parts and employs the surface micromachining technique (*i.e.*, sacrificial layer and epilayer growth) to form the joint parts. The new process provides up to 20 microns mechanical structure with higher production yield.

4.2 LIGA Process

To date, most of the micromachined devices are made using either surface or bulk micromachining. A micromachining technology called the LIGA (an acronym for the German term for lithography, electroforming, and plastic modeling) process is less compatible with integrated circuit manufacturing, but is showing great promise for the development of microstructures. The technique provides well-defined, thick microstructures, that have extremely flat and parallel surfaces. These characteristics are particularly useful for fabricating motors, gear trains, and generators having spinning parallel parts that come in contact.

In a typical LIGA process [9], polyamide is first deposited on a silicon wafer as a sacrificial layer. The shape of the layer is then defined by passing ultraviolet light waves through a mask. Next the substrate is covered with a plating base made of a thin film of titanium under a layer of nickel. A photoresist layer over the plating base is deposited, this usually consists of polymethyl methacrylate, a photopolymer similar to Plexiglas.

A mask is placed over the photoresist material and X rays are radiated through it. The X rays are powerful enough to penetrate a photopolymer as thick as 300 microns. The X-ray mask is a membrane, typically of silicon nitride, which is only 1 to 1.5 microns thick. It is mostly transparent to allow the X rays through, but gold is deposited in defined regions of the mask to absorb certain X rays.

X-ray radiation changes the polymer's molecular weight in selected areas, and chemical etchants dissolve away the regions of material that have been exposed. The resulting geometry is a mold into which nickel is electroplated. In subsequent steps, chemicals are used to remove the mold, plating base, and sacrificial layer. Use of a sacrificial layer allows us to fabricate structures that are semi-suspended or free of the substrate. These components are assembled into parts that are attached to the substrate to make interacting microstructures. In a motor, for example, the shaft and the rotor are fully attached parts, while the spinning rotor is machined free of the substrate and mounted to the shaft.

4.3 Materials and Reliability Issues

The research issues related to materials must start with the application of "traditional materials" in demanding applications. The application of silicon and GaAs single crystals are many and include mechanical structures (membranes, beams, seismic masses *etc.*) as well as transducer structures (piezoresistor, piezoelectric, photonics). Electrochemical properties are critical such as that of porous silicon and etch stops. Traditional thin film materials are in the form of mechanical structures and chemical structure such as silicon nitride for etching masks, protective layers and silicon oxide sacrificial layers.

New materials, which require concentrated research are: piezoelectric layers (PZT, PLZT and ZNO) and shape memory alloys based on titanium nickel alloys. Other aspects of new materials include low stress silicon nitride, glass layers, doped glass layers with high optical index of amplifying properties, and chemical protective layers. Research in polymers includes low stress polymers and glues for enlarged chemical resistance. From a reliability point of view, the compatibility of materials is critical. This area is divided into mechanical effects, chemical effects, thermal effects and optical effects. Mechanical effects include the bimetallic effects and the coupling of external forces into sensitive parts inside the component.

New material analysis techniques are required for microsystems and further analytical is needed for the determination of the mechanical properties of thin films as well as for the chemical properties of protective layers.

REFERENCES

1 O'Connor, L., "MEMS: microelectromechanical systems," Mechanical Engineering, February 1992, 40-47.
2 Wise, K.D., "Integrated microelectromechanical systems: a perspective on MEMS in the 90s," IEEE MEMS Workshop, 1991, 33-38.
3 Middelhoek, A.S. and Audet, S.A., Silicon Sensors, Academic Press, 1989, pp. 118-122.
4 Roylance, L.M. and Augell, J.B., "A batch fabricated silicon accelerometer," IEEE Trans. Electron Dev., ED-26(12), December 1979, 1911-1916.
5 Shajii, J., Ng, K.Y. and Schmidt, M.A., "A microfabricated floating-element shear stress sensor using wafer-bonding technology," J. Microele(-ti-omechanical System, 1(2), June 1992, 89-93.
6 Kim, C.J., Pisano, A.P. and Muller, R.S., "Silicon-processed overhanging microgripper," J. Microelectromechanical System, l(l), March 1992, 31-35.
7 Arimatsu, K., Hashimoto, I., 00ishi, S., Tanaka, S., Sato, T. and Gejyo, T., "Development of large scale ion beam milling machines," Nuclear Instruments and Methods in Phys. Research, B37/38, June 1989, 833-837.
8 Suzuki, K. and Tanigawa, H., "Single crystal silicon rotational micromotors," IEEE MEMS Workshop, January 1991, 15-20.
9 Hannening, M. *et al.*, "Modeling of three dimensional microstructures by the LIGA Process," IEEE MEMS Workshop, 1992, 202-207. and Sorensen, Inc., Providence, RI.

Chapter 14

SURFACE CHARACTERISTICS OF INTEGRATED MEMS IN HIGH-VOLUME PRODUCTION

Jack Martin
Analog Devices, Inc., Micromachined Products Division
Cambridge, MA 02139

1 BACKGROUND

In the early 1970's, silicon pressure sensors were considered to be the leading edge of a micromachining technology that would revolutionize industrial instrumentation. This new technology did grow. However, commercialization was slower than anticipated - delays arose due to quality, cost, lack of standardization and difficulties related to electronic integration. Many of the quality problems were caused by surface instabilities.

A step change in the MEMS industry occurred in 1993 when Analog Devices shipped the first ADXL-50 micromachined accelerometers for automotive air bag control systems. This was a pivotal moment because the ADXL50 incorporated integrated electronics, was produced in a state-of-the-art facility and was sold into a high volume market that would drive both cost and standardization.

Subtle effects are often not recognized until one attempts to produce large quantities. Every new technology stumbles a few times as it progresses up the learning curve, and the ADXL was no exception. Its problems were often rooted in surface effects. In retrospect, this not surprising because:
1. MEMS devices have high surface areas and closely spaced surfaces,
2. The ADXL circuit is fully passivated. However, like most polysilicon microstructures, the MEMS surfaces were unpassivated.

This paper summarizes the surface-related lessons that were learned as ADXL production increased to over 500,000 units/week. It should be clearly understood that the high volume air bag market is allowing ADI to establish a much broader product-enabling capability. The graphic in Fig. 1 illustrates some of the new product areas being spawned by this technology. An understanding of the lessons learned from the ADXL experience helps to predict the future needs of integrated inertial sensors as well as other MEMS products that are barely on the horizon today.

Single and dual-axis accelerometer applications include game controllers, joysticks, virtual reality headsets, shock, vibration, machinery and medical monitors and disk drive head controllers.

This technology can also be used to make low cost gyroscopes, thus further extending its impact in the marketplace.

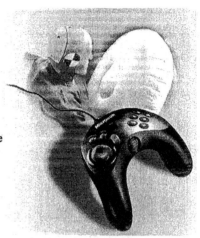

Figure 1. Commercial and industrial applications of low cost accelerometers.

2 DISCUSSION

The initial stumble occurred during ADXL50 reliability tests in late 1992. The data showed that this device exhibited a very slow drift in electrical output during high temperature (125°C) testing. The drift was traced to polysilicon surface characteristics. Initial reaction focused on surface contaminants. However, the sensor surface was extremely clean. Solution of this unanticipated problem led to a successful product launch into the air bag sensor market.

> **LESSON #1:** An understanding of surface science is critical to MEMS. In an inert dry gas, devices that have ultra-clean polysilicon microstructure surfaces can exhibit long-term drift in electrical output when the poly terminals are held at different potentials. The drift became evident when powered units were held for weeks/months at 125°C.

The ADXL50 was extremely reliable and offered on-board integrated electronics. However, it was packaged in headers – an expensive package that customers could not handle with automated equipment. Therefore, the next generation ADXL75 was designed for use in cerdips.

Cerdips have been used for hermetic packaging of semiconductor devices for decades. They are dual-in-line ceramic packages that are sealed with glass. Cerdips were particularly attractive for the ADXL75 because they are a relatively low cost cavity package that can be handled with automated equipment. Assembly includes several furnace passes at 430–500°C in air so they seldom contain organic materials.

Initial lots passed the production qualification without incident. However, two problems became evident soon after volume production began:
1. Handling of assembled devices often caused stiction.
2. Device-to-device initial output (no acceleration) was quite variable.

This differed from the anomalous behavior observed in the ADXL50 headers where the output drifted with time.

The ADXL50 and ADXL75 sensors were designed to measure acceleration with a full-scale output near 50-g's. Stiction was not expected to be a problem because the microstructures should not touch, even at 2000-g overloads. However, as illustrated in Fig. 2, these sensors have closely spaced polysilicon "beams." The beams (2-μm thick in the ADXL75) are electrically connected to form differential capacitor cells with 1.2 μm air gaps. Two of the beams in each cell are rigidly mounted 1.6 μm above the silicon surface. The third beam in each cell is centered between these fixed beams. It is part of a larger mass that is suspended by very compliant springs. An applied acceleration moves the mass (and the center beams) relative to the fixed beams. This relative motion creates an imbalance in the differential capacitor – thus allowing acceleration to be converted into an electrical signal.

It is extremely difficult to generate an acceleration above 2000-g's in air bag modules that are typically mounted near the middle of the car under the dashboard. However, high-g shock waves can be generated when discrete ceramic packages bang into each other in automated part handlers. Clean inorganic parts have very high surface energies so it is not surprising that stiction problems arose during the handling of ADXL75 cerdips.

The solution[1] uses a phenyl-methyl silicone liquid that has unusually high resistance to temperature and oxidation. It also has a boiling point over 400C. In production, a small amount of this liquid is dispensed into each cerdip before it is sealed. As the temperature rises in the furnace, adsorbed gases are driven off of all surfaces. Removal of adsorbed species increases the MEMS surface energy and makes it more chemically reactive.

Further increase in temperature causes the liquid to evaporate. Most of the vapor escapes but not before contacting all surfaces in the cerdip cavity, including the MEMS sensor. The molecules that initially contact the hot MEMS surface chemically react and bond to it. The result is a surface that is enriched in organic groups. Organic-rich surfaces have low energies so

---

[1] U.S. Patent 5,694,740

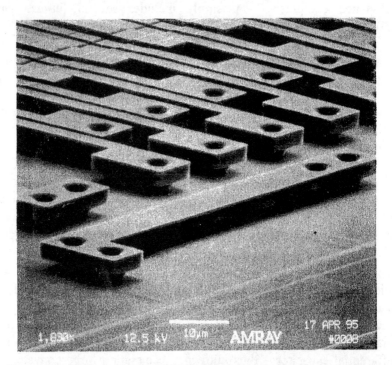

Figure 2. Polysilicon beams rigidly anchored 1.6 microns above the silicon surface in an ADXL accelerometer. The beams entering from the top left of the photo are part of a larger mass that is suspended by springs that have high mechanical compliance.

they are non-reactive. Any additional vapor that contacts the surface immediately volatilizes. The result is a monolayer coating with a high organic content.

Some degradation of the phenyl-methyl silicone does occur in the harsh furnace environment. However, silicones are organic silicon oxides so the degradation by-product is harmless – a few additional angstroms of silicon oxide on top of the native oxide.

> **Lesson #2A:** Even in cerdips, organic-rich coatings that are only a few angstroms thick suppress stiction with high reliability. Monolayer SAM coatings described in the literature do not survive this harsh process. Organics with high thermal and oxidative stability can be used if their volatilization, by-product and purity characteristics are fully considered.
>
> **Lesson #2B:** Gases adsorbed onto MEMS surfaces affect the electrical output of sensitive devices.

The third generation air bag sensor (the ADXL76 and its two-axis counterpart the ADXL276) solved the problems caused by MEMS surface

characteristics in the ADXL75. These products are sold in cerpacs. Cerpacs are similar to cerdips. However, the leads are not straight. Rather, they are formed in order to create a surface mountable package. The ADXL76 solution required a system level approach with three major elements:
1. Continued use of U.S. Patent 5,694,740 anti-stiction technology.
2. Re-design of the MEMS sensor to ensure adequate mechanical recovery force if microstructures come into contact.
3. Circuit re-design that substantially reduced susceptibility to surface electrical effects.

Lesson #3: High volume production of MEMS devices with integrated electronics is practical. The result, an extremely reliable air bag sensor, was achieved only after considerable development and experience.

In 1998, ADI introduced the ADXL202 accelerometer for industrial and commercial applications. This 2-g full-scale product measures acceleration in two axes with one sensor (Figure 3). The sensor design increases stiction susceptibility because shock loading can cause rubbing contact of sensor surfaces. (Even when shock loading causes the surfaces to contact during handling of single-axis air bag sensors, the direction of impact is usually

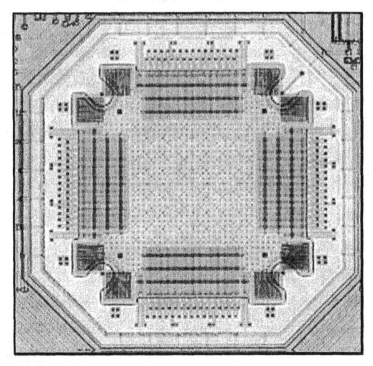

Figure 3. XL202 sensor. This sensor (400 microns square) has beams an all four sides of the center mass so it measures acceleration in both x and y axes.

normal to the surfaces.) The ADXL202 design also required package changes that were difficult to integrate with the package level anti-stiction process. In summary, we had reached the limits of the package level anti-stiction process.

The solution combines a new wafer level anti-stiction process with a slightly re-designed ADXL202 sensor. As noted above, a phenyl-methyl silicone is used in the package level anti-stiction process. The "weak links" in this silicone molecule are the methyl groups – they are more susceptible to thermo-oxidative degradation than the phenyl or siloxane portions. Therefore, the material used in the wafer level process is a diphenyl siloxane – all methyl groups are eliminated so the coating is more robust and is capable of surviving the cerpac assembly process.

The wafer level anti-stiction program examined films formed on silicon from several diphenyl siloxanes. The investigation included a wide variety of vapor deposition conditions and two different types of deposition equipment. One surprising result was the difference in anti-stiction performance observed between classes of diphenyl siloxanes – a difference that was essentially unexplained when coatings were compared using sophisticated surface analysis techniques, including ToF-SIMS and wavelength scanning ellipsometry. The program concluded with a process based on a custom synthesized anti-stiction material[2] that has impressive characteristics. For example, 100 wafer coating runs have a coating thickness uniformity of +/- one angstrom. There is no practical way to measure coatings on MEMS surfaces with this level of precision. Therefore, specially prepared wafers are placed in each furnace boat to monitor thickness. Figure 4 shows typical results. Run to run uniformity is equally impressive (Fig. 5).

Stiction failures occur when the mechanical restoring force is insufficient to overcome stiction forces. There are two classes of stiction forces:

1. Surface forces. The above discussion described techniques for reducing surface forces. The result is adequate for current products, but insufficient for low spring force designs that characterize lower g accelerometers.

2. Electrostatic forces. These forces arise when adjacent structures have different electrical potentials – a situation that is common in MEMS. The solutions to electrostatic capture generally combine both MEMS and circuit design concepts.

---

[2] Patent Application filed.

Surface Characteristics of Integrated MEMS

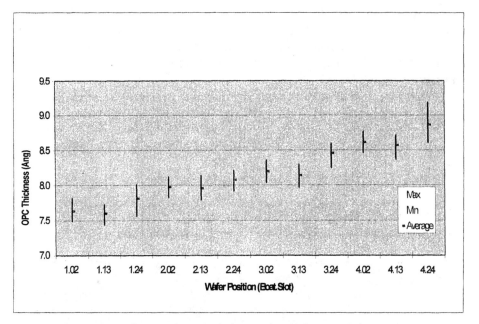

Figure 4: Wafer-to-wafer anti-stiction coating thickness variation (angstroms). 100 wafers.

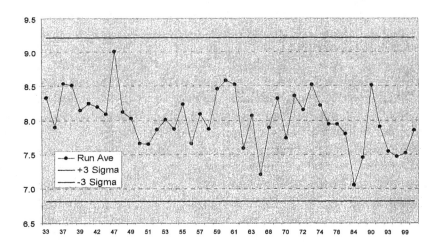

Figure 5. Run to run variation in thickness (angstroms) of anti-stiction coating in 100 wafer coating process.

Competitive market price and application requirements are both driving chip sizes down. As microstructures shrink (Fig. 6), the mechanical restoring force is reduced. However, the stiction forces remain unchanged. Thus, stiction has become a fundamental constraint to further MEMS miniaturization.

Lesson #4: In high volume applications of integrated MEMS products with inertial sensors, stiction is a fundamental performance limitation. New anti-stiction technologies are required. These will likely require lower temperature and lower cost MEMS-friendly packaging technologies.

XL50
10.8 sq mm

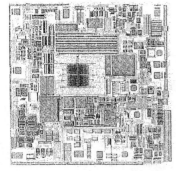

XL76
5.4 sq mm

XL78
2.7 sq mm

Figure 6. Die shrink in single axis air bag sensors. Each device has one half the area of the earlier generation device.

3 CONCLUSIONS

This discussion uses the rapid growth of surface micromachined accelerometers with integrated electronics to illustrate the effect of MEMS surfaces on product performance and reliability. Examples included both electrical characteristics and stiction. Each advance required a combination of scientific insight, practical engineering and considerable investment. Further advances will require even greater resource commitment as smaller and lower range accelerometers as well as low cost gyroscopes are brought to market.

High surface-to-volume ratio and the close proximity of microstructures make an understanding of surface characteristics essential to successful commercialization of future MEMS products. Each new type of device will have unique problems. However, the experience gained with inertial microstructures is invaluable. These hard-earned lessons will guide the development and implementation of new MEMS products, including those designed for optical communications and for microrelays.

Chapter 15

TRIBOLOGICAL ISSUES IN THE IMPLEMENTATION OF A MEMS-BASED TORPEDO SAFETY AND ARMING DEVICE

Wade G. Babcock
Indian Head Division, Naval Surface Warfare Center
Indian Head, Maryland

Abstract NSWC-Indian Head Division is designing and prototyping a MEMS-based Safety and Arming device for deployment in the next generation of small, multi-platform torpedoes. Designs under consideration utilize either LIGA-Nickel, deep reactive ion etch (DRIE) or thin optical lithography nickel barriers, actuators and other moveable, spring supported structures. Critical issues for these devices will be sticking and sliding friction, wear during use, and surface morphology induced "snagging" and friction. In both the nickel-based and the silicon-based structures, there will be a need to characterize and quantify surface roughness and surface films (both intentional films deposited during fabrication and unintentional films resulting from environmental exposure). Another issue will be the long term effects of transportation and storage on the these structures, such as how they might suffer from vibration-induced fretting corrosion or surface wear, and how these tribological phenomena might degrade performance when the time comes for the device to be put to use.

1 INTRODUCTION

The Naval Surface Warfare Center, Indian Head Division (IHDIV/NSWC) is designing and prototyping a MEMS-based Safety and Arming (S&A) device for deployment in the next generation of small, multi-platform torpedoes. This paper presents an overview of the NSWC MEMS program and a summation of tribological topics that will need to be analyzed as the project transitions from prototyping to engineering and manufacturing development.

The S&A's main function is to ensure weapon warhead safety and only allow detonation of the warhead when specific deployment criteria are met. NSWC's goal is to design a mass-produced, lower cost, "smarter" MEMS S&A with increased reliability and safety. The modular approach of this S&A is consistent with the long-term Navy goal of "building block" S&A architecture more easily adaptable to numerous weapon systems. It would require less modification and provide significant development and production cost savings over S&As specifically tailored for each weapon.

Designs under consideration utilize structures fabricated from either LIGA-Nickel (LIGA is the German acronym for electroplating metal into a mold)[1,2], thin optical lithography nickel [3], or deep reactive ion etching (DRIE) processes [4,5]. The structures in the design (Figure 1) consist of:

- A barrier to prevent transmission of shock energy to the high explosive material
- Actuators to physically move the barriers or lock and unlock its motion
- Other moveable, spring-supported structures that interact with the external environment either through water pressure, water flow, or acceleration.

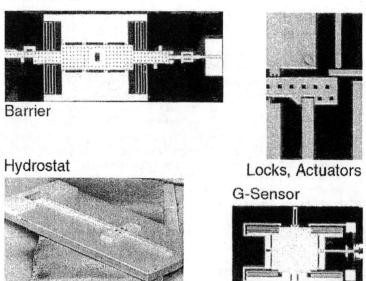

Figure 1. Sampling of structures being integrated into the NSWC MEMS Safety and Arming device. All are X-ray LIGA nickel on 400 μm silicon base wafer. Typical thickness of nickel plate is ~100 μm.

The MEMS S&A may be viewed as a system of checks which confirm that the weapon meets a series of criteria prior to it being able to energize its firing system and detonate. If only some, but not all of the criteria are met, the weapon will not be able to detonate, thus limiting the potential for accidental detonation.

The structures in our design are typically on the order of 5 to 15 millimeters long (at most), 20 to 100 microns wide and 50 to 100 microns thick. In the case of the nickel structures, they are fabricated by electroplating from nickel sulfamate into mold structures (the LIGA process) created by X-ray or optical lithography processes. The plating base and mold are then etched away leaving spring supported structures floating less than 3 microns above the silicon substrate. Our DRIE structures are created by etching through a top layer of silicon down to a sacrificial layer of silicon oxide, which resides on a silicon substrate. The thickness of the top layer of silicon determines the thickness of the structures and the thickness of the separating oxide dictates the clearance of the structure from the substrate.

Since these structures are so compliant out-of-plane (literally paper-thin), the S&A will be packaged with a delimiter top-chip to prevent out-of-plane motion of the structures. (Figure 2) The clearance between the top-chip and the moveable structures is on the order of 15 microns, ±5 microns. The top-chip above the MEMS device will have to allow for at least one of our structures to move out-of-plane, which will be accomplished by etching a "raised ceiling" into specific areas of the delimiter chip.

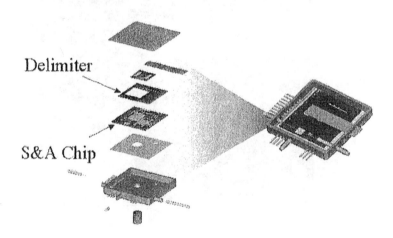

Figure 2. Exploded schematic of the S&A chip and its packaging components. Structures on the chip must be constrained in all three axes during storage and use. Constraint in x and y is accomplished by fixed LIGA structures on the chip, while z motion is constrained by a delimiter chip bonded to the base chip during storage and use. The packaging also allows the chip access to electrical, water pressure, and fiber optic connections.

2 CURRENT TECHNOLOGY ASSESSMENT

Obviously the NSWC device does not fall into the category of MEMS devices which have to actuate thousands or millions of cycles. The S&A

barrier must move about 400 to 500 microns in one direction to arm, and move back into the initial position to re-safe. The current scheme to accomplish this involves two driving actuators which engage and push the barrier forward, and a third which then engages the barrier to prevent it sliding back while the drivers reset to repeat the forward push. This stepping process will be repeated about 25 times to move the barrier to the armed position.

So far in the development process, we have studied the mechanical resistance to be overcome from the spring-supported structures. This has proven to be problematic for a number of reasons, primarily due to the non-adherence of micro-scale nickel to the mechanical properties of bulk, macro-scale nickel.[6] The NSWC development team has been aware of the need for static and sliding friction data, but has only recently built a fully integrated prototype S&A assembly which will allow this type of study.

Figure 3. SEM view of typical LIGA nickel structure surface morphology. Bright layer on the top is a conformal gold electroplated coating. Top surface roughness is primarily due to termination of columnar grain growth at end of plating step.

IHDIV/NSWC has characterized the surface topology of the structures through visual examination from very early in the production process. (Figure 3) The effort was part of our reliability and quality assurance initiative to develop a rigorous understanding of the nascent technology of LIGA fabrication.

During the development process we have compiled information on the plating height variation inherent in the X-ray LIGA process. This has impact on our packaging effort because the height of plated structures varies across the plated wafer. Much effort has gone into controlling plating height to tighter tolerances and designing packaging schemes to allow for plating variations. (Plating height variation in X-ray LIGA nickel has been the main driving force in our pursuit of alternative fabrication techniques such as optical LIGA and DRIE.)

Most work in X-ray LIGA has been with plating heights ranging from less than 50 microns up to 1 mm. Some of the plating height variation that our work has revealed can be compensated for with planarization, but that induces residual stress issues that often cause the structures to "bow" or curve like a potato chip. Our corporate partner in this work (Cronos Integrated Microsystems, Inc., Research Triangle Park, NC) has looked into the viability of optical lithography-based LIGA processes with typical plate heights between 20 and 40 microns. Cronos has had success to the point of beginning fabrication of marketable actuators made with this process.

DRIE fabrication alleviates some of these design problems by controlling height of finished structures in an entirely different way. The SOI (silicon on oxide) wafers used for DRIE are made by oxidizing a standard silicon wafer, creating a layer of a very controllable and uniform thickness oxide. Then another silicon wafer is bonded to this surface and milled to the desired thickness.

The structures are created by etching down to the oxide layer and then undercutting selected structures to make them moveable. They are thickness-controlled by the initial thickness of the bonded silicon wafer. Thus the main issue for tribological study in DRIE structures is the etch-created surfaces between structures and guides and top and bottom wafer.

3 TECHNOLOGY NEEDS

Critical issues for these devices will be static and sliding friction, snagging, wear during use, wear during shipping and storage, and surface films. The barrier and an acceleration sensor (or G-sensor) have fairly large footprints and therefore are most susceptible to sticking and sliding friction. The actuators are long and slender with less total contact surface area and therefore friction is also of concern, but not as much as with the larger structures. Some of the structures also contact the "sharp" edges of etched-through windows in the base wafer and fixed guides which serve to limit their in-plane motion and also increase the risk of snagging during motion. In some cases the gold plating applied to the nickel to lessen oxidation and allow wire-bonding directly to the structures has appeared to delaminate which could pose friction interference with the top-chip. Long-term deleterious effects of storage such as oxidation or corrosion are also of concern.

3.1 Static and Sliding Friction

During fabrication of X-ray and optical LIGA parts, a plating base is deposited onto the base wafer, and then the mold material is applied and patterned. Once the unwanted mold material is removed and the plating base is exposed, nickel (in our case, but in general most any metal) is plated onto the exposed plating base and up to target heights within the mold area. The remaining mold material is then removed, and the plating base is etched away to release the structures. Careful design ensures that fixed structures are large enough to prevent the plating base from being etched from underneath them, while floating structures have their plating base completely undercut and removed.

The bottom surface of the nickel is therefore a relief of the plating base, while the surface morphology of the top of the nickel is produced by the growth of the metal crystals during plating. (Figure 3) The characteristics of the structures' edges are dependent on the mold surface created during patterning.

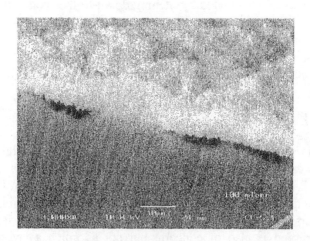

Figure 4. SEM view of side and top of LIGA nickel structure. Vertical striations on the sidewall are due to pixelation effects in the X-ray mask leaving relief in the sidewalls of the polymethylmethacrylate (PMMA) mold. The columnar grain terminations can easily be seen in the top of this structure.

For all the moveable structures, we will need to characterize the top, bottom and side surfaces of the plated nickel. (Figure 4) While the bottom moves in close proximity and possibly in contact with the silicon base wafer, the top of the structure may contact the delimiter chip. Where fixed nickel guides bound the structures, the edges of two plated structures will be in contact. All of these surfaces will have to be studied for their effect on static and sliding friction. This data will be used to assure that the design of the S&A will account for and overcome resultant frictional effects.

3.2 Snagging

Surface morphology resulting from the items discussed in 3.1 will also impact the ability of structures to move past fixed items on the base wafer. These include etched-through windows in the base wafer and the guides, which limit in-plane motion of the S&As structures.

The S&A design includes a window etched into the base wafer to bring a high-explosive pellet in close proximity to a detonator chip. (Figure 5) The barrier also has a window, which when moved in line with the detonator chip and pellet, allows transmission of kinetic shock energy to begin detonation of the warhead. Thus the barrier has to move over a relatively sharp edged window directly etched into the base wafer.

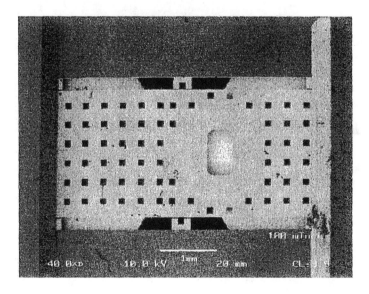

Figure 5. SEM view from the bottom of the S&A chip, showing the "picture frame" opening etched into the underside of the silicon base wafer. The bottom of the LIGA nickel slider/barrier is visible in the center.

At risk here is the possibility of surface roughness on the barrier's underside snagging on the edge of the silicon window.

The edges of the nickel structures are relief images of the mold walls created during patterning (clearly evident in Fig. 4). Roughness here is largely the result of pixelation in creating the mask, and produces a continuous series of top-to-bottom striations along the edge of the structures. As two nickel structures attempt to slide past one another, they would encounter something akin to two pieces of corduroy sliding past one another. The DRIE-fabricated structures' edges are created in an etch process. These surfaces and their interactions are largely unknown.

All of the large moving structures are bound in-plane with fixed guides, creating contact areas where sliding occurs. In addition, friction effects play a role in the case of locking structures attempting to disengage. Here a structure will have to move out of a slot, much like a deadbolt retracting from a door jam. (Figure 6) The drive actuators also have gear teeth, which engage similar teeth on the barrier. At risk here would be binding due to surface roughness and a resultant inability of the actuator teeth to disengage.

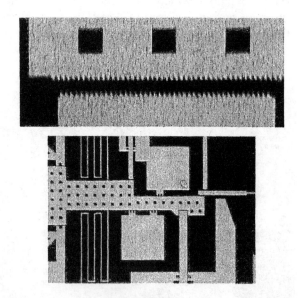

Figure 6. (Top) Close-up view of the teeth patterned on the edge of LIGA structures to facilitate better actuator grip. (Bottom) General view of a portion of the S&A showing actuators (to the right), a lock (center-bottom) and spring-supported slider/barrier (center, left to right).

All of these surfaces would need to be characterized and correlated to physical device malfunctions, then resultant design considerations employed.

3.3 Wear during Use

The NSWC MEMS device will not suffer from wear issues inherent to devices that operate for thousands or millions of cycles. We anticipate wear of a frictional nature common to objects sliding past one another not to be a factor. Of critical importance however is the ability of the S&A's structures to stand up mechanically to the process of tooth engagement or lock disengagement.

It is not unreasonable to presume that during just 10 or so cycles there is a possibility of the actuator teeth becoming dulled or damaged due to misalignment on engagement. This could affect performance in the latter stages of arming. Since this phenomenon is most affected by mechanical properties of the nickel, it is not directly of interest to the tribological community.

3.4 Wear during Storage and Shipping

One of the aspects unique to MEMS devices like the S&A and automobile airbag accelerometers is the necessity for potentially long-term storage and transportation in a complex mechanical environment where vibration is commonplace. This creates the possibility that in-situ wear from very minute vibration-induced motions could impair the device's ability to function as designed and therefore must be addressed during the early development phases of such devices.

Factors that would need to be addressed include:
- Lock Degradation – structures in physical contact and exposed to vibrational wear may disengage due to increased gaps or tolerances.
- Lock Binding – contact between structures could create wear patterns, which might prevent a lock from being able to disengage when needed.
- Debris – particulate accumulation in the S&A package resulting from long term frictional wear.
- Tolerances – if locks and other structures degrade, their tolerances could be increased to the point of functional impairment.

Study of prototype S&As and other MEMS devices already in use for a period of time should elucidate this problem.

3.5 Surface Films

As mentioned before, the nickel structures are plated with a protective layer of gold during fabrication. (Gold plated coating may be seen in Figure 4.) What would be of interest is the effects of this conformal coating and other coatings intentionally applied during fabrication on the aforementioned tribological issues.

The NSWC device has exposure to its environment in the form of water pressure and flow, and the chemical compounds present in the high explosive. Exposure of this type could lead to unintentional surface films, which could have unpredictable properties and implications. During design the need to hermetically seal the S&A has been studied but is not resolved, and the level of isolation the S&A needs would depend on surface characterization and tribological issues discussed in this paper. The need to seal the S&A raises manufacturing cost and could be avoided if it is unnecessary.

3.6 Two Notes on Surface Characterization

Surface characterization might be done with white light interferometry or similar processes, with the main caveat being that we are in need of studying large (at least in terms of MEMS) surface areas. Thus techniques such as AFM and SEM have no or only limited application.

The long shelf life of these devices raises important issues of how to simulate age and storage conditions. Surface characteristics could be drastically affected by large temperature ranges and exposure to environmental conditions. Chemical reactions between the MEMS materials and the compounds present in the explosive train are a completely virgin area of study, which may have serious effects on the surfaces in question.

4 CONCLUSION

In both the nickel-based and the silicon-based structures, there will be a need to characterize and quantify surface roughness and surface films (both intentional films deposited during fabrication and unintentional films resulting from environmental exposure). Another issue will be the long term effects of transportation and storage on the these structures, such as how they might suffer from vibration-induced fretting corrosion or surface wear, and how these tribological phenomena might degrade performance when the time comes for the device to be put to use.

These areas are largely uncharted and present formidable challenges (or opportunities!) for fundamental contributions to the MEMS industry.

ACKNOWLEDGMENTS

The author would like to thank Donald Garvick and Lawrence Fan for assistance on this paper. The Indian Head Division, Naval Surface Warfare Center's MEMS Fuze/Safety and Arming program is funded by the Office of Naval Research and the Defense Advanced Research Projects Agency.

REFERENCES

1. Madou, M., *Fundamentals of Microfabrication*, Washington, D.C.: CRC Press, 1997 (p. 275).
2. Spiller E, Feder R, Topalian J, Castellani E, Romankiw L, Heritage M. X-ray Lithography for Bubble Devices, Solid State Technology, April 1976; 62-68.
3. Madou, M., *Fundamentals of Microfabrication*, Washington, D.C.: CRC Press, 1997 (pp. 303-304).
4. Pandhumsoporn T, Feldbaum M, Gadgil P, Puech M, Maquin P. High Etch Rate, Anisotropic Deep Silicon Plasma Etching for the Fabrication of Microsensors. Micromachining and Microfabrication Process Technology II. Austin, Texas, USA, 1996 (pp. 94-102).
5. Brodie I, Murray JJ. The Physics of Microfabrication. New York: Plenum Press, 1982.
6. Last HR, MEMS Reliability, Process Monitoring, and Quality Assurance. Proceedings of the MEMS Reliability for Critical and Space Applications Symposium; 1999 August. Society of Photo-Opitcal Instrumentation Engineers, 1999.

Chapter 16

CHALLENGES FOR LUBRICATION IN HIGH SPEED MEMS

Kenneth Breuer
Division of Engineering
Brown University
Providence RI 02912

Frederic Ehrich, Luc Fréchette, Stuart Jacobson, C-C Lin, D.J. Orr, Edward Piekos, Nicholas Savoulides and Chee-Wei Wong
Gas Turbine Laboratory
Massachusetts Institute of Technology
Cambridge, MA 02139

Abstract Design considerations for lubrication in high speed rotating MEMS devices are discussed, using the MIT Microengine program as the primary example. This device relies on the rotation of a microfabricated silicon rotor at speeds well in excess of one million RPM, and uses gas bearings to support the rotor both in the axial and radial directions. MEMS fabrication places unique and novel constraints on the design of the gas bearing system and these lead to a relatively unusual design space for high-speed lubrication systems. The overall architecture for the microengine lubrication system is described using both experimental and computational data to illustrate the effects of geometric and manufacturing variations on the overall system performance.

1 INTRODUCTION

As microengineering technology continues to advance, driven by increasingly complex and capable microfabrication and materials technologies, the need for more and more sophistication in MEMS design will increase. For the most part, existing MEMS devices tend to be very low energy density machines, rarely pushing material or thermodynamic limits. This is now beginning to change as scientists and engineers are attempting to achieve more and more complex functionality from MEMS devices and to increase the power density of the devices towards their practical maximum. At the macroscopic scale, gas turbine engines and aerospace vehicles are examples of high power density devices operating at the limits of materials, structures and fluid capabilities. Similar developments are underway at the microscopic scale and will no doubt be of great importance as micro and nanoengineering develop in the years to come.

An example of such "Power-MEMS" development is provided by a project initiated at the Massachusetts Institute of Technology in 1995 to demonstrate a fully functional microfabricated gas turbine engine [3]. The baseline engine, illustrated in Fig. 1, consists of a centrifugal compressor, fuel injectors (hydrogen is the initial fuel, although hydrocarbons are planned for later configurations), a combustor operating at 1600K and a radial inflow turbine. The device is constructed from single crystal silicon and is fabricated by extensive and complex fabrication of multiple silicon wafers, which are fusion bonded in a stack to form the complete device. An electrostatic induction generator may also be mounted on a shroud above the compressor to produce electric power instead of thrust [8]. The baseline MIT Microengine has at its core a "stepped" rotor consisting of a compressor with an 8 mm diameter and a journal bearing and turbine with a diameter of 6mm. The rotor spins at 1.2 million RPM.

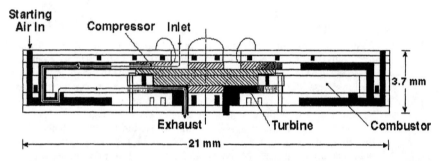

Figure 1. Schematic of the MIT Microengine, showing the air path through the compressor, combustor and turbine. Forward and Aft thrust bearings located on the centerline hold the rotor in axial equilibrium, while a journal bearing around the rotor periphery holds the rotor in radial equilibrium.

Key to the successful realization of such a device is clearly the ability to spin a silicon rotor at high speed in a controlled and sustained manner, and clearly the key to spinning a rotor at such high speeds is the demonstration of efficient lubrication between the rotating and stationary structures. The lubrication system needs to be simple enough to be fabricated and yet with sufficient performance and robustness to be of practical use both in the development program and in future devices. To develop this technology, a micro-bearing rig has been fabricated and is illustrated Fig. 2. Here, the core of the rotating machinery have been implemented, but without the substantial complications of the thermal environment that the full engine brings. The rig consists of a radial inflow turbine mounted on a rotor and embedded inside two thrust bearings, which provide axial support. A journal bearing located around the disk periphery provides radial support for the disk as it rotates.

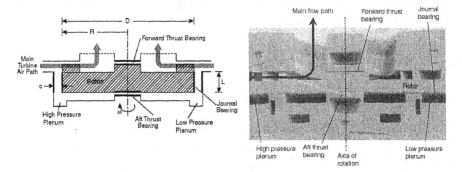

Figure 2. Illustrating schematic, and corresponding SEM of a typical microfabricated rotor, supported by axial thrust bearings and a radial journal bearing.

1.1 Prior State of the Art

Prior to this work, little thought or effort was placed in lubrication issues for micromachined devices. The reason was clear - most MEMS operated at low speeds (particularly low surface speeds) and at very low power density. In addition, in the interests of research most devices were not required to operate for extended periods of time. For all these reasons little effort was placed on lubrication and dry-contact "lubrication" has been used for a variety of rotating devices, most notably a series of electrostatic micromotors first fabricated in the 1980's by research teams at MIT and Berkeley [1,7], and more recently at Sandia [15]. This solution was acceptable for the low surface speeds (300,000 RPM on a 100-μm rotor corresponds to a rim speed of only 1.5 m/s). However, for the high power density application such as the MIT Microengine, a more robust solution was required and this need initiated a research program on developing the necessary tools and appreciation for gas bearing technology as applied to high performance MEMS devices.

1.2 Objectives and Outline

The objective of this paper is two-fold. Firstly we wish to report on a particular device – the MIT Microengine - which requires a high-integrity lubrication system and for which we have successfully designed, fabricated and tested a micro-scale gas bearing system. The second objective is somewhat broader, and that is to try to identify endemic issues that face all microbearing systems as a result of their size, the existing state-of-the-art in microfabrication technology and, perhaps most importantly the rigors of multidisciplinary design in which every decision is constrained by the fundamental laws of physics, the ability to manufacture a given design and the interactions that every aspect of the lubrication sub-system has on the successful operation of the entire device.

2 FUNDAMENTAL SCALING ISSUES

2.1 The Cube-Square Law

The most dominant effect that changes our intuitive appreciation of the behavior of microsystems is the so-called "cube-square law" which simply reflects the fact that volumes scale with the cube of the typical length scale while areas (including surface areas) scale with the square of the length scale. Thus, as a device shrinks, surface phenomena become relatively more important than volumetric phenomena. The most important consequence of this is the observation that the device mass (and inertia) becomes negligibly small at the micro- and nano-scales. For a hydrodynamic bearing, this is illustrated by the non-dimensional load parameter:

$$\zeta = \frac{F}{pLD}, \qquad (1)$$

where F is the applied load, p the ambient pressure, L the bearing length and D the bearing diameter. Thus, the load parameter compares the applied force to the pressure force acting on the projected bearing area. If we allow F to be the weight of the journal:

$$F = \rho_s \pi R^2 L, \qquad (2)$$

(where ρ_s is the density of the rotor, typically silicon), then we find that the load parameter has the following dependence:

$$\zeta = \frac{\rho_s \pi R}{p}, \qquad (3)$$

from which it can clearly be seen that the load parameter decreases linearly as the device shrinks. In particular, the load parameter (due to the rotor mass) for the MIT Microengine is approximately 10^{-3}. This has advantages and disadvantages. The benefits are that orientation becomes effectively irrelevant and that unloaded operation is easy to accomplish, should one desire to do so. The primary disadvantage of the low natural loading is that unloaded operation is often undesirable (in hydrodynamic lubrication where a minimum eccentricity is required for journal bearing stability), and thus a scheme for applying an artificial load needs to be developed. This is discussed in more detail later on in the paper.

2.2 Applicability of the Continuum Hypothesis

A common concern in microfluidic devices is the appropriateness of the continuum hypothesis as the device scale continues to fall. At some scale, the typical inter-molecular distances will be comparable to the device scales and the use of continuum fluid equations becomes suspect. For gases, this is measured by the Knudsen number (Kn) - the ratio of the mean free path to the typical device scale. Numerous experiments (*e.g.*, Arkilic, *et al.* [1]) have determined that non-continuum effects become observable when Kn reaches approximately 0.1, and that continuum equations become meaningless (the "transition flow regime") at Kn of approximately 0.3. For atmospheric temperature and pressure, the mean free path (of air) is approximately 70 nm. Thus, atmospheric devices with features smaller than approximately 0.2 μm will be subject to non-negligible non-continuum effects. However, in the case of the microengine, such dimensions are not present, and thus the fluidic analysis for the microengine design can use the standard Navier-Stokes equations.

Nevertheless, in applications where viscous damping is to be avoided (for example in high-Q resonating devices such as accelerometers or gyroscopes) the working structure is often packaged at low pressure. In such cases non-continuum effects do need to be fully considered. This regime, however, is not discussed here.

3 CONSTRAINTS ON MEMS BEARING GEOMETRIES

3.1 Device Aspect Ratio

Perhaps the most restrictive aspect of microbearing design is the fact that MEMS devices are limited to rather shallow etches, resulting in devices with low aspect ratio. Even with the advent of Deep Reactive Ion Etchers, or DRIE (in which the ion etching cycle is interleaved with a polymer passivation step), the maximum practical etch depth that can be achieved while maintaining dimensional control is about 500 microns (even this has an etch time of about nine hours which makes its adoption a very costly decision). This is in comparison with typical rotor dimensions of a few millimeters. The result is that microbearings are characterized by very low aspect ratios (length/depth, or L/D). In the case of the MIT microturbine test rig, the journal bearing is nominally 300 microns deep while the rotor is 4 mm in diameter, yielding an aspect ratio of 0.075. To put this in perspective, commonly available design charts [6] present data for values of L/D as low as 0.5 or perhaps 0.25. Prior to this work there was no data for lower L/D. The implications of the low aspect ratio bearings are that the task becomes supporting a *disk* rather than a *shaft*.

In fact, the low aspect ratio bearings do not have terrible performance by any standard. The key features of the low L/D bearings are:

- The load capacity is reduced compared to "conventional" designs. This is because the fluid leaking out of the ends relieves any tendency for the bearing to build up a pressure distribution. Thus, for a given geometry and speed, a short bearing supports a lower load (per unit length) than its longer counterpart.
- The bearing acts as an incompressible bearing over a wide range of operation. The reason for this is as above – pressure rises, which might lead to gas compressibility, are minimized by the flow leaking out the short bearing. The implication of this is that incompressible behavior (without, of course, the usual fluid cavitation that is commonly assumed in incompressible liquid bearings) can be observed to relatively high speeds and eccentricities.

3.2 Minimum Etchable Clearance

It is reasonable to question why one could not fabricate a 300-µm long "shaft," but with a much smaller diameter, and thus greatly enhance the L/D. For example, a shaft with a diameter of 300 µm would result in a reasonable value for L/D. However, this raises the second major constraint on bearing design raised by current microfabrication technologies – that of the minimum etchable clearance.

In the current microengine manufacturing process, the bearing/rotor combination is defined by a single deep and narrow etch, currently 300 microns deep and about 12 µm in width. No foreseeable advance in fabrication technology will make it possible to significantly reduce the minimum etchable clearance, and this has considerable implications for bearing design. In particular, if one were to fabricate a bearing with a diameter of 300 µm (in an attempt to improve the L/D ratio), it would result in a bearing with a clearance to radius, c/R, of 12/300 or 0.04. For a fluid bearing, this is *huge* (two orders of magnitude above conventional bearings) and has several detrimental implications.

The most severe of these is the impact on the dynamic stability of the bearing. The non-dimensional mass of the rotor depends on (c/R) raised to the *fifth* power [12]. Thus, by bringing the bearing into the center of the disk and raising the c/R by a factor of 10 results in a mass parameter increasing by a factor of 10^5. This increase in effective mass has severe implications for the stability of the bearing, which are difficult to accommodate.

All of these reasons, and others not enumerated here (such as the issue of where to locate the thrust bearings) make the implementation of an inner-radius bearing less attractive and thus, the constraint of small L/D is, in our opinion, unassailable as long as one requires that the micro device be fabricated in-situ. If one were to imagine a change in the fabrication process such that the rotor and bearing could be fabricated separately and subsequently assembled (*reliably*), this situation would be quite different. In such an event, the bearing gap is not constrained by the minimum etch dimension of the fabrication process, and almost any "conventional" bearing geometry could be considered and would probably be superior in performance to the bearings discussed here. Such fabrication could be considered for a "one-off" device, but does not appear feasible for mass production, which relies on the monolithic fabrication of the parts. Lastly, the risk of contamination during assembly – a common concern for all precision-machined MEMS – effectively rules out piece-by-piece manufacture and assembly and again constrains the bearing geometry as described.

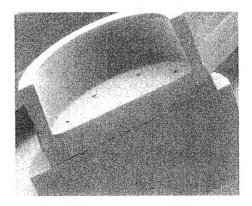

Figure 3. Close-up cutaway view of micro thrust bearing showing the pressure plenum (on top), the feed-holes and the bearing gap (faintly visible). SEM courtesy of C. C Lin.

4 THRUST BEARINGS

Because of the low aspect ratio of the rotor and bearing system, high performance thrust bearings are critical to contain both the axial forces generated by the turbine (or compressor) as well as any tilting tendencies as the disk spins. For simplicity the initial test articles of the MIT rotors use hydrostatic thrust bearings of a rather conventional design in which a pressurized plenum drives air through a series of small restrictors and radially outward through the thrust bearing film. This is illustrated in Fig. 3, which shows an SEM of the fabricated device cut though the middle to reveal the plenum, restrictor holes and the bearing lubrication gap, which is approximately 1-μm wide. Key to the successful operation of hydrostatic thrust bearings is the accurate manufacture of the restrictor holes, maintenance of sharp edges at the restrictor exit and carefully controlling the dimension of the lubrication film. In an initial fabrication run, the restrictor holes were fabricated 2 μm larger than specified. While the bearing operated, its performance was well below its design peak, due to the off-design restrictor size. Current specifications of the fabrication protocols control the restrictor size carefully, ensuring close to optimal operation.

One implication of the cube-square law is that surface-generated lubrication forces will become more and more effective as the device gets smaller. For this reason, hydrodynamic thrust bearings ("spiral groove" bearings) become more attractive for MEMS bearings.

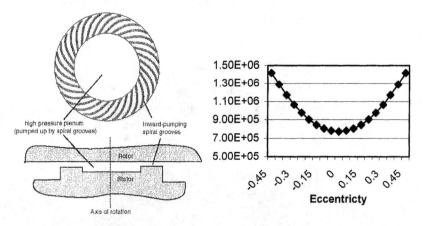

Figure 4. Schematic and predicted performance (stiffness, in N/m, versus axial eccentricity) for a typical spiral groove thrust bearing for use in a high speed MEMS rotor.

Spiral groove bearings, or SGBs (illustrated in Fig. 4) operate by using the rotor motion against a series of spiral grooves (etched in the bearing) to viscously pump fluid into the lubrication gap and thus create a high-pressure cushion on which the rotor can ride. The devices typically have relatively low load capacity, which has limited their use in macroscopic applications. However, the load capacity becomes adequate at microscales due to cube-square scaling. Thus, they gain considerable advantage when compared with conventional hydrostatic thrust bearings as the scale (and Reynolds number) decreases. In addition, the fabrication of the multitude of shallow spiral features (which is an expensive task for a traditional SGB) – is ideally suited for lithographic fabrication technologies such as MEMS. Figure 4 illustrates the bearing stiffness for a single-point design for the MIT micro-rotor rotating at design speed (2.4 million RPM) and supported by matched forward and aft spiral groove bearings. The stiffness at full speed is quite impressive (superior to comparable hydrostatic bearings), but the SGB do suffer at lower speeds since the bearing stiffness is roughly proportional to rotational speed. However, for this design, the lift-off speed (the speed at which the film can support the weight of the rotor and the pressure distribution associated with the turbine flow) is only a few thousand RPM, and so the dry rubbing endured during startup will be minimal. They also have the strong advantage that the two matched spiral groove bearings (forward and aft) naturally balance each other (no supply pressures to maintain or adjust) and most importantly, that the removal of the thrust bearing plena and restrictor holes considerably simplifies the overall device fabrication, and can allow for the use of two fewer wafers in the wafer-bonded stack – a considerable advantage from the perspective of manufacturing process cost and yield.

5 JOURNAL BEARINGS

As discussed earlier, the journal bearing resides on the outer rim of the rotor and is used to support the radial motion of the disk as it spins. Unlike almost any other journal bearing in common usage, the MEMS journal bearing is dominated by its very low aspect ratio and its relatively large gap, both dictated by etching constraints. Journal bearings can operate in two distinct modes - hydrodynamic and hydrostatic, although typically any operating condition will contain aspects of both modes.

5.1 Hydrodynamic Operation

Hydrodynamic operation occurs when the rotor is forced to operate at an eccentric position in the bearing housing. As a consequence of this, a pressure distribution develops in the gap to balance the viscous stresses that arise due to the rotor motion. This pressure distribution is used to support the rotor both statically (against the applied force) and dynamically (to suppress random excursions of the rotor due to vibration, *etc.*). Hydrodynamic operation has the advantage that it requires no external supply of lubrication fluid. However, it has two distinct drawbacks: Firstly, it requires a means to load the rotor to an eccentric position and secondly, insufficient eccentricity results in instability (the so-called "fractional speed whirl") and likely failure. Both of these issues are particularly difficult in the case of MEMS bearings.

5.1.1 Static Journal Bearing Behavior

The static behavior of a MEMS journal bearing is shown in Fig. 5, which presents the load capacity, ζ, and the accompanying attitude angle (the angle between the applied load and the eccentricity vector) as functions of the bearing number and the operating eccentricity. The bearing number is defined as:

$$\Lambda = \frac{6\mu\omega}{p}\left(\frac{R}{c}\right)^2, \qquad (4)$$

Where μ is the fluid viscosity, ω, the rotation rate, p, the ambient pressure and R/c, the ratio of the radius to clearance. For a given geometry, Λ can be interpreted as operating speed.

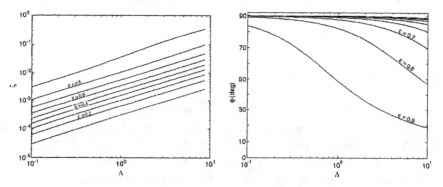

Figure 5. Static performance (eccentricity and attitude angle vs. bearing number) for a journal bearing with $L/D = 0.075$. Notice that the load lines are almost constant (linear), indicating the absence of compressibility effects. This is also indicated by the attitude angle which remains close to 90 degrees except at very high eccentricities [11].

Several aspects of these results should be noted. Firstly, the load capacity is quite small when compared with bearings of higher L/D. As mentioned before, this is due to the fact that, for very short bearings, the applied load simply squeezes the fluid out of the bearing ends, which has difficulty developing significant restoring force. The same mechanism is responsible for the fact that the load lines are straight, indicating that little compressibility of the fluid is taking place (which usually results in a "saturation" of the load parameter at higher values of the bearing number). Again, this is because any tendency to compress the gas is alleviated by the fluid venting at the bearing edge. This point is driven home by the behavior of the attitude angle, which maintains a high angle (close to $\pi/2$) over a wide range of bearing numbers and eccentricities. This value of the attitude angle corresponds to the analytic behavior of a Full-Sommerfeld incompressible short bearing [9] which clearly is a good approximation for such short bearings at low to moderate eccentricities, so long as the eccentricity remains below approximately 0.6, at which point compressibility finally become important. This incompressible behavior is much more extensive than "conventional" bearings of higher aspect ratio and has profound ramifications, particularly with respect to the dynamic properties of the system.

5.1.2 Journal Bearing Stability

The stability of a hydrodynamic journal bearing has long been recognized as troublesome and is foreshadowed by the static behavior shown in Fig. 5. The high attitude angle suggests that the bearing spring stiffness is dominated by cross-stiffness as opposed to direct stiffness. Thus, any perturbation to the rotor will result in its motion perpendicular to the applied force. If this reaction is not damped, the rotor will enter a whirling motion. This is precisely what is observed and gas bearings are notorious for their susceptibility to fractional-speed whirl. The instability is suppressed by the generation of more damping and/or increased direct stiffness, both of which are obtained by increasing the loading and the static eccentricity of the rotor.

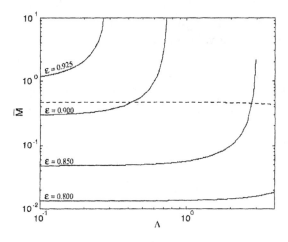

Figure 6. Stability boundaries for a typical microbearing, plotted versus bearing number (speed for a fixed geometry). The dotted line represents an operating line for a microbearing which has almost constant $\overline{M}$ (varying only due to centrifugal expansion of the rotor at high speeds [11].

Figure 6 shows a somewhat unusual presentation of the stability boundaries for a low-aspect ratio MEMS journal bearing. The vertical axis shows the non-dimensional mass of the rotor, $\overline{M}$, which is defined as

$$\overline{M} = \frac{mp}{72L\mu^2}\left(\frac{c}{R}\right)^5. \quad (5)$$

This is the "mass" which appears in the non-dimensionalized equations of motion for the rotor and it is fixed for a given geometry (Close inspection of Fig. 6 indicates that $\overline{M}$ does changes very slightly with speed. This is due to the expansion of the rotor due to centrifugal forces, the variation in the ambient pressure at different speeds and temperature effects on viscosity). The horizontal axis of Fig. 6 shows the bearing number, which can be interpreted as speed, for a fixed bearing geometry. The contours on the graph represent the stability boundary at fixed eccentricity. Stable operation lies above each line. Thus, for a fixed $\overline{M}$, at low bearing number (*i.e.*, speed), a minimum eccentricity must be obtained to ensure stability. As the speed increases, this minimum eccentricity remains almost constant (the lines are horizontal), until a particular speed at which point the lines break upwards and the minimum eccentricity required for stability starts to drop (Note that as Λ increases, the load required to maintain a fixed eccentricity increases linearly due to the stiffening of the hydrodynamic bearing, as indicated by Fig. 5). The key feature of this chart is that the minimum eccentricities are *very* high and suggest that stable operation requires running very close to the wall – this is troublesome. The high eccentricities are driven by high values of the mass parameter, $\overline{M}$, which is due to the relatively high value of the clearance-to-radius ratio (c/R), and the short length, L. The low aspect ratio (L/D) also contributes to high minimum eccentricities. At high speeds, the problem becomes less severe, since the high speed has allowed the bearing to generate sufficient direct stiffness. However, even at these points, the eccentricity is very high and may not be manageable in practical operation.

Orr [9] has demonstrated, on a scaled-up experimental rig that matches the microengine geometry, that stable high-eccentricity operation is possible for extended periods of time. His experiments achieved 46,000 RPM, which when translated to the equivalent speed at the microscale correspond to approximately 1.6 million RPM. However, in order to accomplish this high eccentricity operation, he noted that the rotor system must (a) have very good axial thrust bearings to control axial and tipping modes of the rotor system, and (b) be well-balanced – a rotor with imbalance of more than a few percent could not be started from rest. Piekos [1999] also explored the tolerance of the microbearing system to imbalance and found that it was surprisingly robust to imbalance of several percent. However, his computations were achieved assuming that the rotor was at full speed and then carefully subjected to imbalance. In practice, the imbalance will exist at rest and so it might well be the case that the rotor is stable at full speed, but unable to accelerate to that point. This "operating line" issue is discussed in some more detail by Savoulides [14] who explored several options for accelerating microbearings from rest under both hydrodynamic and hydrostatic modes of operation.

A convenient summary of the trade-offs for design of a hydrodynamic MEMS bearing is shown in Fig. 7. This presents the variation of the low-speed minimum eccentricity asymptote (or worst-case eccentricity, as demonstrated in Fig. 6) as functions of the mass parameter, $\overline{M}$, and other geometric factors (L/D, clearance, c, etc.). Notice, that the worst-case eccentricity improves as the L/D increases and the $\overline{M}$ decreases. However, it is interesting to note that the physical running distance from the wall is actually increased (slightly) by running at a higher eccentricity, but with a larger bearing gap. In all cases, however, the stable eccentricity is alarmingly high and other alternatives need to be sought for simpler stable operation.

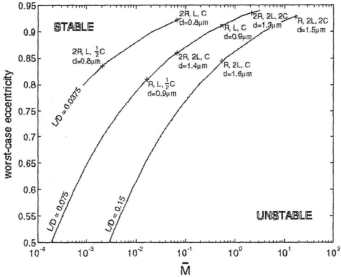

Figure 7. Tradeoff chart for microbearing design. For a given length-to-diameter (L/D) and a given $\overline{M}$, the worst-case (*i.e.*, low-speed) eccentricity is shown for a variety of geometric perturbations. In general lower eccentricities are preferred [11].

5.2 Advanced Journal Bearing Designs

One prospect for further improvement in the journal bearing performance is the incorporation of "wave bearings" [2] as illustrated in Fig. 8. These bearing geometries have been demonstrated to suppress the sub-synchronous whirl due to the excitation of multi-synchronous pressure perturbations imposed by the bearing geometry. As an added attraction, the geometric complexity of the wave bearing is no problem for lithographic manufacturing processes such as are used for MEMS, thus alleviating many of the reservations and cost that might inhibit their adoption. However, since the MEMS constraint is the minimum gap dimension, the wave bearing in a MEMS machine can only be implemented by selectively enlarging the bearing gap. Piekos [11] has analyzed the performance of the wave bearing for the microengine geometry and has found (Fig. 9) that, while the load capacity is diminished, the stability is enhanced (as expected) and that the load required to maintain stable operation (*i.e.*, to achieve the minimum stable eccentricity) is reduced considerably with the introduction of a wave geometry. In microbearings the load capacity is usually quite sufficient (the cube-square law to our rescue!) and thus the wave bearing looks very attractive as a stabilizing mechanism.

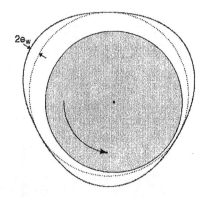

Figure 2. Geometry of a wave bearing, with the clearance greatly exaggerated for clarity [11].

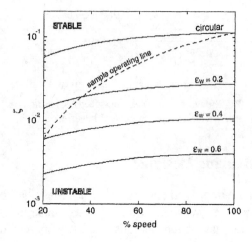

Figure 3. Effect of wave bearing amplitude on journal bearing stability as a function of rotor speed. The dotted line shows a typical operating line for a microengine [11].

Having said that, rotor imbalance, which is increasingly becoming a first-order issue, can only be contained with excess load capacity, and so this tradeoff is by no means clear. The adoption and testing of wave bearing geometries are scheduled to be explored in the near future, as part of our development program.

5.3 Side Pressurization

Due to the small mass of the rotor in a MEMS device, any eccentricity required to enable stable hydrodynamic operation must be applied using some other means. Typically, this requires the use of a pressure distribution introduced around the circumference of the bearing, which serves to load the bearing preferentially to one side. Such a scheme is illustrated in Fig. 10 for the MIT Microengine. Here, the aft side of the rotor is divided into two plena, isolated by seals. Each plenum can be separately pressurized. The pressure in each plenum forces an axial flow through the journal bearing to the forward side (which is assumed to be at a uniform pressure), and thus establishes two differing pressure distributions on the high- and low-pressure sides of the rotor. Of course, nothing comes without its consequences, and the axial flow through the bearing has consequences of its own - the generation of a hydrostatic stiffness mechanism and an associated hydrostatic critical frequency. This is discussed in the following section.

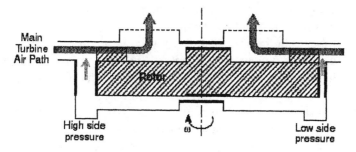

Figure 10. Schematic of the pressure-loading scheme used in the microengine to provide a side-load to the rotor during hydrodynamic operation. The side load is developed by applying a differential pressure to the two plena located on the aft side of the rotor.

5.4 Hydrostatic Operation

Although hydrodynamically lubricated bearings with low aspect ratio are predicted to operate successfully and have now been demonstrated on a scaled up level [9], there are a number of issues that make them undesirable in a practical MEMS rotor system. The primary difficulty is that, in order to satisfy the requirements of sub-synchronous stability, the rotor needs to operate at high eccentricity. For a MEMS device this means operating 1 to 2 μm from the wall. This is hard to control, particularly with the limited available instrumentation that plagues MEMS devices. An alternative mode of operation is to use a hydrostatic lubrication system. In this mode, fluid is forced from a high-pressure source through a series of restrictors, each of which imparts a fixed resistance. The fluid then flows through the lubrication passage (the bearing gap). If the rotor moves to one side, the restrictor/lubrication film act as a pressure divider such that the pressure in lubrication film rises, forcing the rotor back towards the center of the bearing. The advantages of using hydrostatic lubrication in MEMS devices are that

- The rotor tends to operate near the center of the housing and thus small clearances are avoided. This is both "safer" and results in lower viscous resistance.
- Since the hydrostatic system is a zero-eccentricity based system, no position information about the rotor is needed. This greatly simplifies instrumentation requirements

However, there are significant disadvantages to a hydrostatic system, including:

- Pressurized air needs to be supplied to the bearing. This requires supply channels, which complicate the fabrication process. It also comes with a system cost – the high-pressure air must come from somewhere (in a turbomachinery application, bleed air from the compressor could be used).
- Since the bearing gaps are relatively large (due to minimum etchable dimensions discussed above), the mass flow through hydrostatic systems can be substantial and might be impractical in anything but demonstration experiments
- Fabrication constraints make the manufacture of effective flow restrictors very difficult. Flow restrictors need to have very well controlled dimensions, sharp edges as well as other specific geometric features. Only the simplest restrictors can be implemented without undue cost and effort, severely limiting the hydrostatic design.

For journal bearings with low aspect ratio, a novel method for achieving hydrostatic lubrication was demonstrated by Orr [9] and Piekos [11]. The mechanism relies on the fact that small pressure differences will inevitably exist between the forward an aft sides of the rotor and that, for a short bearing of the kind seen in MEMS devices, the flow resistance to that pressure differences is small enough such that an axial flow will ensue.

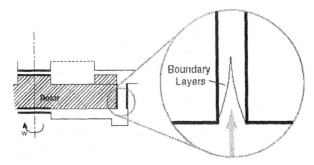

Figure 11. Schematic illustrating the origin of the axial-throughflow hydrostatic mechanism.

As the flow enters the bearing channel, boundary layers develop along the wall, eventually merging to form the fully developed lubrication film. This boundary layer development (Fig. 11) acts as an inherent restrictor so that if the rotor moves off the centerline and disturbs the axisymmetric symmetry of the flow, a restoring force is generated. This source of hydrostatic stiffness acts to support the bearing at zero eccentricity and is effective even when the rotor is not moving. The stiffness is also enhanced by the conventional inherent restriction of the flow entering the lubrication channel. The stiffness, coupled with the rotor mass defines a natural frequency which was measured by Orr [9,10] (indeed the presence of this frequency led to the discovery of the axial-through-flow mechanism). Simple theory [9,10,11] was also able to predict the frequency with reasonable accuracy (Fig. 12). Unfortunately, experiments [9] and computations [14] suggest that the hydrostatic stiffness might not support the rotor up to high speeds and that a stability boundary at some loosely defined supercritical speed limits the top speed that the rotor can achieve using only axial-through-flow hydrostatic support. The accurate prediction of the hydrostatic frequency and the maximum achievable speed under hydrostatic support are still un-resolved issues and are in need of further study. In addition, since the bearing gaps are reasonably large due to minimum etchable dimensions, the mass flow required to generate the hydrostatic support can be substantial and might make the mechanism unfeasible for machines with a tight efficiency budget (like almost any gas turbine engines). Nevertheless, the axial through-flow mechanism has proved to be of tremendous value in the microrotor test program due to its relative simplicity and robustness. To date all high-speed runs (up to 1.4 million RPM with a 4-mm rotor, as of this date) have taken place under axial hydrostatic conditions.

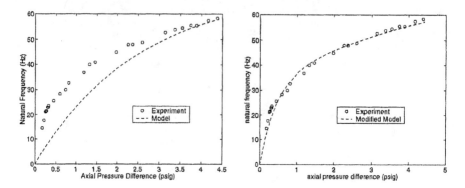

Figure 12. Prediction and measurement of natural frequency associated with axial through-flow in a low-aspect ratio microbearing. The left-hand frame shows the measurements (from a 26:1 scaled-up experimental rig) along with the theoretical predictions based on the assumed geometry while the right-hand frame shows the same measurements compared with the same model, but using slightly modified geometric parameters [11]

6 FABRICATION ISSUES

A key challenge to the successful operation of a high-speed microbearing is the accurate fabrication of bearing geometries. Two aspects of this need to be considered, the first being the need to hold design tolerances in any given fabrication process, and the second being the ability to manufacture multiple devices with good uniformity in a single fabrication run.

The first issue – that of achieving design tolerances – is a matter of process maturity and the attention paid to the maintenance of tight tolerances and small details is the hallmark of a well-established fabrication process. The microengine process is *very* complex and continually advances the state of the art in micromachining complexity. Almost any fabrication run that results in a freely rotating turbine should deservedly be considered a manufacturing triumph [5]. However from the standpoint of the success of the *system*, we must have much more stringent manufacturing requirements. The bearing designs are very sensitive to critical dimensions such as the bearing-rotor gap and the size of restrictor holes for hydrostatic injectors, and the failure to hold these dimensions within a specified tolerance can make the difference between a device that operates with a lubricated film and one that grinds the rotor and stator surface until failure. The very first version of the microbearing rig ran in just this mode, with very occasional demonstrations of lubricated operation. However, subsequent designs and builds have paid extreme attention to dimensional accuracy, and the fabrication protocols at this stage are quite mature so that this precision is ensured from one build to the next.

6.1 Cross-Wafer Uniformity

Typically, multiple microengines are fabricated in parallel on a single silicon wafer. Thus, in addition to the accurate manufacture of critical dimensions on a single microengine die, the importance of manufacturing uniformity from one die to the next on a single wafer is vital, and has revealed itself to be a major obstacle to device yield. It is very common for a given process to exhibit cross-wafer variations. For example, a shallow plasma etch into a silicon substrate might show a variance of as much as 10% from one side of the wafer to the other due to variations in the plasma that are intrinsic to the fabrication tool. All fabrication processes will exhibit such variations to one degree or another, and any microfabrication process needs to identify and accommodate these variations. Should they be unacceptably large, either the fabrication tool needs to be improved, or a different processing path needs to be considered. This need is a common driver throughout both the MEMS and microelectronics industry (which also desires greater process uniformity as feature size diminishes and processing moves to larger and larger wafers).

As discussed above, for microbearing design there are several critical etches that need to be controlled to a high degree of precision. The difficulty in maintaining cross-wafer uniformity thus results in some operational devices on the wafer (typically from the center of the wafer, where the process was initially honed to precision), while many devices (from the wafer edge) are out of spec. and will not operate satisfactorily. At this stage, most of the uniformity issues have been addressed; however, two items are still troublesome. Firstly, the Deep Reactive Ion Etcher, being a relatively new tool (perhaps 5 years in maturity) exhibits a fairly significant variation in etch rate between the center and edge of a wafer. This variation results in a gradient in etch depth that is particularly severe on devices lying on the wafer periphery (perhaps 3 microns variation across a 4 mm rotor wheel). This gradient may not seem important, but it contributes to a mass imbalance of as much as 25% of the bearing gap, rendering the bearing inoperable at high speed (the imbalance force increases with the square of the rotational speed).

The second continuing difficulty is that of front-to-back mask alignment during fabrication. It is common during the fabrication process for a single silicon wafer to be patterned on both the front and back surfaces. For example, the rotor has the turbine blades patterned from one side and the bearing gap patterned from the other side. Any slight misalignment between the lithography on the front and back surfaces of the wafer will result in an offset of front and back features which, as with the etch-depth gradient, leads to a rotor imbalance. At the present, mask alignment of critical features (primarily the rotor blades and the bearing gap) must be maintained to within 0.5 μm or better in order to ensure operable rotors from every die on the wafer. This is an extremely tight, but achievable, tolerance and work continues to improve the alignment even further and to improve process design to minimize imbalance.

6.2 Deep Etch Uniformity

The last issue for fabrication precision is that of deep-etch uniformity. Any high-speed bearing depends critically on the straightness and parallelism of the sidewalls that constitute the bearing and rotor surfaces. This is particularly true for hydrodynamic operation at high eccentricity. Unfortunately, in the drive to generate deep trenches so that the bearing aspect ratio is minimized, the quality of the bearing etch is often compromised. These two issues – the etch depth and the etch quality – constantly pull against each other and their relative advantages need to be weighed against each other in any final design.

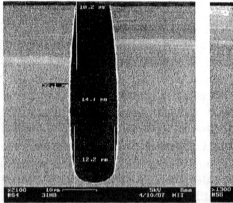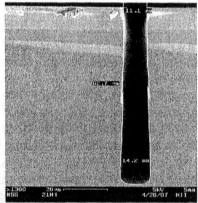

Figure 13. SEMS of bearing etches, illustrating typical manufacturing non-uniformities. The left-hand SEM shows an etch with a bow in the center, while the right-hand SEM shows an etch with a taper (SEM courtesy of A. Ayon).

For DRIE (Deep Reactive Ion Etching), typical non-uniform etch profiles are shown in Fig. 13, illustrating two common phenomena – etch bow (where the trench widens in the middle) and etch taper (where the trench widens, usually at the bottom). The effects of these non-uniformities have been analyzed computationally [11]. As one might expect, the static performance of the bearing (load capacity) is degraded by the blow-out, particularly in the case of the tapered bearing where fluid pressure cannot accumulate in the gap but rather leaks out the enlarged end. The bowed bearing, due to its concave curvature tends to "hold" the pressure a bit more successfully and the loss in load capacity is typically less severe. However, as mentioned earlier, load capacity is less of an issue in microbearings, and it is the effect on hydrodynamic stability that is of most interest. The effects of bow and taper on hydrodynamic operation are summarized in Fig. 14, which shows the minimum stable eccentricity as a function of bearing number (*i.e.*, speed, for a fixed bearing geometry) for different levels of both bow and taper. Clearly, the effects of taper are most severe and considerable effort has been placed in the fabrication process design to minimize bearing taper

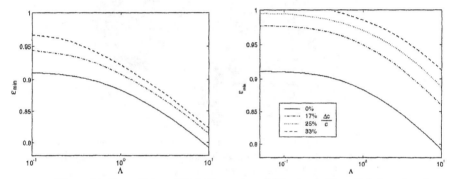

Figure 14. Degradation of hydrodynamic stability due to Bow (left frame) and Taper (right frame), as indicated by the minimum eccentricity required for hydrodynamic stability at a given bearing number (speed) [11].

6.3 Material Properties

One of the key benefits that can be realized at the microscale is the improvement in strength-related material properties. This is particularly true in silicon-based MEMS where the baseline structural material is single-crystal silicon, which can be fabricated to have very good mechanical properties. The strength of brittle materials is controlled primarily by flaws and to some extent by grain boundaries, both of which become smaller or non-existent in a single-crystal material with surfaces defined by microfabrication processes. In addition, the device size becomes comparable with the flaw distribution, such that the incidence of "super-strong" devices increases in microscale systems. Lastly, silicon is a light material with a density (2330 kg/m^3) lower than that of aluminum (2700 kg/m^3). For these reasons, the strength-to-weight ratio of silicon micromachined structures is un-paralleled, which is a key for high speed rotating machinery. However, notwithstanding its high specific strength silicon is still a very brittle material. For a high speed rotating system, such as a turbine, this can be problematic since an impact or touchdown at any appreciable speed is likely to result in a catastrophic failure rather than an elastic rebound or more benign plastic deformation. This is illustrated in Fig. 15, which shows a photograph of a microturbine rotor, which crashed during a high-speed test run. The importance of robust bearings is emphasized, as the material is extremely unforgiving.

Figure 15. Photograph of the remains of a silicon rotor after experiencing a high-speed crash. The instant fracture of the rotor (largely along crystallographic planes) is clearly visible.

7 CONCLUSIONS

In this paper, we have tried to focus on some of the more practical issues that face us in the design, manufacture and operation of micro devices that need to operate at high speed. Even with dramatic advances in lubricant and surface treatment technologies, high speed operation will likely require gas bearings which have always offered high speed, low wear operation with the attendant cost of a narrow and treacherous window of stable operation. Many of the commonly held assumptions and design rules that have guided previous fluid film bearings in conventional machinery have had to be re-visited due to both the consequences of scaling as well as the current (and foreseeable) limitations in microfabrication technology.

There are many areas where future research needs to focus attention. On the manufacturing side, the single largest obstacle to trouble-free production of gas bearings for high-speed rotors and shafts is the issue of precision microfabrication. Macroscopic systems, with typical scale of 1 meter, need precision manufacturing in places with sub-millimeter tolerances. So too, microdevices with typical dimensions of 1mm will need tolerances of 1 µm or less. The ability to manufacture with such precision will require much improved understanding of micromachining technologies such as etching and deposition so that cross-device and cross-wafer uniformities can be improved.

At the level of lubrication technologies, the new parameter regimes that are exposed by microfabricated systems (very low aspect ratios, relatively large clearances, insignificant inertial properties, *etc.*) need to be further explored and understood. Despite the low Reynolds numbers, inertial losses are critically important for hydrostatic lubrication mechanisms and need to be better understood and predicted. Similarly, the coupled fluid-structure interactions at high eccentricities, and the interactions between hydrodynamic and hydrostatic mechanisms need to be more fully explored. Last, and by no means least, fundamental issues of fluid & solid physics need to be addressed as the scale continues to shrink. Gas surface interactions, momentum and energy accommodation phenomena and the effects of surface contamination (whether deliberate or accidental) all need to be rigorously studied so that the macroscopic behavior can be predicted with some certainty. These issues are only going to become more important as manufacturing scales decrease further and as continuum assumptions become less and less applicable.

A final word needs to be said about the roles of computation and experiments in this emerging science. With the decrease in system scale, the possibility of direct molecular simulations becomes both feasible and attractive. At the same time, experiments at the micro- and nano-scale become very difficult, as many interfering factors need to be controlled and compensated for. It must, however, be realized that *both* of these approaches need proceed hand in hand and that neither should be neglected. Increasing computational power will enable more and more complete simulations of micro- and nano-systems, but only the "full analog simulation" that an experiment provides can ensure that we are including all the right terms and relevant physics.

When all is said and done, the "proof is in the pudding." A MEMS-based gas bearing system has been designed and fabricated and the rotor spins at speeds (as of this writing) approaching 1.4 million RPM (300 m/s tip speed). This achievement is remarkable (and serves as testament to thousands of hours of hard work by many people in multiple disciplines) and the work continues to both increase the speed and to incorporate the bearing technology into fully operational and useful micro-turbomachinery.

ACKNOWLEDGEMENTS

This work represents the efforts of a large group of people, with whom we have been most fortunate to work during the past several years. In particular, the contributions of Arturo Ayon, Alan Epstein, Reza Ghodossi, Ravi Khanna, Jon Protz, Marty Schmidt, Mark Spearing and Xin Zhang should be noted, and the assistance of the entire Microengine team (too many to list individually) is gratefully acknowledged. The work has been supported by the Army Research Office, Grant #DAAH04-95-1-0093, monitored by Dr. Richard Paur and by DARPA-ARO, Contract #DAAG55-98-1-0365, monitored by Dr. Robert Nowack.

REFERENCES

1. E.B. Arkilic, M.A. Schmidt, & K.S. Breuer. "Gaseous slip flow in long microchannels." *J. MicroElectroMechanical Systems,* **6** (2). June 1997.
2. S.F. Bart, T.A. Lorber, R.T. Howe, J.H. Lang and M. F. Schlecht. "Design Considerations for Micromachined Electric Actuators." *Sensors and Actuators.* Vol. **14**. 269-292. 1988.
3. F. Dimofte. "Wave Journal Bearing with Compressible Lubricant. Part 1: The Wave Bearing Concept and a Comparison to the Plain Circular Bearing." Tribology Trans. Vol **38**. No. 1. pp153-160. 1995.
4. A.H. Epstein *et al.* "Micro-Heat Engines, Gas Turbines and Rocket Engines - the MIT Microengine Project." AIAA Paper 97-1773. Snowmass, CO. June 1997.
5. B.J. Hamrock. *Fundamentals of Fluid Film Lubrication.* McGraw-Hill, New York 1984.
6. C-C. Lin, R. Ghodssi, A.A. Ayon, D-Z. Chen, S. Jacobson, K.S. Breuer A.H. Epstein & M.A. Schmidt. "Fabrication and characterization of micro turbine/bearing rig." Proceedings, MEMS99. January 1999.
7. D.F. Wilcox (Editor) *Design of Gas Bearings.* M.T.I. Inc. Latham, NY 1972.
8. M. Mehregany, S.D. Senturia and J.H. Lang. "Measurement of Wear in Polysilicon Micromotors." IEEE Transactions on Electron Devices, Vol. **39**, No. 5, 1136-1143, May 1992.
9. S. F. Nagle and J. H. Lang; "A Micro-Scale Electric-Induction Machine for a Micro Gas-Turbine Generator;" Presented at the 27th Meeting of the Electrostatics Society of America, Boston, MA; June 1999.
10. D.J. Orr. *Macro-Scale Investigation of High Speed Gas Bearings for MEMS Devices.* PhD Thesis. Department of Aeronautics and Astronautics. December 1999.
11. D.J. Orr, E.S.Piekos, & K.S. Breuer "A Novel Mechanism for Hydrostatic Lubrication in Short Journal Bearings." (Manuscript in Preparation). 2000.
12. E. S. Piekos. *Numerical Simulations of Gas-Lubricated Journal Bearings for Microfabricated Machines.* Ph.D. Thesis. Department of Aeronautics and Astronautics. February 2000.
13. E.S. Piekos & K.S. Breuer. "Pseudo-Spectral orbit simulation of non-ideal gas-lubricated journal bearings for microfabricated turbomachines." *J. Tribology,* 1998.
14. N. Savoulides. *Low Order Models for Hybrid Gas Bearings.* M.S. Thesis. Department of Aeronautics and Astronautics. February 2000.
15. J. Sniegowski and E. Garcia, "Surface-micromachined Geartrains Driven by an On-Chip Electrostatic Microengine," IEEE Electron Device Letters, vol. 17, no. 7, p. 366. 1996.

Chapter 17

ELECTRICAL PHENOMENA AT THE INTERFACE OF ROLLING-CONTACT, ELECTROSTATIC ACTUATORS

Cleopatra Cabuz
Honeywell Technology Center
Plymouth, MN 55441
e-mail: cleo.cabuz@honeywell.co

1 INTRODUCTION

We live today in a hyper-dynamic social, economical and technological environment that calls increasingly for portable and wearable systems. People want to be, at all times, informed and in control of the events affecting their lives. The cell phone, the pager, the laptop and palmtop computers or the implantable drug delivery systems and peacemakers are some of the first wearable/portable devices to be widely used. New, wearable, biological and chemical analysis systems are now being developed, and they will provide continue monitoring of the environment and/or of the individual's health condition.

Actuators for flow control are a significant part of any chemical or biological microsystem. Actuators size and power requirements are, in many cases, the barrier toward miniaturization and portability. Honeywell has championed the development of low power electrostatic microactuators for flow control, with extremely good performance in terms of power consumption, output work capability, size and weight. These actuators have a rolling contact configuration and are working in a pull-in mode. As it will be shown in more detail below, the rolling-contact configuration gives the best displacement/force product that can be obtained with an electrostatic actuator. However, the superb advantages of the rolling contact actuators are coming, unfortunately, with a bag of reliability issues, related to the intimate contact between the rolling surfaces of the actuator plates. While surface energy is certainly an issue, the paper will focus on a less investigated aspect, related to electrical phenomena going on at the interface between the actuator plates. It will be shown that the electrical characteristics of the surfaces of the different materials used in the rolling-contact configuration are in fact controlling the operation of these devices. A new electrical model

NANOTRIBOLOGY

of such actuators together with extensive experimental results will be presented trying to explain some of the observed behavior. The research in materials and surface science needed to support this type of devices will be emphasized.

2 BACKGROUND

The basic structure of the rolling-contact actuator is illustrated in Fig. 1c, together with two other commonly used configurations of electrostatic actuators: the comb-drive actuator (Fig. 1a), and the parallel plate actuator (Fig. 1b).

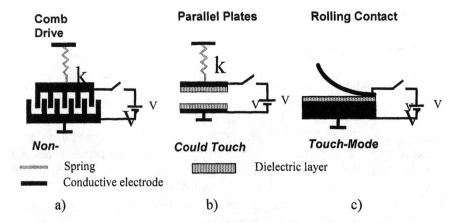

Figure 1. Commonly used configurations of electrostatic actuators. In all these configurations, the upper plate of the actuator is movable and spring-loaded, while the lower plate is generally fixed to a substrate.

A spring-loaded plate moves under the attractive electrostatic force generated by a distribution of opposite charges placed on the two electrodes of the actuator. The charges are generated by a power supply, V, connected to the electrodes.

The attractive electrostatic force for a parallel plate actuator in the open state can be described with Eq. (1),

$$F_{el} = \frac{QE}{2} \cong \frac{\varepsilon A}{2d^2} V^2, \qquad (1)$$

where Q is the charge accumulated on the electrodes, E is the electric field in the gap separating the actuator plates, A is the electrode area, d is the spacing between the two electrode plates and ε is the dielectric constant of the medium filling the actuator gap. The scaling of the electrostatic force with distance (the force increases at smaller spacings) and the simple structure of the actuator makes electrostatic force an adequate actuation force for microscale actuation.

In the comb-drive actuator, the displacement of the moving comb depends linearly on the applied driving voltage. In this structure, large displacements but very low electrostatic pressures can be achieved (fractions of an atmosphere). The small output work capability of such actuators limits their use to micro-world applications such as the control of precision xyz stages for AFM microscopes or the driving of resonant micromechanical sensors.

In the parallel-plate actuator the displacement depends in a highly non-linear manner on the driving voltage and is limited by the initial gap between the electrodes. Large gaps translate in very high actuation voltages, as will be shown in the example below.

> **Example:** Consider a clamped, circular diaphragm, for which a hydrostatic pressure of 0.2 psi applied on one side of the diaphragm produces a center deflection of 100 μm. If this diaphragm is placed in a parallel-plate, electrostatic actuator configuration, the voltage required to produce a similar deflection would be of about 1000 V. The voltage required to produce the same center deflection in a rolling-contact configuration would be of about 50V!

This example shows that the rolling contact configuration provides the same overall displacement as a parallel plate device at considerably lower driving voltages. Or, for similar driving voltages, the rolling contact configuration provides actuation forces at least one order of magnitude higher.

Other advantages deriving from the simple structure, the low power and the high output work of the rolling-contact, electrostatic actuators are given below:
- low-mass of actuating elements:
  - electrodes: about 10-nm aluminum (mass: 2×10^{-2} g/m^2)
  - dielectric: about 2-μm polymer: (mass: 3 g/m^2)

  Total specific mass of actuating elements: 3 g/m^2.
- high power efficiency:
  - about 50% power efficiency in cycled actuation; essentially zero power in hold;
- scalability:
  - electrostatic actuators are *the only ones* that can be economically used for large, 2D and 3D arrays of actuators for large area shape control, modular systems, redundant systems;
- versatility:

- a variety of substrates can be used, as requested by specific applications:
  - membranes with thickness limited by handling capability (typical 1mil);
  - embedded electrodes and optical coatings on top of actuating elements;
  - molded plastic substrates for rigid structures.

Figure 2 shows a SEM photograph of a silicon micromachined electrostatically actuated microvalve array based on individual rolling-contact actuators.

Figure 2. Honeywell's silicon micromachined electrostatic microvalve array.

Table 1 presents figures of merit (defined as output power per volume or output power per weight of actuator) for different actuation mechanisms and systems. The data in Table 1 shows that in terms of output power per weight, the rolling contact actuator used in a linear-actuator configuration comes third, after the 777 jet engine and the high energy PZT transducer. Other performances, however, such as the actuation speed and the maximum displacement makes it a more useful actuator than the PZT, making the rolling contact, linear actuator configuration, a high performance actuator.

The large output-power/weight provided by the rolling contact electrostatic actuators as well as their compatibility with array configurations recommend them for use in space systems, micro air vehicles and portable and wearable systems. There is increasing interest in such structures at major institutions such as Xerox Corp. [15], Germany's Karlsruhe Research Center [16], Honeywell/DARPA [14, 19], Lawrence Livermore Lab [18], University of Nagoya [17]. Devices in individual or array configurations have been demonstrated and showed the predicted characteristics: low power consumption, low voltage, and increased displacement. These devices use a variety of dielectric materials, (from silicon based dielectrics to aluminum oxides or to organic dielectrics), and

Table 1. Comparison of figures of merit for different actuators and/or machines.

	Output	Displcement	Speed	Size	FOM (2W/cm^3)	FOM (W)	Ref.
Model airplane	2 hp (1 in^2 displ.)			200 cm^2	7.5	0.1	
777 jet aircraft	90,000 lb (thrust)	700 mph		10 ft Φ 10 ft long	3		
Switched reluctance motors	4000 Nm/m^3		3600 rpm		1.5	0.02	3
PZT high-energy transducer	0.05 J/cm^3		100 Hz		3		4
Human muscle	400 W (20-m duration)			8×10^3 cm^3	0.05	0.05	9
Molecular motors (myosin)	5 pN	10 mm	0.05 s	10 nm Φ 15 nm long	1	1	6
Nanomotor (pieze gripper/slider)	0.03 N	5 mm	1 mm/s	3.5 mm Φ 15 mm long			10
Electrostatic linear actuator	4.4 N (1.6W)		1.4 m/s	7 g (4.5 cm^3)	0.15	0.1	11
Permanent magnet DC electric motor	20 Nm (torque)		20,000	3 mm Φ 8 mm long	2.2	0.3	13
Permanent magnet DC electric motor w gear box	2m Nm (torque)		350 rpm	5 mm Φ 21 mm long	0.15	0.02	13
Rolling contact polyMEMS	2 N/cm^3	3 mm	0.01 s	1 cm^3	0.7	0.7	

of driving approaches, resulting in widely different capabilities. While all devices are showing promising results, the operation consistency and the reliability is widely different and it is obvious that not all the interface effects are understood and addressed. The technology based on rolling-contact, electrostatic actuators, has high potential but has yet to mature.

3 CHALLENGES IN ROLLING-CONTACT ELECTROSTATIC ACTUATORS

In the non-touching electrostatic actuators, such as the comb-drive structure shown in Figure 1a, the two electrodes of the actuator will not touch during a normal operation cycle. The spring-suspended comb moves toward the fixed comb with the fingers of the comb sliding along each other, at very close spacing (~2 μm) but without touching each other. Such actuators have no problems related to surface interactions. They are extremely reliable and are already used in commercial products. However, as it has been mentioned, the force/unit-area generated by such actuators is extremely small, being of limited applicability in micro-to-macro-world interactions.

In the touch-mode actuators, such as the parallel plate actuator shown in Fig. 1b or the rolling-contact actuator shown in Fig. 1c, the two plates of the actuator come in touch at some point of the actuation cycle. For such actuators, the electrodes (one or both) are covered with insulating layers, to prevent electrical shorting during the contact phase of the actuation cycle. These actuators show several failure modes, all being related to the contact phase of the actuation cycle and to the presence of the dielectric material.

The most common failure modes in TMEA are:
- Actuator's failure to close or to remain closed;
- Actuator's failure to open, also called stiction;
- The catastrophic breakdown of the dielectric, preventing actuator's driving.

Involved in the above failure modes are:
    a. The dielectric material through its characteristics:
- Dielectric constant;
- Dielectric strength vs. thickness;
- Surface properties:
  - surface energy ;
  - electrical properties: surface conductivity, dielectric constant, density of surface states.
- Mechanical properties (Young's modulus, mechanical strength).

b. The electrode material through its characteristics:
      - Injection properties into the specific dielectric;
      - Surface energy (when exposed).
   c. The mechanical properties of the actuator, of which the most important is the spring constant of the spring loaded plate.

As far as the catastrophic dielectric breakdown is concerned, it can generally be avoided by maintaining the maximum electric field in the actuator below the dielectric strength of the material and, in addition, by using electrodes with self-healing properties.

The stiction and the actuator's failure to close or remain closed are more complicated issues and they are intimately related to the interface interactions between the contacting surfaces of the two actuator plates.

Given the increased surface/volume ratio of MEMS devices, such devices are generally more prone to stiction than their macroscopic counterparts. The MEMS community has dedicated significant efforts to reducing the surface energy in MEMS actuators. Low-surface energy coatings have been developed specifically for the silicon based actuators [21]. However, little attention has been paid to the electrical phenomena responsible for actuator's failure to close or for the electrically induced stiction. This paper will focus on electrical interactions in the touch-mode, electrostatic actuators.

4 MODELING AND SIMULATION OF INTERFACE INTERACTIONS IN THE TOUCH-MODE, ELECTROSTATIC ACTUATORS

For ease of representation, a parallel-plate configuration will be considered in this presentation. However, qualitatively, the results apply to a rolling-contact actuator as well.

Figure 3 shows a schematic representation and the equivalent electrical circuit for an ideal parallel-plate electrostatic actuator in both the open and the closed state. In this ideal case, the actuator plates are both covered with the same dielectric material, which has no surface layer. In the closed position, this actuator becomes a simple parallel-plate capacitor with uniform dielectric. The charge responsible for actuation in the closed position is given by Eq. (2),

$$Q = CV_{DC}; \quad C = \frac{\varepsilon_0 \varepsilon_{SiN} A}{2 d_{SiN}}, \quad (2)$$

and the electrostatic field between the actuator's electrodes is given by Eq. (3),

$$E = \frac{V_{DC}}{2d_{SiN}}. \qquad (3)$$

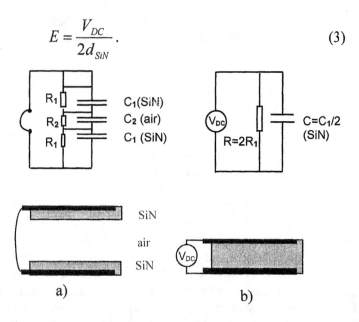

Figure 3. Schematic representation and equivalent electrical circuit for an ideal, parallel plate, touch-mode electrostatic actuator in the open (a), and in the closed state (b).

In the closed state, such an ideal actuator has no dielectric/dielectric interface and, consequently, no unexpected electrical phenomena associated with it. In the real cases, however, the dielectric layers covering the electrodes have a more complex structure. Sometimes the dielectric is produced as a stack of dielectric layers with different properties, to negotiate different aspects of the fabrication process. In all cases, the dielectric materials have surface properties different from the bulk, creating a structure with multiple electrical interfaces, as illustrated in Fig. 4.

Figure 4 shows that, in the real structures, the actuator plates, even though covered with the same dielectric material, do not form, when brought together, a simple capacitor. This is the first step toward the understanding and the improvement of the touch-mode electrostatic actuators. Knowing what the properties of the surface layer are, and predicting the behavior of the actuator in different environmental conditions, is a more complicated endeavor, and this paper will teach through examples.

The tool used to complement the experimental results in the study of the TMEA was SPICE simulation [22]. The closing and opening of the actuator under an applied electrostatic field was simulated with a voltage-controlled switch with hysteresis, as shown in Fig. 5.

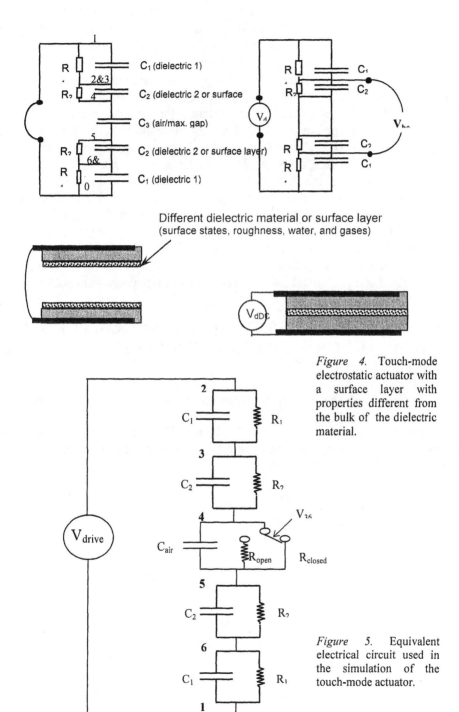

Figure 4. Touch-mode electrostatic actuator with a surface layer with properties different from the bulk of the dielectric material.

Figure 5. Equivalent electrical circuit used in the simulation of the touch-mode actuator.

The key elements in the operation of the electrostatic actuator are the electric field and the charge responsible for the actuation. While the driving voltage, V_{drive}, is ultimately responsible for initiating the actuation process, the actual movement of the actuator is determined by the electrostatic field across the air gap and by the electric charges at the nearest interfaces. The nearest interfaces in the case of the equivalent circuit in Fig. 5 are the nodes #4 and #5 (the surfaces of the actuator plates) for the open state, and the nodes #3 and #6 for the closed state. However, it should be noted that, when the actuator is opened, the impedance of the air gap is much larger than the impedance of the dielectric layers. The voltage across the air gap, V_{4-5}, is essentially the same as the voltage between the nodes #3 and #6. With this simplifying assumption, it is reasonable to consider that the actuator is actually driven by the voltage between the nodes #3 and #6 in both the opened and the closed state. This allows the simulation of actuator's motion with a switch controlled by the voltage V_{3-6}. In order to simulate the non-linearity of the electrostatic force, the switch has a hysteresis, i.e. it closes at about 15V and opens at about 5 V (experimentally determined values).

5 CASE STUDIES FOR INTERFACE INTERACTIONS IN THE TOUCH-MODE, ELECTROSTATIC ACTUATORS

Case 1: Stack of Dielectric Layers with Different Resistivities

In this case, the dielectric #1 and the dielectric #2 are different dielectric layers, deposited to accommodate certain fabrication constraints. Dielectric

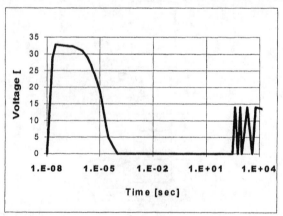

Figure 6. Simulated voltage between the nodes #4 and #5 of the actuator's equivalent circuit, for Case 1: a stack of dielectric layers with different resistivities. The material with lower resitivity is next to the air gap.

#1 is made out of 1000Å of a type of sputtered silicon nitride, with a resitivity of about 10^{14} ohm-cm. This nitride is lower stress but also less resistant to alkaline solutions used to etch silicon. Dielectric #2 is made out of 500 Å of a different type of sputtered nitride, with a resistivity of about 10^{13} ohm-cm. This nitride is more resistant to alkaline etchants and was used as a protective layer in several devices. Both nitrides have relative dielectric constants of about 9.

Figure 6 shows the time evolution of the voltage between nodes #4 and #5, following the application of a 50V driving-voltage step.

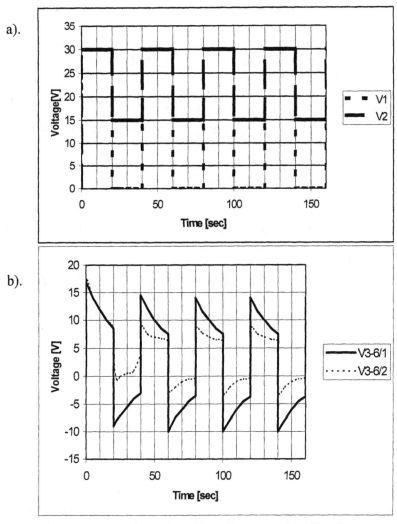

Figure 7. Simulated actuation voltage V_{3-6} versus time (b) under two types of periodical DC pulses: 30 V-0 (V1) and 30V-15 V (V2) (a).

Immediately after the application of the DC step, V_{4-5} increases and produces the actuation. As the two plates move toward each other, the voltage across the air gap decreases, and when the plates touch, it gets to zero. Figure 6 shows that after staying in zero for about 100 sec, the voltage V_{4-5} starts increasing again, indicating the fact that the actuator is opening. When V_{4-5} reaches 15 V (closing voltage), the actuator closes again, and the cycle gets repeated. During that time, the driving voltage is constant and equal to 50V. It is therefore obvious that the actual movement of the actuator is not determined by the applied DC driving voltage alone, but by the field across the air gap, which depends on the actual structure of the dielectric stack. The simulated results are in full agreement with experimental observation in devices with a structure as that shown in Fig. 5.

Figure 7 shows the time evolution of V_{3-6}, which can be considered the actual driving voltage for the actuator, in response to a sequence of 20-s long, 30-V high DC pulses intended to close and open the actuator.

As the driving voltage, V1, increases from zero to 30 V (Figure 7a), the voltage V_{3-6} increases from zero to about 18V. On the plateau of V1, V_{3-6} decreases as a result of the voltage redistribution in the dielectric stack with non-uniform conductivities. On the negative front of V1 (from 30 V to zero), V_{3-6} follows with a negative jump of about 18 V, from +9 V to about – 9 V, keeping the actuator closed (the electrostatic force depends on V^2, i.e. it is not dependent on the sign of the applied voltage). After a while the charge will leak and the actuator will eventually open. Opening times from several seconds to few minutes have been experimentally measured on such actuators, in full agreement with the simulated data.

Figure 7 also shows that if the actuator is driven with a 30-V DC step, V2, and then the voltage is decreased to 15 V instead of going to zero, V_{3-6} will remain below the holding voltage of |5 V| and the actuator will open. All these effects have been repeatedly observed on actuators with stack of dielectrics with different conductivities used in a dry environment.

Case 2: Dielectric Material with Hydrophilic Surface

In this case the dielectric layer#1 is a sputtered silicon nitride ($k = 9$, resistivity $=10^{13}$ ohm-cm). The dielectric layer #2 is the native oxide layer, hydrated in the presence of environmental humidity ($k = 80$, resistivity lower than 10^{10} ohm-cm and depending on surface cleanliness). Figure 8 shows the simulated voltage V_{4-5} for the hydrated surface case. A 30V DC step is applied at time $t = 0$. The voltage V_{4-5} drops to zero after about 0.01 s, meaning that the actuator plates get in touch. However, even though the driving voltage is maintained at 30 V, as soon as the contact is established the actuator plates separate again, as shown by the increase in V_{4-5}. When

V_{4-5} reaches 15 V, the actuator closes again, and it starts to vibrate at a frequency of about few Hz. For the hydrated surfaces case, the time-to-open is much shorter than in the case of the dielectric stack and the vibration frequencies are much higher. The simulated results are in full agreement with the experimental observations for this type of actuators.

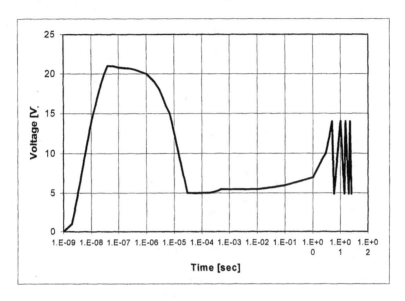

Figure 8. The simulated voltage V_{4-5}, across the actuator's gap, for a touch-mode electrostatic actuator with hydrated dielectric surfaces.

Case 3: Dielectric Material Coated with a Self-Assembled-Monolayer (SAM)

To eliminate the hydration of the dielectric surface under the effect of environmental humidity, hydrophobic self assembled monolayers were developed and applied on test actuators. Two types of coatings were used: a fluorinated coating (per-fluoro-decyl-trichloro-silane-PFTS) and a non-fluorinated coating (octadecyl-tricholorosilane-OTS) [2, 22]. Both coatings were able to essentially remove the effect of the environmental humidity.

Figure 9 shows experimental data on the dependence of the generated electrostatic pressure on the relative humidity in a non-coated and an OTS-coated rolling-contact actuator. It can be seen that the non-coated actuator stops generating any useful force at environmental humidities greater than about 35%, while the coated actuator can be used at RH as high as 98% whithout changes in the generated force/pressure.

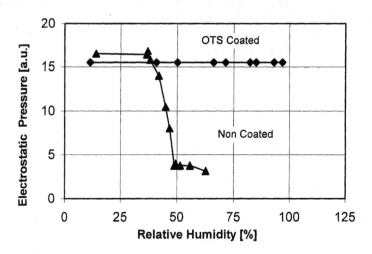

Figure 9. The dependence of measured electrostatic pressure generated in a non-coated and an OTS coated rolling-contact electrostatic actuator under an applied square-wave, AC driving voltage of $25V_p$

Our simulation and the experimental data show that, in rolling contact actuators, the hydrophobic surfaces are required not (only) for reducing stiction, but mainly for preventing the loss of actuation force due to the high dielectric constant and the low resistivity of the hydrated interfaces. Without a hydrophobic coating, a rolling-contact actuator used in a humid environment will actually not close, instead of getting stuck. These effects have been experimentally observed and are now fully supported with a theoretical model.

A variety of SAM coatings have been proven to reduce the surface energy of the coated surfaces [21, 24-26]. The electrical properties of these layers are not discussed in literature. For the operation of touch-mode actuators, such properties are, however, crucial.

Figure 10 shows the simulated voltage across the air gap (V_{4-5}) for two types of SAMs with different resistivities (greater than 10^{15} ohm-cm and lower than 10^{14} ohm-cm respectively). It can be seen that the two actuators behave initially the same (the traces overlap). However, after a few seconds, even though the driving voltage stays constant, the actuator with a SAM of a lower resistivity opens while the actuator with a higher resistivity SAM stays closed. Comparing the simulated data with the experimental results, it has been concluded that the fluorinated PFTS coating has a resistivity of less tha 2×10^{14} ohm-cm, resulting in eventual opening of the actuators coated with it, while the OTS coating has a resistivity greater than 10^{15} ohm-cm.

The higher resistivity of the OTS layer prevents the opening of the actuator but produces an accumulation of charges at the OTS/nitride interface resulting in a newly observed type of stiction: an electrostatic stiction. We consider that this effect is responsible for the higher stiction observed in the OTS-coated devices as compared to the non-coated devices.

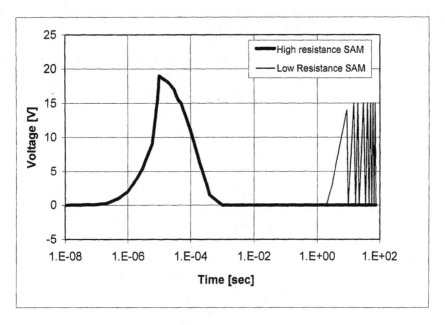

Figure 10. Time evolution of the voltage across the actuator's air gap, V_{4-5}, for two SAM coated actuators: a high resistivity SAM layer ($\rho > 10^{15}$ ohm-cm) and a low resistivity SAM layer ($\rho < 10^{14}$ ohm-cm).

6. DISCUSSIONS AND CONCLUSIONS

Through experimental evidence and theoretical modeling and simulation we have shown that the rolling-contact electrostatic actuators are very sensitive to the detailed electrical properties of the dielectric layers covering the electrodes. While bulk properties of the dielectric material such as dielectric strength, dielectric constant and resistivity are important for the operation of the device, the behavior of the actuator is, in most cases, controlled by the surface properties of the materials involved: the hydrophobic/hydrophilic character, the resistivity of the surface layer, the relative properties of the layers in the dielectric stack. It becomes evident that the ideal material for such applications would be a material "without surface" (the surface has the same properties with the bulk). In addition, such a material would have to be stable in a humid environment, to have a reduced surface energy in order to prevent van der Waals stiction and to be free of charge trapping interfaces. In order to maximize the electrostatic force produced by a certain driving voltage, the dielectric used would also have to have a relatively high relative dielectric constant (8...9). Also, in order to allow application in the narrow, intricate shapes of the MEMS devices, such a material would have to be conformally applied in very thin layers (fraction of a micron), preferably in a dry phase process. Organic dielectrics are, probably, best candidates for such applications, but they have reduced dielectric constant. It is the hope of the MEMS community that the researchers in the materials science, interface science and tribology will come together to develop the material that would allow the development of the rolling-contact electrostatic actuators to their full potential.

ACKNOWLEDGEMENTS

The work reported here is the result of a large team, whose contributions are greatly acknowledged: Masayuki Oneda (Yamatake, Japan) and Eugen Cabuz for the SPICE simulation and experimental determinations, David Zook and Bob Horning for assembling the data in Table 1, Jim Neus and Dawn Murphy for processing the tests samples, Tom Ohnstein for proposing and building the first rolling-contact electrostatic actuators.

The author is very grateful to David Arch from Honeywell's Management for supporting the publication of this study.

REFERENCES

1. T. Ohnstein et al. (Honeywell) " Micromachined Silicon Microvalve", MEMS'90, IEEE Press. 95-98, (1990)
2. C. Cabuz, "Dielectric Related Effects in Micromachined Electrostatic Actuators", 1999 Conference on Electrical Insulation and Dielectric Phenomena, IEEE press. 327-332, (1999)
3. M.R.Harris et al. "Limitations of Reluctance Torque in Double-Salient Structures", *Proc. Int. Conf. On Stepping Motors and Systems,* Univ. of Leeds, July 1994, pp. 158-168 Reprinted in "Variable Reluctance Drives" 1988
4. D.A.Berlincourt, IEEE Trans. Sonics and Ultrasonics, SU-15, 89 (1968)
5. J. A. Spudich, "How Molecular Motors Work", *Nature 372,* 515-518 (8 December 1994)
6. J.T. Finer, et.al., "Single Myosin Mechanics: Piconweton Forces and Nanometer Steps", *Nature 368,* 113-119 (1994)
7. J. Howard, " Molecular Motors: Structural Adaptations to Cellular Functions", *Nature 389,* 561-167 (1997)
8. H. Fujita, K. Gabriel, "New Opportunities for Microactuators", MEMS'91, IEEE Press, 14-20 (1991).
9. F.R. Witt and D.G. Wilson, "Bicycling Science", The MIT Press, (1982)
10. V. Klocke and S. Kleindiek GbR, Soerser Weg 37, D-52070 Aachen, Germany, (1996 brochure)

11 T. Niino, et al, "High Power and High Efficiency Electrostatic Actuator", MEMS'93, IEEE Press, pp. 236-241 (1993)
12 R. Perline et al, MEMS'97, IEEE Press, pp.238-243
13 RBM Miniature Bearings, Inc., 29 Executive Parkway, Ringwood, NJ 07456 (1997 Brochure)
14 C. Cabuz et al, "Mesoscopic Sampler Based on 3D Array of Ellectrostatically Activated Diaphragms", Transducers'99, Sendai, Japan, (June 1999)
15 D. Biegelsen et al, (Xerox Corp) "High Performance Electrostatic Air Valves Formed by Thin-Film Lamination", ASME-MEMS 1999 IEEE Press 163-168, (1999)
16 C. Goll et al, (Forschungszentrum Karlsruhe, Germany) "An Electrostatically Actuated Polymer Microvalve Equipped with a Movable Membrane Electrode", J-Micromech and Microeng. 7, 224-226, (1997),
17 F. Isono et al, (Nagoya Univ. Japan), "Fabrication Process of an Electrostatic Microvalve Using Platingand Electrodeposition Photoresist", 16th Sensor Symposium, 177-180, (1998)
18 A.P. Lee et al, (LLL), " A Low Power, Tight Seal, Polyimide Electrostatic Microvalve", ASME MEMS, 345-349, IEEE Press (1996)
19 C. Cabuz et al, A Polymer Micro-Actuator Array with Macroscopic Force and Displacement", US Patent Application #09/223 368
20 Mike Philpott et al. UIUC, Integrated Mesoscopic Cooler, DARPA/DSO Contract,
21 R. Maboudian and R. Howe, Critical Review: Adhesion in surface micromachined structures, *J. Vac. Sci. Technol. B 15(1), Jan/Feb 1997.*
22 Roberts and Sendra, SPICE, 2nd Edition, Oxford University Press, 1997.
23 C. Cabuz et al, Factors Enhancing the Reliability of Touch-Mode Electrostatic Actuators, Sensors and Actuators 79(2000) 245-250.
24 B-H. Kim et al, A New Class of Surface Modifiers for Stiction Reduction, Proceedings of the 12th Conference on Micro Electro Mechanical Systems, pp. 189-193, January 1999, Orlando, Florida, USA.
25 Y. Matsumoto et al, The Property of Plasma Polymerized Fluorocarbon Film in Relation to CH_4/C_4F_8 Ratio and Substrate Temperature, pp. 34-37, The 10th International Conference on Solid-State Sensors and Actuators, June 7-10, Sendai, Japan.
26 J. Sakata et al, Anti-Stiction Silanization Coating to Silicon Microstructures by a Vapor Phase Deposition Process, pp. 26-29, The 10th International Conference on Solid-State Sensors and Actuators, June 7-10, Sendai, Japan.

Chapter 18

THE USE OF SURFACE TENSION FOR THE DESIGN OF MEMS ACTUATORS

Chang-Jin "CJ" Kim
Mechanical and Aerospace Engineering Department, University of California
Los Angeles, CA 90095-1597
cjkim@ucla.edu; http://cjmems.seas.ucla.edu

1 INTRODUCTION

This paper attempts to shine light to a fundamental shift of mechanics as the scale of mechanical elements goes down far below millimeters. While the usual dynamics of motion, the knowledge essential to design any moving elements, most typically deals with the balance between forces and inertial and frictional resistances, the role of inertia becomes negligible in nanometer scale by all practical means. Instead, surface forces come under the limelight when analyzing and, more importantly, designing nano- and micro-mechanical elements for MEMS or microsystems.

Surface tension, a line force, is a unique type of force, which scales directly to length. When the dimension of interest shrinks down to sub-millimeter range, the size range typical for MEMS devices, surface-tension force becomes dominant over most other forces, such as those based on pressure (surface force) or mass (body force), *e.g.*, see [1,2]. This dominance of surface tension has most typically been the source for a serious hindrance against successful fabrication and operation of microdevices, calling for elaborate analyses and fabrication techniques in MEMS [3-6].

A completely different approach would be to design microdevices which *use* surface tension. Unfortunately, however, there has been little knowledge on how to make a machine run by this unfamiliar force. This review reports our on-going effort to establish surface tension as a useful force for microdevices and eventually a main driving force for micromachines. We explain several examples of using surface tension for microdevices, including an exploration to implement continuous electrowetting (CEW) phenomenon for MEMS devices in order to control surface tension and drive microdevices.

2 MECHANICS IN MICRO AND NANOMETER SCALE

Manifestation of the scale effect is abundant in nature, as we see ants not collapsing from the loads they carry but helplessly trapped in a water droplet, small creatures climbing up the wall, or a water strider running on water surface. It is interesting that biologists reported by empirical observation of nature that surface force becomes more important than body or muscle force in the world below 1 mm scale [1], the domain we did not experience building mechanical systems in until the advent of MEMS. By the way, be reminded that typical MEMS devices are even smaller than ants by orders of magnitude.

If human beings were creatures of sub-millimeter in size, the main force of interest to us would have been surface tension. If a thousand times smaller, we would fear being stuck to many liquid surfaces around us but never worry about getting hurt falling. In this scale, it is interesting to suspect that we might have invented many machines that are driven by surface tension.

Indeed there have been a few attempts in MEMS to utilize surface tension, some passively using surface tension to block unwanted movements and some actively controlling it to induce movements. Let us start with passive examples.

3 PASSIVE USE OF SURFACE TENSION

Many examples can be found that use surface tension to stabilize small objects, such as picking up a contact lens with a finger tip. Those with an engineering flavor include such examples as keeping ink from leaking through the micro-nozzles in the popular inkjet printers. This stabilizing effect can be used much more elaborately for the microdevices made by MEMS technologies.

3.1 Example 1: Microinjector

Figure 1 shows a novel droplet ejection mechanism that uses the flow-blocking effect of bubbles in microchannels. Due to the large surface tension in microscale, the bubble in microchannel functions as a check valve, which can be created thermally by a patterned resistor and elminated by conductive heat dissipation. With a bubble in place (*i.e.*, virtual check valve closed), a droplet can be ejected with high pressure without the problem of chamber-to-chamber cross talk. With the bubble collpased (*i.e.*, valve open), ink refills the microchamber quickly without much restriction in the flow path. The result is a high frequency droplet ejection without any microvalve fabricated. The device demonstrated an ejection frequency several folds higher than commercial inkjet printers at the time of publication (*e.g.*, from 5 to 35 kHz), not to mention the complete elimination of the so-called satellite droplet, with no added fabrication complexity [7,8].

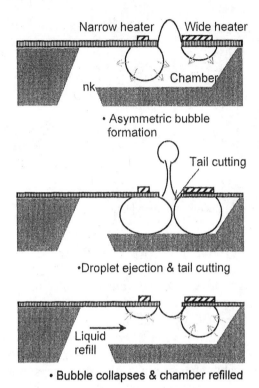

Figure 1. Principle of bubble valve for micro injector [7,8]. Nozzle size is 5-20 μm. Drawn not to scale.

3.2 Example 2: Bubble Pump

It has been shown that sequential generation of thermal bubbles can pump the liquid in microchannel [9]. In Fig. 2, the bubble on the left side functions as a check valve while a new bubble is formed and expands on the right side. The effect is a non-symmetric expansion of the new bubble and flow of liquid to the right. The device can also be operated with one bubble by using the temperature gradient that creates asymmetric vapor pressure and surface tension between the two ends of the bubble. This valveless pumping mechanism, demonstrated with 2 μm high, 10 μm wide microchannels made of SiO_2, has produced about 200 μm/s of flow velocity and pressure head of 800 Pa.

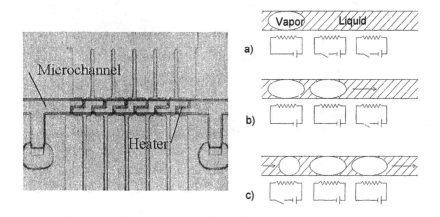

Figure 2. Valveless micropump using bubbles [9].

3.3 Example 3: Micromechanical Switch

Taking advantage of the strong physical stability of droplets in microscale, even a droplet-based micromechanical switch can be made, as shown in Fig. 3. A droplet of liquid metal (*e.g.*, mercury), 5–20 μm in diameter, closes the circuit between the reference line and one of the two electrodes. The bistable droplet can be moved to switch by electrostatic attraction. Yet, it is stable against physical disturbances even at thousands of G's, thanks to the high surface tension and negligible inertia effect in microscale. Since the switch contact is between metal and liquid metal, many of the typical contact problems in solid-solid contact switches can be eliminated. The droplet being the only moving part, as opposed to the long beams found in typical microswitches in MEMS, an array of microswitch as densely populated as LSI circuits can be envisioned.

4 ACTIVE USE OF SURFACE TENSION

More ambitious goal is the active usage of surface tension, such as driving liquids directly by surface tension. Surface-tension-induced motion is possible by creating surface tension difference in fluid-fluid or fluid-solid interface. There are several ways to control the surface tension, such as chemical (*i.e.*, use of surfactant), thermal (*e.g.*, Marangoni force), and electrical methods (*i.e.*, electrocapillary, [10,11]). Among them, use of electrocapillary appears most promising due to the energy efficiency (vs. thermal) and potentially simple and long-lasting operation (vs. chemical), as well as the simplicity in realizing by microfabrication.

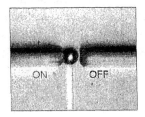 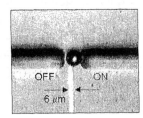 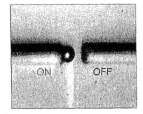

Figure 3. Microswitch with a droplet of liquid metal (switching at 70 V).

4.1 Continuous Electrowetting (CEW) Principle

While the electrocapillary, in general, refers to a relationship between surface tension change of liquid metal due to applied voltage, continuous electrowetting (CEW) refers to a principle of *moving* a lump of liquid metal using electric potential [10]. See Fig. 4 below. In brief, a drop of liquid metal (*e.g.*, mercury) travels along an electrolyte-filled channel or tube when electric potential is applied across the length of the channel. The movement of a mercury slug in electrolyte is due to the local change of surface tension.

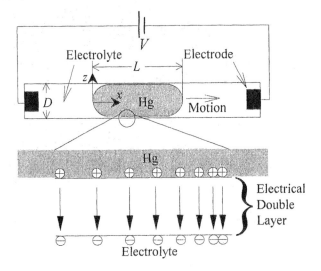

Figure 4. Continuous electrowetting (CEW) effect.

Figure 4 shows a state of electric charge distribution at the mercury-electrolyte interface, which is responsible for surface tension gradient according to Lippman's equation [10], and the motion induced by it. The electric potential across the electrical double layer becomes higher as x-coordinate increases. The surface tension decreases as the electric potential increases and the motion occurs to the right where the surface tension is lower. Movement of the liquid-metal droplet to the lower surface-tension area can be interpreted as the tendency to minimize the energy by wetting low surface tension region more than the higher one.

4.2 Example: Liquid Micromotor

Using the enabling technologies developed to make microscale mercury droplets on lithographically defined spots and to encapsulate electrolyte in lithographically defined region [12,13], CEW effect has been realized with MEMS [14,15]. Although the moving distance of the liquid metal is found limited for microscale devices due to the severe polarization effect caused by the unusually small size of the electrode, a long-range movement can still be obtained by alternating the polarity to the overlapping electrodes. The application of long-distance travel strategy to a loop of channel results in indefinite circular motion of a liquid metal droplet, which we call a *liquid micromotor*.

Figure 5 is the liquid micromotor. The picture on the right is a frame image of the device under operation. The loop diameter is 2 mm, channel width is 200 μm, and the deep-RIE depth is 20 μm. This liquid micromotor demonstrated successful operation with maximum speed of 420 RPM at only 2.8 V and an average current around only 10 μA. This is in contrast to electrostatic actuation, which consumes little current but typically requires high voltage, and electromagnetic or thermal actuation, which uses low voltage but consumes high current. Being a liquid motion, the movement of

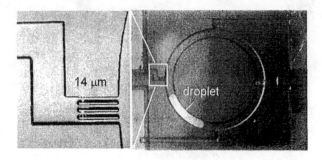

Figure 5. Liquid micromotor under testing [15]. Diameter of the channel loop is 2 mm.

the liquid-metal slug by CEW is very smooth and free from dry friction and wear from which many microactuation mechanisms are suffering. We thus anticipate a long operating life cycle and reliability with this type of devices. Numerous devices in different application areas are expected, such as micro inertia sensors and motors (*e.g.*, momentum wheel), micro switches, and micro optical switches

4.3 Other Electrowetting Mechanisms

The CEW described above is one of the several actuation mechanisms based on electrowetting. Other mechanisms include the so-called electrowetting-on-insulator as well as normal electrowetting. Although more attractive due to the absence of liquid-metal, these electrowetting mechanisms pose further challenges in realizing as working devices with the planar-based MEMS technologies. Nevertheless, recent new development in micromachining technologies and the success of the CEW devices above has led us to launch the development of MEMS devices based on regular electrowetting principle [16, 17].

5 SUMMARY

Surface tension dominates almost all other forces in micro and nanometer scale, but there is little knowledge on how to make use of it in designing mechanical devices. Summarized was our on-going effort to establish surface tension as a useful force for MEMS devices and eventually a main driving force for micromachines that feature many nanomechanical aspects in their key operations. We are entering the new world of machine design where the disciplinary boundaries between conventional chemistry and mechanical engineering become blur.

ACKNOWLEDGEMENTS

Several past and current members in the Micromanufacturing Lab. of UCLA, including J. Lee, F.-G. Tseng, J. Kim, and T. Jun have carried out the work summarized in this paper. The reported work has been supported by ONR and NSF.

REFERENCES

1. F.W. Went, "The Size of Man," *American Scientist*, Vol. 56, pp. 400-413, 1968.

2. W.S.N. Trimmer, "Microrobots and Micromechanical Systems," *Sensors and Actuators*, Vol. 19, No. 3, pp. 267-287, Sep. 1989.
3. R. Maboudian and R. T. Howe, "Critical Review: Adhesion in Surface Micromechanical Structures", *J. Vac. Sci. Technol.*, Vol. B15, No. 1, Jan/Feb. 1997, pp. 1-20.
4. C.-J. Kim, J. Y. Kim, and B. Sridharan, "Comparative Evaluation of Drying Techniques for Surface Micromachining", *Sensors and Actuator*, Vol. A64, Jan. 1998, pp. 17-26.
5. C.H. Mastrangelo and C.H. Hsu, "Mechanical Stability and Adhesion of Microstructures under Capillary Forces - Part I : Basic Theory," *J. MEMS*, Vol. 2, No. 1, pp. 33-43, Mar. 1993.
6. U. Srinivasan, M.R. Houston, Roger T. Howe, and R. Maboudian, "Self-Assembled Fluorocarbon Films for Enhanced Stiction Reduction," *Int. Conf. Solid-State Sensors and Actuators (Transducers '97)*, Chicago, IL, pp. 1399-1402, Jun. 1997.
7. F.-G. Tseng, C.-J. Kim, and C.-M. Ho, "A Novel Microinjector with Virtual Chamber Neck", *Proc. IEEE MEMS Workshop*, Heidelberg, Germany, Jan. 1998, pp. 57-62.
8. F.-G. Tseng, C.-J. Kim, and C.-M. Ho, "A Microinjector Free of Satellite Drops and Characterization of the Ejected Droplets", *Proc. MEMS*, ASME IMECE, Anaheim, CA. Nov. 1998, pp. 89-95.
9. T. K. Jun and C.-J. Kim, "Valveless Pumping using Traversing Vapor Bubbles in Microchannels", *J. Applied Physics*, Vol. 83, No. 11, June 1998, pp. 5658-5664.
10. G. Beni and S. Hackwood, "Electro-wetting displays," *Applied Physics Letter*, Vol. 38, No. 4, pp. 207-209, Feb. 1982.
11. H. Matsumoto and J. E. Colgate, "Preliminary Investigation of Micropumping Based on Electrical Control of Interfacial Tension," *Proc. IEEE MEMS '90*, Napa Valley, CA, pp. 105-110, Feb. 1990.
12. J. Simon, S. Saffer, and C.-J. Kim, "A Liquid-Filled Microrelay with a Moving Mercury Micro-Drop", *J. MEMS*, Vol. 6, No. 3, Sept. 1997, pp. 208-216.
13. C.-J. Kim, "Microgasketing and Adhesive Wicking for Fabrication of Micro Fluidic Devices", Proc. Vol. 3515, Microfluidic Devices and Systems, *SPIE Symp. Micromachining and Microfabrication*, Santa Clara, CA, Sept. 1998, pp. 286-291.
14. J. Lee and C.-J. Kim, "Liquid Micromotor Driven by Continuous Electrowetting", *Proc. IEEE MEMS Workshop*, Heidelberg, Germany, Jan. 1998, pp. 538-543.
15. J. Lee and C.-J. Kim, "Microactuation by Continuous Electrowetting Phenomenon and Silicon Deep RIE Process", *Proc. MEMS*, ASME IMECE, Anaheim, CA. Nov. 1998, pp. 475-480.
16. C.-J. Kim and S. Nelson, "Microactuation by Electrical Control of Surface Tension", NSF XYZ-on-Chip Program.
17. C.-J. Kim and R. Garrell, "Integrated Digital Microfluidic Circuits Operated by Electrowetting-on-Dielectrics (EWOD) Principle", DARPA/MTO Bio-Fluidic Chips (BioFlips) Program.

Chapter 19

MESOSCALE MACHINING CAPABILITIES AND ISSUES

Gilbert L. Benavides, David P Adams and Pin Yang
Sandia National Laboratories
P.O. Box 5800, MS 0958
Albuquerque, NM 87185

Abstract Mesoscale manufacturing processes are bridging the gap between silicon-based MEMS processes and conventional miniature machining. These processes can fabricate two and three-dimensional parts having micron size features in traditional materials such as stainless steels, rare earth magnets, ceramics, and glass. Mesoscale processes that are currently available include, focused ion beam sputtering, micro-milling, micro-turning, excimer laser ablation, femtosecond laser ablation, and micro electro discharge machining. These mesoscale processes employ subtractive machining technologies (*i.e.*, material removal), unlike LIGA, which is an additive mesoscale process. Mesoscale processes have different material capabilities and machining performance specifications. Machining performance specifications of interest include minimum feature size, feature tolerance, feature location accuracy, surface finish, and material removal rate. Sandia National Laboratories is developing mesoscale electro-mechanical components, which require mesoscale parts that move relative to one another. The mesoscale parts fabricated by subtractive mesoscale manufacturing processes have unique tribology issues because of the variety of materials and the surface conditions produced by the different mesoscale manufacturing processes.

1 INTRODUCTION

Sandia National Laboratories has a need to machine mesoscale features in a variety of materials. In the past, Sandia has developed precision miniaturescale electro-mechanical components. Presently, Sandia has been developing functionally similar electro-mechanical components using technologies such as silicon based MEMS and LIGA. The authors recognized that there was a void in our ability to fabricate mesoscale parts and features. There is also a need to machine mesoscale features in traditional engineering materials like stainless steels, ceramic and rare earth magnets. Examples of mesoscale features are fillets, spherical radii, contours, holes, and channels. Figure 1 is an illustration of the relative size of critical dimensions for miniature, meso-, and micro-machining. In general, meso-machining processes should be capable of machining feature sizes of 25 microns or less. Unlike LIGA which is an additive technology, the meso-machining technologies that are being developed are subtractive in that material is removed to fabricate a part. These subtractive technologies are, focused ion beam machining, micromilling, micro-turning, laser machining, and micro electro-discharge machining. Sandia is driven to develop micro- and mesoscale fabrication technologies to meet the needs of the nuclear weapons stockpile.

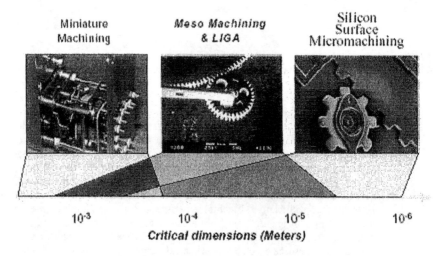

Figure 1. Perspective of miniature machining, meso-machining, and micro-machining.

The focused ion beam (FIB) machines metals by bombarding the work piece with a nanometer scale diameter beam of gallium ions. The material removal rate for focused ion beam machining is very low, on the order of 0.5 μm^3 per second. Given the low material removal rate, the effort is placed upon fabricating tools that can be used repetitively to remove material at much faster rates. Examples of these tools are 25-μm diameter end mills, masks for photolithography and masks for laser machining. Sandia has successfully milled square channels having a cross section of 25 μm by 25 μm in PMMA, aluminum, brass, and 4340 steel using a high precision milling machine. The work related to using the FIB to fabricate hard tooling, has been a joint effort between Sandia National Laboratories and Louisiana Tech University.

The two laser machining processes that are being developed are nanosecond excimer ablation and femtosecond Ti-sapphire ablation. The excimer laser, which has a nanosecond pulse width, can readily machine mesoscale holes and channels in polymers and ceramics. A mask projection technique can be introduced in the expanded portion of the excimer laser beam to project a complex de-magnified replica of the mask onto the workpiece. The femtosecond Ti-sapphire laser can readily machine microscale holes and channels in metals. The femtosecond laser machining process can fabricate a one micron diameter, high aspect ratio hole in metal with minimal debris. Laser machining can be used to create three dimensional features because depth of cut is very well correlated to exposure time.

Sandia's Agie Compact 1 micro-sinker electro-discharge machine (EDM) is being used to machine features as small as 25 microns in difficult materials such as stainless steels and kovar. This class of EDM technology employs a micro-generator that is capable of controlling over-burn gaps to as little as three microns. LIGA technology is being employed to fabricate small intricate copper electrodes. These are mounted to the micro-sinker EDM to machine the complementary shape into these difficult materials.

These subtractive mesoscale machining technologies generate issues in regards to cleanliness, assembly, and tribology. Some issues are unique to mesoscale machining while other issues can be regarded as an extension of similar macro-scale issues. Cleanliness is important because mesoscale critical dimensions can easily be exceeded by dirt particle size or debris created during the machining process. Mesoscale milling and turning can create chips and burrs that can block holes or create a mechanical interference. Surface morphology and surface finish conditions vary greatly depending upon the mesoscale machining technology. The great variety of materials and surface conditions create a complex parameter set for characterizing tribological phenomenon. Mesoscale parts are difficult to handle and align which makes assembly a challenge.

2 MESO-SCALE MACHINING PROCESSES

2.1 Focused Ion Beam, Micromilling and Microturning

The focused ion beam (FIB) sputters material from a workpiece by gallium ion beam bombardment. An illustration of this sputtering process is shown in figure 2. The workpiece (*e.g.*, a ground rod) is mounted to a set of precision stages and is placed in a vacuum chamber underneath the source of gallium (see figure 3). The two translation stages and one rotation stage in the vacuum chamber make various locations on the workpiece available to the beam of gallium ions. A tunable electric field scans the beam to cover a pre-defined projected area. A high voltage potential causes a source of Gallium ions to accelerate and collide with the work piece. The collisions strip away atoms from the work piece. In the example shown in figure 2, the result of the FIB machining process is to create a near vertical facet. Our FIB has either a 200 or 400 nanometer beam diameter, although some FIBs have beam diameters as small as 5 nanometers, making the FIB a true micro-scale capable machine. Sandia National Laboratories has teamed with Louisiana Tech University to further develop the FIB machining technology.

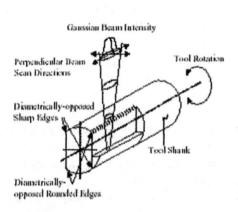

Figure 2. Beam of Gallium ions machining a facet on a cylindrical work piece.

Figure 3. FIB system showing source of Gallium sitting over target.

The hexagonal tool shown in figure 4 was fabricated in the FIB by using the rotational stage to rotate a ground rod to six equally spaced angular positions. This hexagonal tool is similar to the micro-milling tool that was mounted on a high precision milling machine (Boston Digital BostoMatic 18) to machine a channel in aluminum as shown in figure 5. The micro-turning tool shown in figure 6 is another example of tool that can be fabricated in the FIB. This micro-turning tool was used on a lathe to fabricate a finely threaded rod.

The bulk of the FIB effort has been to machine hard tooling instead of directly machining features onto the end workpiece. The slow material removal rate has rendered the FIB as impractical for direct machining large features (see table 1). The hard tools, however, can remove material at an impressive rate and are durable enough for several hours of machining time (one tool was used for six hours). Nevertheless, the FIB is practical for directly machining complex three dimensional shapes that do not require a

Table 1. Comparison of Meso-scale machining technologies.

Technology / Feature Geometry	Minimum feature size / Feature tolerance	Feature positional tolerance	Material removal rate	Materials
Focused Ion Beam / 2D & 3D	200 nanometers / 20 nanometers	100 nanometers	.5 cubic microns/sec	Any
Micro milling or micro turning / 2D or 3D	25 microns / 2 microns	3 microns	10,400 cubic microns/sec	PMMA, Aluminum, Brass, mild steel
Excimer laser / 2D or 3D	6 microns / submicron	submicron	40,000 cubic microns/sec	Polymers, ceramics and metals to a lesser degree
Femto-second laser / 2D or 3D	1 micron / submicron	submicron	13,000 cubic microns/sec	Any
Micro-EDM (Sinker or Wire) / 2D or 3D	25 microns / 3 microns	3 microns	25 million cubic microns/sec	Conductive materials
LIGA / 2D	submicron / 0.02um--0.5 um	-0.3um nom. across 3"	N/A	Electroformable: copper, nickel, permalloy (see note)

Note: LIGA can also be used to fabricate parts in polymers, pressed powders, ceramics, and rare-earth magnets with a little degradation in machining performance specifications.

Figure 4. 25-m end mill. (M. Vasile, Louisiana Tech University).

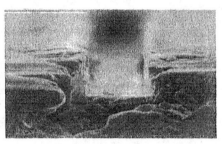

Figure 5. 27×25 µm channel machined in aluminum.

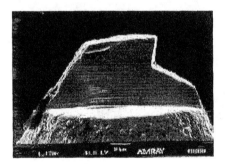

Figure 6. 30-m width turning tool.

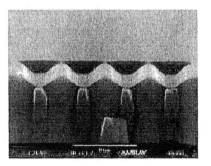

Figure 7. A sine wave directly machined by a FIB. Wavelength = 7 µm (M. Vasile, Louisiana Tech University).

substantial material removal rate. The sinusoidal profile shown in figure 7 was directly machined into silicon by the FIB. Length of exposure and angle of incidence can greatly affect the geometry of directly machined features.

2.2 Laser Machining

Sandia has an excimer laser (Lumonics Hyperex-400, 248nm) set up for mesoscale machining. The excimer laser machines material by pulsing it with nanosecond pulses of ultraviolet light. The workpiece is mounted to precision translational stages. A controller coordinates the motion of the workpiece relative to the stationary UV laser beam and coordinates the firing of the pulses. Figure 8 is a schematic of a mask projection technique that can be used to define machining geometries. The mask is inserted into the expanded part of the beam where the laser fluence is too low to ablate the mask. The mask geometry is de-magnified through the lens and projected onto the work piece. This approach can be used to machine multiple holes simultaneously. Figure 9 is an image of an array of 48 μm holes simultaneously machined into alumina. Sandia's excimer laser has been used to machine polymers, ceramics, glass and metals having feature sizes as small as 12 μm. Figure 10 is an image of a 25 μm by 25 μm channel machined by the excimer laser into PZT. The vertical walls in the channel are a result of good coupling between the UV wavelength (248 nm) and the PZT material. Figure 11 is an image of a 12-μm hole machined by the excimer into kovar using the mask projection technique. The debris on the entrance side of the kovar sheet is a result of thermal melting of the material. The mask projection technique has been used successfully with YAG lasers as well; however, re-solidified thermal debris is still problematic (figure 12c).

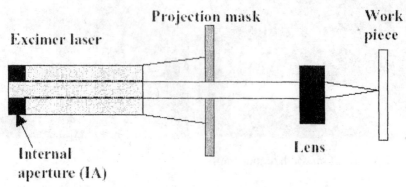

Figure 8. Schematic of excimer laser and mask projection technique.

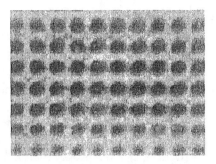

Figure 9. An array of 48-μm holes machined into 275-μm-thick alumina using excimer laser mask projection.

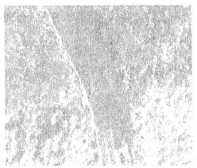

Figure 10. 25-μm-wide by 25-μm-deep trench in PZT using excimer laser.

Figure 11. 12-μm-diameter hole by 25-μm deep in kovar using excimer laser.

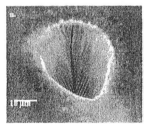 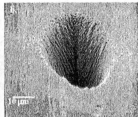 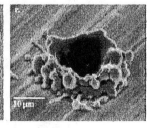

Figure 12. Collaboration with Pulsed Power (SNL): Micro-holes machined in kovar using a Ti:sapphire system (120 fs) in air (a.) and in vacuum (b.) These are compared with a hole (c.) drilled by a Nd: YAG laser ($\lambda = 1.06$ μm, pulse width = 100 ns, power = 50 mW, 2 kHz). All images are taken from the entry side of the foil.

A cleaner laser machining approach is to use a Ti-sapphire femtosecond laser (pulse width on the order of 10–15 s). Figure 12 shows three images of laser machined holes into identical sheets of kovar, each hole having a similar diameter. The first two holes (a & b) were machined with the Ti sapphire femtosecond laser while the third hole (c) was machined with a YAG laser. The hole shown in Figure 12a, which was machined in air, is about as clean as the hole in figure 12b, which was machined in a vacuum. The detectable debris that can be observed in images a & b, are nanosize particles. These nanosize particles resemble a thin film deposition coating. The minimal debris resulting from femtosecond machining has been observed by Clark MXR and the University of Nebraska (figures 13 & 14). Figure 14 is also an illustration of a deep one micron-size feature that can be fabricated using the femtosecond laser. The femtosecond laser ablation process is unique in that it breaks atomic bonds instead of thermally ablating material. The femtosecond laser machining process is appealing for several reasons. It is cleaner, it is micron capable, and it is not material specific.

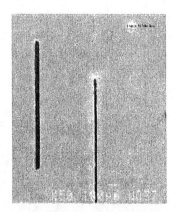

Figure 13. The slit on the left was machined with ultrafast (fs) laser pulses while the slit on the right was machined with long pulses (ns). Material: Invar, 1mm thick. (Clark MXR Inc.)

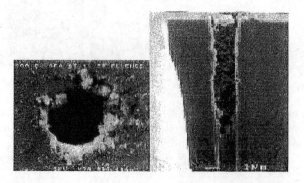

Figure 14. Top view and cross-sectional view of 1 μm diameter hole in silicon. Aspect ratio: 10 to 1. (University of Nebraska, Center for Electro-Optics)

2.3 Micro Electro-Discharge Machining

Electro-discharge machining removes material through a spark erosion process. The micro-EDM machines can machines features as small as 25 μm because the micro-generator needed to create the spark has the necessary fine control for the smaller features. For either the sinker or the wire micro-EDM machine, the two major considerations for determining feature size are the electrode size and the over-burn gap. Sandia has used electrodes as small as 13 μm in diameter and over-burn as little as 3 μm. Currently, Sandia has an Agie Compact 1, micro-sinker EDM machine (see figure 15) and will soon possess an Agie Excellence 2F microcapable wire EDM machine (see figure 16). The advantage to the sinker EDM process is that an electrode having a complex three-dimensional geometry can be sunk into a workpiece creating the conformal geometry in the workpiece. A disadvantage to the sinker EDM is that the electrode also erodes during the EDM process (although at a much slower rate). The advantage to the wire EDM process is that unused wire can be circulated to the work piece during the EDM process thereby presenting to the work piece an electrode having a known geometry. The disadvantage to the wire EDM process is that the feature cuts are of a simple geometry (two and a half dimensional).

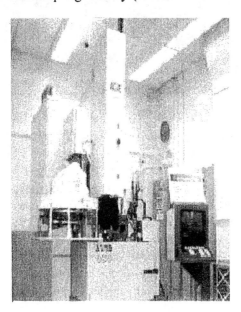

Figure 15. Agie Compact 1, microsinker EDM machine.

Figure 16. Agie Excellence 2F micro capable wire EDM machine.

Figure 17. Micro-EDM electrode in copper made with the LIGA process.

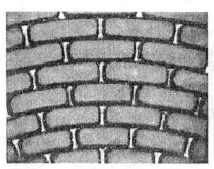

Figure 18. Entrance side of kovar part.

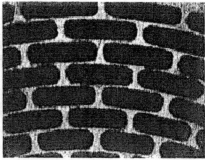

Figure 19. Exit side of kovar part.

Creating an electrode having a complex geometry for the sinker EDM machine is not trivial. Although, graphite is a very desirable electrode material because it machines easily and erodes slowly, copper is also very common. One approach to fabricating a complicated sinker EDM electrode for a mesoscale part is to use the LIGA process. Copper, which is a good performing electrode material, can be plated into LIGA molds. The copper LIGA electrode can then be mounted onto the sinker EDM machine to fabricate a part in a different material such as stainless steel or kovar. This hybrid process is a method of extending the material base for LIGA. Figure 17 is an image of a small intricate copper electrode fabricated by the LIGA process. This electrode was then used to EDM a 0.006 inch thick kovar sheet. If the electrode can be described as the bricks then the machined workpiece would be the "mortar" that fits between the bricks. Figures 18 and 19 show the entrance side and the exit side of the kovar part. The most challenging feature on the kovar part is the portion of the "mortar" that is 0.002 inch thick by 0.006 inch deep (measured into the page).

3 SUMMARY OF CAPABILITIES

As in the macro world, no one mesoscale machining process can do it all. Some mesoscale processes are more encompassing than others, but each process has its niche. As in the macro world, designers usually require a variety of materials to optimize performance of mechanical components. For example hermeticity and corrosion resistance may be important characteristics for housing (or packaging) materials but wear and friction characteristics may be important for gears internal to the housing. Table 1 is an attempt by the authors to summarize mesoscale machining processes. The data in the table is meant to be a representation of the technology but does not represent any particular machine manufacturer. The first column in the table lists the technology and also whether the technology can fabricate 3D features or 2D (actually 2.5D) features only. The second column indicates minimum feature size and the tolerance associated with that feature. For example the minimum feature size for micro-milling is 25 µm channel plus or minus 2 µm. The source for the tolerance assigned to micro-milling is a result of the radial run out of the tool when mounted in the collet. The third column lists the feature positional tolerance, which is mostly based upon the quality of positional stages used on the machine. In the example of micro-milling, the 25-µm channel can be positioned on the work piece to plus or minus 3 microns. The fourth column lists the material removal rate, which is an indicator of how quickly parts can be machined. The FIB has a very poor material removal rate, which was a driver for using the FIB to fabricate micro-tools. Although it can take a long time to make a micro-tool, the micro-tool has a high material removal rate and can be used for a significant duration (up to six hours). A feature tolerance and feature positional tolerance of about 3 µm does not compare favorably with LIGA. Nevertheless, a 3 micron profile tolerance on ratchet teeth (for example) on a 6 mm diameter part is an equivalent ratio to a .001 inch profile tolerance on a 2 inch diameter part. The point is, 3 µm is probably plenty good enough.

4 MIRCOSCALE MACHINING ISSUES

Designers are comfortable with traditional materials (*e.g.*, stainless steel) because these materials have a long history and have been very well characterized through the years. Mesoscale machining processes allow the designer to use traditional materials. On the other hand, tribological issues for mesoscale parts may or may not emulate what is already known. Subtractive mesoscale machining technologies expand the material base and increase the combinations of materials that can come into contact. Galling may be anissue with some material combinations. Each particular mesoscale machining process uniquely affects the surface roughness and morphology. Micro-milling and micro-turning may generate burrs and particles that can cause mechanical interference. Micro-EDM may leave a recast layer that can have particular wear and friction characteristics. Friction effects of mesoscale parts sliding with other parts may have limited points of contact and are not accurately modeled by surface contact models. Some mesoscale machining technologies, such as micro-EDM, are fairly mature, while others, such as femtosecond laser machining, require additional development. Many issues have yet to be identified.

ACKNOWLEDGEMENTS

Part of this work was performed at Sandia National Laboratories and supported by the United States Department of Energy under Contract No. DEAC04-94AL85000. Sandia is a multiprogram laboratory operated by Sandia Corporation, a Lockheed Martin Company, for the United States Department of Energy. We thank Professor M. J. Vasile, Louisiana Tech University for his technical contributions to this paper.

REFERENCES

1   D.E. Bliss, D.P. Adams, S.M. Cameron, T.S. Luk, "Laser Machining with Ultrashort Pulses: Effects of pulse-width, frequency and energy," Material Science of Microelectromechanical Systems (MEMS) devices, Mat. Res. Soc. Symp. Proc. vol 546 (1999).
2   D.P. Adams, G.L. Benavides, "Micrometer-scale Machining of Metals and Polymers Enabled by Focused Ion Beam Machining," Material Science O Microelectromechanical Systems (MEMS) Devices, Mat. Res. Soc. Symp. Proc. vol 546 (1999)
3   H.C. Moser, B. Boehmert, "Trends in EDM," Modern Machine Shop Feb. 2000.
4   P. Bado, "Ultrafast pulses create waveguides and microchannels," Laser Focus World, April 2000.

Chapter 20

AN OVERVIEW OF NANO-MICRO-MESO SCALE MANUFACTURING AT THE NATIONAL INSTITUTE OF STANDARDS AND TECHNOLOGY (NIST)

Edward Amatucci, Nicholas Dagalakis, Bradley Damazo, Matthew Davies, John Evans, John Song, Clayton Teague and Theodore Vorburger
National Institute of Standards & Technology
Gaithersburg, MD 20899

Abstract The future of nano-, micro- and meso-scale manufacturing operations will be strongly influenced by a new breed of assembly and manufacturing tools that will be intelligent, flexible, more precise, include in-process production technologies and make use of advanced part design, assembly and process data. To prioritized NIST efforts in nano, micro and meso manufacturing, several visits to industrial and government research laboratories and two workshops were organized. The identified needs at the meso and micro Scale include: dimensional and mechanical metrology, assembly and packaging technology and standards, and providing a science base for materials and processes emphasizing materials testing methods and properties data. Nanocharacterization, nanomanipulation, nanodevices and magnetics industry support have been identified at the nanoscale. Nanometrology and nanomanipulation have a substantial base from which to expand within the Manufacturing Engineering Laboratory (MEL). Therefore, MEL is initiating a new long-term Strategic Program in nano-manufacturing to conduct research and development, to provide the measurements and standards needed by industry to measure, manipulate and manufacture nanoscale discrete part products. The program will address the measurement and standards issues associated with the manufacture of both nanoscale products themselves, as well as with the development of the production systems required for their manufacture.

1 INTRODUCTION

Nano-Micro-Meso (NMM) scale manufacturing measurement capabilities, standards development, and performance measures are emerging as critical needs for various manufacturers. Leveraging a history of dimensional metrology, nanotechnology, sensor calibration and development, controls and automation, and systems integration for manufacturing enterprises, the Manufacturing Engineering Laboratory (MEL) of the National Institute of Standards and Technology (NIST) is poised to meet these new challenges while working with our partners in government, academia and industry. This overview will review the MEL exploratory project and share with you our findings of the industrial needs and motivations in micro and meso manufacturing. Furthermore, strategic programs being organized to meet new challenges in nano scale manufacturing will be discussed.

Micro- and meso-scale devices are a very exciting new arena for many research groups and companies around the world. Meso-scale devices, defined as things between the size of a sugar cube and the size of one's fist, include small chemical process systems, air and water purifiers, refrigerators and air conditioners that weigh only grams, small robotic devices for military uses, and all manner of electronics and sensors and mechanisms that are tiny and light weight. This is a domain where electrostatic actuation works well and there are significant advantages in heat transfer, kinetics, and high surface to volume ratios in building highly parallel systems from simple unit processes. The Defense Advanced Research Projects Agency (DARPA) is pushing this area very hard, for small, lightweight devices for the individual soldier to carry. A huge new consumer electronics industry deriving from this work is just over the horizon.

Manufacturing is constantly pushing toward smaller and smaller scales. This is not a consistent continuum, and in fact micro-manufacturing which includes semi-conductor manufacturing, wafer and chip manufacturing processes, *etc.* is a larger field than meso-manufacturing because of exploitation of microelectronics fabrication technologies and the large markets for data storage devices. There is activity at all scales and there is a pressing need for NIST and particularly MEL services at all scales. Making devices at increasingly smaller scales is one of the most exciting and challenging frontiers of manufacturing.

The future of meso-scale manufacturing operations will be strongly influenced by a new breed of assembly and manufacturing tools that will be intelligent, flexible, and precise will include production technologies with in-process control and sensor capabilities to produce parts quicker and more accurate; and will make use of advanced part design, assembly and process data.

Nano-manufacturing is a growing area for MEL and we have initiated a strategic program that will provide the measurements and standards needed by industry to measure, manipulate and manufacture nano-scale part products.

2 MANUFACTURING ENGINEERING LABORATORY EXPLORATORY PROJECT

The needs of U.S. manufacturing industries in the area of meso-, micro-, and nano-manufacturing were identified by [Reference 1]:
- Visiting some 20 companies and laboratories and asking about the technology needs and about the specific measurements, standards, and data needs;
- Running two workshops, one co-sponsored with DARPA and one with National Science Foundation, to establish the technology base and the prioritized needs and opportunities for NIST efforts in this area;
- Attending courses and lectures and conferences in MEMS (microelectromechanical systems) and nanotechnology;
- Organizing a NIST-wide coordinating group to inventory base efforts and needs; and
- Carrying out a literature search and review.

The *prioritized needs* for NIST efforts in meso and micro manufacturing were identified *by industry* during our visits and at our workshops to be:
- Dimensional and Mechanical Metrology
- Assembly and Packaging Technology and Standards
- Providing a Science Base for Materials and Processes, emphasizing materials testing methods and properties data.

The needs for NIST efforts at the nanoscale were grouped into four categories:
- Nanocharacterization
- Nanomanipulation
- Nanodevices
- Magnetics industry support

A key issue at the nanoscale for MEL, not highlighted through the study yet where MEL has a substantial base effort, is nanometrology. This in part fits into nanocharacterization and in part fits into nanomanipulation. MEL sees this as eventually addressing a continuum of scales from macro to nano, and MEL has to be involved at all scales to meet industry needs.

3 TECHNICAL OPPORTUNITIES FOR NIST PROGRAM

The needs identified above and prioritized by industry at MEL workshops are specifically NIST mission roles: *measurement, standards, data and infrastructure technology*. More than 100 issues that NIST and NSF should address were rank ordered; the above are a composite of the top ranked issues that are specifically NIST issues. Specific needs related to these areas to be addressed by MEL are discussed below:

3.1 Metrology

Maintaining U.S. leadership in global markets requires conformity with existing standards (*e.g.*, International Standards Organization or ISO) as well as addressing faster transfer of technology from laboratories to production, and higher quality products. The manufacturing industries we are addressing here produce approximately $20 billion output per year (which in turn leverages much larger portions of the gross national product (GNP) by providing key components for much more expensive macro-scale discrete part products). Current growth is 20–25% per year, several times faster than overall discrete part manufacturing ($1.1 trillion growing about 7% per year). So MEL is addressing a real and important need with high leverage for impact over the next decade.

Figure 1. Demonstrates large size of probes used on CMMs compared to small meso-scale components needed to be checked.

New and innovative approaches to metrology need to be address as represented by Fig. 1. The size of the critical features on the fiber optic connector needed to be checked are much smaller then the size of conventional probes on the coordinate measuring machine (CMM). These parts have small diameter holes and the tolerances between the holes are critical for optical alignment of components. How manufacturers measure these internal features and the distance between these features accurately and reliably is a major concern. New tools and methods need to be developed and expanded from the current capabilities at the macro and in the micro and nano scales. These meso components fall within a gap in the most current expertise.

It should be noted that Japan and Germany are investing very heavily in these areas. One indicator, cited by Al Pisano, formerly of DARPA and currently at Berkeley, is the number of organizations working in the MEMS field. In 1994, the US was well in the lead; in 1997 the US was behind both Germany and Japan (150 in the United States, 270 in each of the others). Before 1990 there were never more than 10 MEMS Patents issued in the world in any year; in 1997 there were 150, of which only 50 were U.S. Patents. The United States must invest to stay in a leadership position; NIST will play a small, but key role in measurement and standards.

3.2 Assembly and Packaging

At the meso and micro level, successful automation in photonics alone would save approximately $7 billion[1] to $12 billion[2] in the next 10 years in the U.S. economy [Reference 1].

Furthermore, assembly and packaging are *the* essential problems for MEMS devices, one subset of meso manufacturing. According to one expert:

System Planning Corporation[3] accomplished a market analysis for DARPA's MEMS effort. They estimated that the market potential for MEMS was ~$13 billion in the year 2000. The market has never developed due to the price of the MEMS products. The current market is mainly automotive and is about $400M/year. This is air bag deployment sensor. The two main players are Analog Devices and Motorola. The main issue is packaging. Plastic packaging is critical for low cost. No one to date has a viable plastic packaging concept or capability. The other issue is more challenging. MEMS devices cannot be handled by normal die handling tools due to the mechanical structure. This makes not just the package cost expensive but the act of packaging very expensive.

---

[1] Estimate by Gordon Day, Chief of NIST Office of Optoelectronics Programs, Boulder.
[2] Estimate by Brian Carlisle, CEO of Adept Technology, leading US manufacturer of robots for assembly and member of ATP funded Precision Optoelectronics Assembly Consortium (POAC).
[3] Any mention of specific companies or their products does not imply endorsement by the National Institute of Standards and Technologies.

Figure 2. Calibration and Control of Micro Positioners – Microstage used for testing to demonstrate concepts to Precision Optoelectronics Assembly Consortium. Fall 1999.

3.3 DARPA

NIST is actively moving to try to address these issues; NIST's microstage work (Fig. 2) for POAC, for example, has been enthusiastically received by the photonics industry. We have the capability to make a significant contribution to positioning and automation of meso scale components. Our interaction with industry during the Advanced Technology Program's (ATP) POAC effort and during the exploratory study has indicated that NIST should have a positive impact in the photonics industry. NIST expects this impact to exhibit itself through information exchange, demonstrations, performance measures for quantitative evaluation of components, development of key microrobotics and microsensing technologies, providing for metrology and calibration support, and proactive pursuit of interim and *de facto* standards. Researchers have even advanced the state of the art in microstage capabilities in pursuit of measurement and standard capabilities, as illustrated in Fig. 3.

3.4 Process Technology

Proper materials test methods (process metrology) will allow transitioning of technology from the laboratory to production. Individual manufacturers are able to establish repeatability but not real process control. Process metrology and standardization of materials test methods are basic NIST mission issues and were identified as needs in every lab we visited. The 20 % to 25 % growth rate cited will not happen without infrastructure technology, standards, and measurement technology.

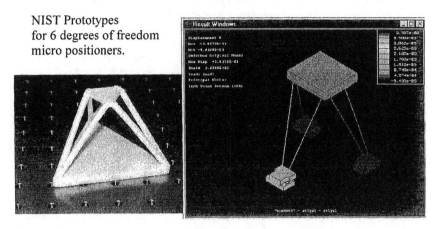

Figure 3. NIST prototype micro positioners.

One specific need is MEMS which is the use of semiconductor fabrication techniques to make sensors and actuators rather than integrated circuits. Several observers have noted that this field has been a major commercial disappointment so far, with only the air bag sensors and deformable mirror devices commercial successes after twenty years of work. This is true, but others point out that it has taken that time to develop an understanding of problems and that there are many new applications that will be major successes within a few years. Another reason for the delay is that MEMS requires better process control than microelectronics. For example, a 5% change in film thickness is within specification for conductivity of an interconnect, but a 15% change in resonance frequency would result if that film were used in a resonant beam (stiffness changes as the cube of thickness). This is totally unacceptable for making RF filters. Similarly, a 0.5% astigmatism in stepper optics is fine for microelectronics but is unacceptable for making gyro rotors that will turn at several hundred thousand revolutions per minute. Discussions with leading organizations in the field indicate significant problems with process metrology, process control, and process control software in trying to build MEMS devices. All agree that NIST should be a major contributor in leading the attack on these problems.

Another important measurement need is for developing much better force measurement capabilities to support calibration and manufacturing of meso and micro scale components used in current products. For example, Fig. 4 shows an early prototype of a force transducer to help investigate force measurement and calibration of suspensions used in electronic disk drives.

NIST fabricated prototype force transducer for calibrating suspensions

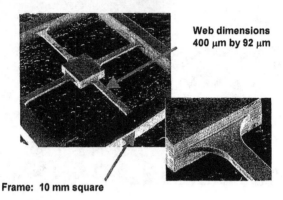

Web dimensions
400 μm by 92 μm

Frame: 10 mm square

Figure 4. Example of research force measurement capabilities.

3.5 Integration Technology

Creating a real, working, distributed manufacturing test bed will dramatically accelerate the development and diffusion of this technology into the marketplace by allowing large numbers of users to access centers of specific process expertise. DARPA has funded work on this concept and has Berkeley, Cornell, Sarnoff, and Stanford under contract, with four industrial applications in negotiation. These organizations and universities believe quality control, and remote supervision or even remote operation of metrology equipment to monitor process operation and quality are the keys to making this concept work from a business standpoint. DARPA sought NIST's involvement, citing problems with interoperability of software and data exchange, problems with process metrology since MEMS fabrication needs better process control than microelectronic fabrication as noted above, and problems with remote testing and monitoring.

4 NANO MANUFACTURING STRATEGIC PROGRAM

MEL is initiating a new strategic program in Nano-manufacturing to provide the measurements and standards needed by industry to measure, manipulate and manufacture nano-discrete part products. The program will address issues associated with the manufacture of both nano-discrete-part products themselves, as well as with the development of the production systems required for their manufacture.

Some key concepts that will be addressed in this program are defined below:

- *Nano-discrete part products* is defined as a product having critical part features with dimensions of <= 100 nm for either a single discrete part or an assembly of discrete parts. Product classes are being drafted that further clarify what is meant by a discrete part. These classes include a single discrete part, assemblies and complex assemblies.
- *Measure* refers to the mechanical properties of product features with accuracy at the sub-atomic scale, *e.g.*, dimensional accuracy better than 0.1 nm.
- *Manipulate* is defined as tools that can grasp, position, and assemble to sub-nanometer accuracy.
- *Manufacture* is defined as the fundamental production and assembly processes of material removal, addition, reshaping, and transformation for producing large lot sizes of nano-molecular discrete parts at and below the nanometer level.

A key point in understanding nano-manufacturing is to realize that at some point nano scale devices have to interface to the macro world. This is a significant technological barrier to adopting nanotechnology to commercial uses.

5 CONCLUSIONS

Micro and Meso-scale devices are a very exciting new arena for many research groups and companies around the world. The future of meso-scale manufacturing operations will be strongly influenced by a new breed of assembly and manufacturing tools that will be intelligent, flexible, more precise, include real time process control production technologies and make use of advanced part design, assembly and process data. The field of Intelligent Integrated Microsystems has a tremendous growth potential expected to reach $30 billion in the first part of this century, with applications in industrial areas like machinery and plant manufacturing, production control, power systems, and home and building control [Reference 1]. And, this is just a projection for the Integrated Circuit Industry. Many more exciting and important discoveries and applications are on the horizon.

Based on our industrial visits and workshops, which provided prioritized inputs, certain minimal elements for research considerations have been defined. These are:

♦ Dimensional Metrology: A suite of measurement technologies at meso- and micro-dimensions to handle metrology requirements as shown in Figure 1.

♦ Force and Torque Measurement and Calibration Services at micro- and nano-Newton levels.

- Assembly and Packaging: information exchange, microrobotics and microsensing technology, performance measures, interim and de facto standards to accelerate commercialization
- Process Technologies and Materials Properties

Research and development is moving rapidly in this field. For example, in basic metrology, the need is chronic. A leading manufacturer of suspension arms for disk drive read heads needs force and torque calibration (at μN and $nN \cdot m$ resolution) now; they will have even more stringent needs five years from now. Current NIST calibration services stop at 44 N [Reference 1].

Another requirement of a major electronic manufacturer is the need to inspect location and form of 125 µm holes in fiber optic connectors that have 1 µm tolerances. Future needs will push the tolerance requirement by 10 fold. This is one of many examples of the need for a micro-CMM that is not one instrument but rather a suite of measurement technologies applied to 3D mechanical parts and devices and systems that fall in the range between classic Coordinate Measuring Machine (CMM) metrology and Scanning Electron Microscopes (SEM), Scanning Tunneling Microscopes (STM), and Atomic Force Microscopes (AFM).

The following summary from our workshop, co-sponsored with NSF and well represented by industry, academia and government, helps to capture the importance and necessity of pushing technology development, measurements and standards at all scales:

"In the not-too-distant future, it is expected that fabrication on an even smaller scale—the nanoscale—will yield functional components, devices, and systems that employ atoms and molecules as their building blocks. The ultimate objective, of course, is to link the enabling technologies across all scales—nano, micro, meso, and macro—accounting for such factors as materials and device characterization methods, fabrication technologies and processes, measurement and test methods, and modeling and simulation tools.

Experts in the broad field of manufacturing research and development agree on the urgent need to address the challenges posed by any factors that might inhibit progress toward further ultraminiaturization. The implications are significant not only for society at large, but also for the commercial sector. Given that significant growth is expected especially in the near-future demand for and availability of micro- and mesoscale applications in medicine, communications, defense, aerospace, and consumer products, it follows that the makers of products in those ranges must develop cost-effective manufacturing techniques. Thus, there is a pressing need to focus research attention on enabling process technologies, packaging and assembly technologies, standards, measurement methods and instruments, and data and tools for engineering" [Reference 3].

ACKNOWLEDGEMENTS

MEL Exploratory Project Team who are Edward Amatucci, Nicholas Dagalakis, Bradley Damazo, Matthew Davies, John Evans, John Song, Clayton Teague, and Theodore Vorburger.

Our colleagues in ATP, CSTL, EEEL, MSEL, PL, BFRL and MEL at NIST who participated and contributed to our findings are too numerous to mention.

DARPA/NIST workshop:

Kevin Lyons, former DARPA Program Manager (currently at NIST)

NIST/NSF workshop organizers:

Dr. Robert Hocken, Director of the Center for Precision Metrology at the University of North Carolina, Charlotte

Dr. Ming C. Leu, Program Director of NSF's Manufacturing Processes and Equipment program

Dr. John Evans, Chief of the Intelligent Systems Division, MEL, NIST

Dr. E. Clayton Teague, Chief of the Automated Production Technology Division, MEL, NIST

Companies/Laboratories we visited include but not limited to: Remmele Engineering, Honeywell, Hutchinson Technology, Professional Instruments, MicroFab Technologies, M-DOT, Sandia Microelectronics Laboratory, UC Berkley, NIST Boulder Optoelectronics Division, Fanuc Berkley Research Center, Adept Technology, Johns Hopkins / Applied Physics Lab, Potomac Photonics, Intuitive Surgical, Inc.

REFERENCES

For additional information, visit http://www.isd.mel.nist.gov/meso_micro/.
1. Teague, E.C., Evans, J., Vorburger, T., et al., Final Report on Micro-Meso Scale Manufacturing Exploratory Project, NIST MEL Internal Report, 1999.
2. Lyons, K., et al., Proceedings of the Manufacturing Technology for Integrated Nano- to Millimeter (In2m) Sized Systems, DARPA/NIST sponsored, March 1999.
3. Hocken, R., et al., Proceedings of the Manufacturing Three-Dimensional Components and Devices at the Meso and Micro Scales, NIST/NSF sponsored with support from University of North Carolina – Charlotte, May 1999.

Chapter 21

LIGA-BASED MICROMECHANICAL SYSTEMS AND CERAMIC NANOCOMPOSITE SURFACE COATINGS

W. J. Meng, L. S. Stephens* and K. W. Kelly
Mechanical Engineering Department
Louisiana State University
Baton Rouge, LA 70803

* *Mechanical Engineering Department*
University of Kentucky
Lexington, KY 40506

Abstract With the X-ray lithography-based microfabrication technique LIGA (Lithographie, Galvanoformung, and Abformung), mechanical components incorporating features ~100 μm in size have been designed and built. Examples include seals with embedded micro-heat-exchangers. Preliminary heat transfer testing showed that micro-heat-exchangers embedded close to the load-bearing surface are effective in controlling the surface temperature during service while wear testing showed that the wear characteristics of electroformed Ni surfaces need to be improved.

A low-pressure, high-density, inductively coupled plasma (ICP) assisted hybrid physical vapor deposition (PVD)/chemical vapor deposition (CVD) tool has been constructed. This deposition tool offers independent control of ion energy and flux, and much higher plasma densities as compared to conventional magnetron glow-discharge plasmas. With such a tool, a wide range of ceramic nanocomposite surface coatings has been deposited. The modulus and hardness of nanocomposite coatings depend systematically on the composite phase fraction, offering the possibility of true coating engineering for specific applications. Ceramic nanocomposite coatings offer much improved wear characteristics as compared to electroformed Ni surfaces. The potential of applying advanced surface coating technologies and materials to tribology needs for LIGA-based micro-mechanical-systems is high and awaits to be fully explored.

1 INTRODUCTION

Contemporary mechanical design puts continuous demands on traditional mechanical systems such as bearing assemblies or automotive transmissions to function at increasingly higher power to mass ratios. This increased power density means higher contact stresses on mechanical components such as bearings and gears. Demands on power density of traditional mechanical systems such as automotive transmissions have reached a point where traditional surface engineering techniques, such as carburizing, induction hardening, and shot peening, will be insufficient to support further contact stress increases without premature contact-induced surface or near surface failures [1]. Figure 1 shows a typical contact-induced pitting failure on a helical gear flank.

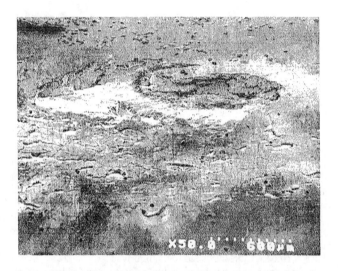

Figure 1. A typical pitting failure on a helical gear flank.

Mechanical seals are another example of a macro-scale component where tribological and thermomechanical issues are of central importance. Seals are present in rotating machinery to limit lubricant loss and isolate the process environment from the operating environment. Figure 2 shows a schematic of a conventional mechanical face seal. The rotating and stationary faces of the active seal experience sliding motion, resulting in wear being the predominant failure mode [2]. Frictional heat generation plays an important role in seal wear, since localized heating causes surface protuberances and leads to thermoelastic instabilities and premature seal failure [3]. The ability to control the surface temperature during seal operation thus offers great potential in improvement of seal performance.

With the X-ray lithography-based microfabrication technique LIGA (Lithographie, Galvanoformung, and Abformung) [4], electroformed Ni load-bearing surfaces with embedded micro-heat-exchangers have been designed and built. A concept drawing for mechanical face seals incorporating such micro-heat-exchangers is shown in Fig. 3. This seal incorporates high-

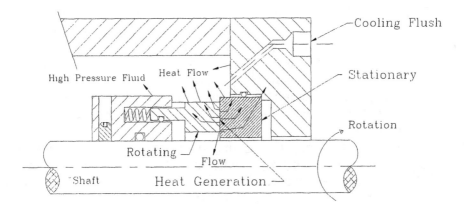

Figure 2. A schematic of a conventional mechanical face seal.

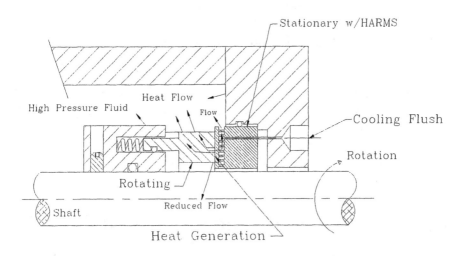

Figure 3. A schematic of a seal with micro-heat-exchangers.

aspect-ratio-microstructures (HARMs), consisting of LIGA fabricated posts and flow channels embedded underneath the stationary seal face. Typical post dimensions are ~150 × 150 × 500 µm. Embedding such HARMs is expected to increase heat conduction and enable better temperature control of the active seal face.

Preliminary testing demonstrated enhanced heat transfer characteristics of seals with embedded HARMs heat exchangers. However, the electroplated Ni surface shows high wear. To improve the overall performance of the HARM-enhanced component, surface modification of electroformed HARMs will be essential.

The application of high-density plasma technology to coating synthesis and surface engineering is in its infancy. Low-pressure, high-density plasmas have the unique capability to apply a wide variety of ceramic surface coatings to impart HARM-enhanced surfaces with desirable tribological properties, including wear resistance, frictional characteristics, and contact fatigue resistance. Ceramic nanocomposite surface coatings synthesized by plasma assisted vapor phase deposition show systematic mechanical property variation as a function of coating composition and microstructure, and offer high potential for coating engineering for specific applications [5].

A variety of schemes for generating low-pressure, high-density plasmas have been studied extensively, including electron cyclotron resonance (ECR) [6], helicon [7], and inductively coupled plasma (ICP) [8]. Among these plasma generation techniques, radio frequency (rf) ICP operates in a non-resonant configuration without external magnetic fields, and has the advantages of ease of operation and simplicity of construction [9].

In this paper, we describe fabrication and testing of prototype face seals with embedded micro-heat-exchangers based on the LIGA technique. Selected results of heat conduction and wear testing on fabricated seals are reported. We describe the construction of a low-pressure, high-density, inductively coupled plasma (ICP) assisted hybrid physical vapor deposition (PVD)/chemical vapor deposition (CVD) tool. Plasma characteristics in such a deposition tool are given. Results on characterization of the mechanical property and microstructure of Ti-C:H surface coatings are summarized and used to illustrate the current understanding on ceramic/ceramic nanocomposite coatings. Directions for future work are discussed.

2 HARM-ENHANCED MECHANICAL SEALS

Masking, X-ray exposure, and polymer sheet development steps of the LIGA fabrication process were carried out at the J. Bennett Johnston Sr. Center for Advanced Microstructures and Devices at Louisiana State University (LSU CAMD) [10]. Prototype seals incorporating micro-heat-exchangers were fabricated onto the end face of annular sleeves. Developed polymer sheets, perforated with high-aspect-ratio holes, were mechanically pressed and clamped onto the end face of the annular sleeve. The assembly was fitted into a polymer plating-jig. The jig was subsequently immersed in a sulfamate Ni-electroplating bath. During the electroplating process, Ni was first deposited into the perforations in the polymer sheet. After filling of the perforations, continued plating (overplating) allowed the electroformed Ni structures to merge and form a new surface parallel to the initial annular sleeve face. At the end of the overplating process, the annular sleeve was removed from the plating-jig. The remaining polymeric material in the seal

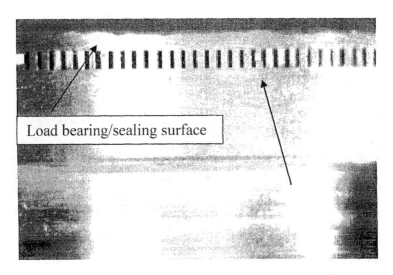

Figure 4(a). View of the outer diameter of an annular sleeve with electroformed Ni micro-heat-exchangers and overplated load-bearing surface.

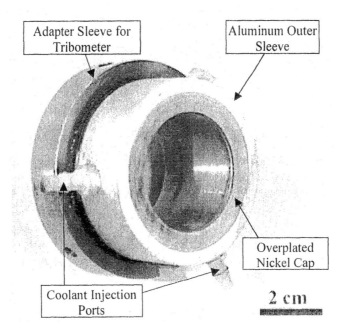

Figure 4(b). An assembled annular sleeve with embedded micro-heat-exchangers.

was dissolved in solvents. Figure 4(a) shows a view of the outer diameter of an annular sleeve with electroformed Ni micro-heat-exchangers and overplated Ni load-bearing/sealing surface. An assembled prototype seal is shown in Figure 4(b). Further details on the LIGA fabrication process and design of micro-heat-exchangers for seals have been presented elsewhere [11].

Electroformed Ni load-bearing/sealing surfaces with embedded micro-heat-exchangers were tested with a thrust washer rotary tribometer, a schematic of which is shown in Fig. 5. Annular seals with embedded micro-heat-exchangers were held stationary, while the mating rings made of brass or stainless steel were rotated at specified loads and speeds. Outputs from the tribometer instrumentation, including load, rotational speed, frictional torque, wear depth, and surface temperature as measured by thermistors embedded into the rotating mating ring in proximity to the active seal surface, were fed into a computer-controlled data acquisition system. Further details on the tribometer design and testing procedures have been reported elsewhere [11].

To assess the cooling capacity and wear resistance of prototype seals with an overplated Ni sealing surface and embedded micro-heat-exchangers, stationary sealing surfaces are tested with the thrust washer rotary tribometer against rotating stainless steel mating rings under a variety of loads and rotation speeds. Seal face temperature and wear were monitored as a function of time. 10W30 engine-oil was used, which provides mixed boundary/hydrodynamic lubrication for the sealing face. Figure 6 shows typical seal surface temperature as a function of testing time. Tests of no cooling and cooling by room temperature air running through the micro-heat-exchangers were conducted. In the air cooling cases, the test was run first without air for 15 minutes and then with air at a constant flow rate for the remaining of the test. Air inlet temperature was 296K. For comparison, seal surfaces with a solid layer of electroformed Ni were also tested using the same procedure. Testing data such as those shown in Figure 6 indicate that embedded micro-heat-exchangers are effective in controlling the temperature of the active sealing surface. Seal faces with embedded micro-heat-exchangers are superior in heat transfer because it provides an axial heat flow path and thus enables a radially more uniform seal face temperature.

Figure 7 shows typical wear as a function of testing time. It is evident that wear is approximately linear versus time. It is also clear that cooling leads to a large reduction in wear rate as compared to the no cooling cases, which may be indicative of oxidative wear mechanisms [12]. Additional heat transfer and wear testing results have been reported elsewhere [13]. From test data such as those shown in Figure 7, a wear coefficient k can be calculated,

$$k = \frac{h}{t} \times \frac{1}{PV}, \qquad (1)$$

where h, t, P, and V are the wear depth, testing time, pressure, and sliding speed, respectively [2]. Wear coefficients for the stainless steel/electroformed Ni seal face pair with and with out air cooling are $\sim 3 \times 10^{-6}$ and 3×10^{-5} mm^3/Nm, respectively. These wear coefficients are higher as compared to that of standard face seal material pairs such as carbon-graphite/cast iron and significantly higher than ceramic face seal materials such as silicon carbide [2].

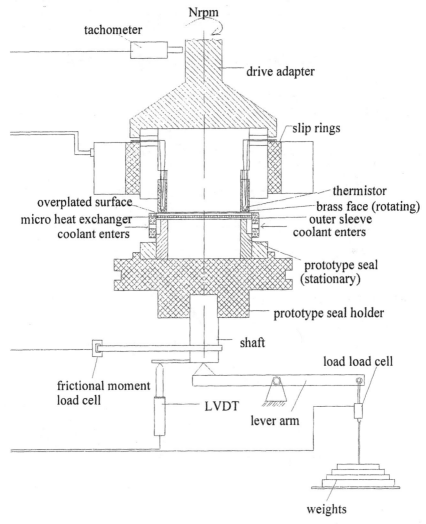

Figure 5. A schematic of the thrust washer rotary tribometer used for testing electroformed Ni seals with embedded micro-heat-exchangers.

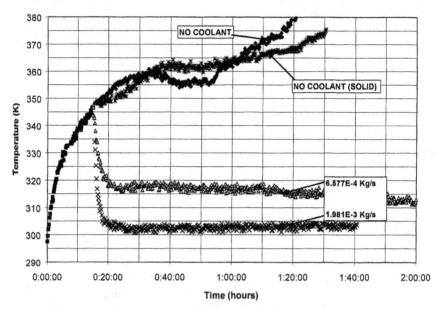

Figure 6. Thermistor temperature versus time during wear testing of electroformed Ni seals with embedded micro-heat-exchangers. Room temperature air with constant flow rates was used as coolant. Testing conditions were seal pressure = 15 psi and rotation speed = 1800 rpm. PV = 14754 psi ft/min.

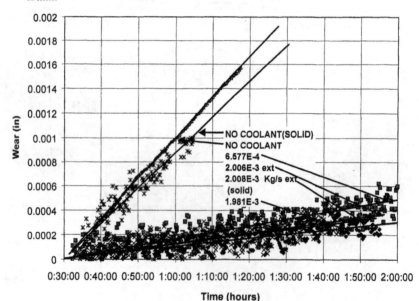

Figure 7. Wear depth versus time. Room temperature air with constant flow rates was used as coolant. Data obtained on solid electroformed Ni layers are also included.

Thus while the heat transfer characteristics of the HARM-enhanced prototype seal is superior, improvement in the tribological characteristics of electroformed Ni surface is critically needed to put these novel micromechanical-systems into actual service.

3 ICP ASSISTED HYBRID PVD/CVD TOOL

Most hard ceramic coating materials of interest to surface engineering applications, such as transition metal nitrides, carbides, and their composites, are refractory with melting temperatures ~3000 K [14]. At deposition temperatures <500 K, surface atoms during growth have low mobility. Ceramic coatings deposited without additional energy input into the surface tend to be porous with grain boundary voids [15]. Energetic ions from plasmas or ion-beams have been widely used for supplying additional energy to the growing surface during deposition [16]. While high-energy ions promote densification of ceramic coatings, they cause atomic displacements and generate interstitials in the deposit, which in turn introduces residual stresses into the coating [17]. In contrast, high-flux bombardment of the growing surface by low-energy ions has been shown to promote coating densification without excessive defect generation [18].

A custom designed, ICP assisted hybrid PVD/CVD tool, with the plasma characteristics of low potential, high density, and independent variation of potential and density, was used to deposit a wide variety of ceramic surface coatings. The advantages of this tool over the standard commercial coating deposition technology of unbalanced magnetron sputtering (UBM) [19] are plasma generation independent of cathode operation, uniform and much higher plasma density within the deposition zone, and independent adjustment of plasma potential and density [20].

A schematic of the deposition zone in this deposition tool is shown in Figure 8 [21]. In a cross configuration, 13.56-MHz rf power was fed into two facing, planar, four-turn induction coils. The induction coils excited a plasma column through dielectric windows via the inductive-coupling mechanism [22]. Two facing balanced magnetron guns, rotated 90 degrees from the induction coils, enable dc sputtering of metal cathodes in different inert/reactive gas mixtures, with or without operation of the induction coils.

Figure 9 shows the total ion flux, as measured by a planar Langmuir probe, in ICP plasmas of pure Ar and Ar/N_2 as a function of the total rf input power into the induction coils. Plasma density increases linearly with the input rf power. At a total input rf power of 2.5 kW, the density of Ar plasma is ~ 13×10^{12} cm^{-3}, about 2 orders of magnitude higher than that of the glow discharge plasma present in conventional magnetron sputtering systems [23].

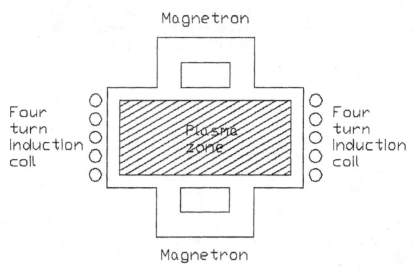

Figure 8. A schematic of an ICP assisted hybrid PVD/CVD tool.

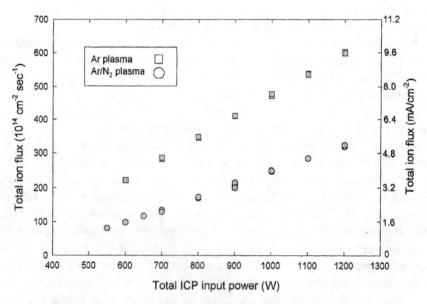

Figure 9. Total ion flux in ICP Ar and Ar/N$_2$ plasmas as a function of total rf input power.

Figure 10 shows the plasma potential, as measured by an emissive Langmuir wire probe, in ICP plasmas of Ar/N$_2$ as a function of the total rf input power. The plasma potential remains approximately constant ~20 V over the entire input power range. By changing the total rf input power, the ion flux can be changed without significant changes in the plasma potential.

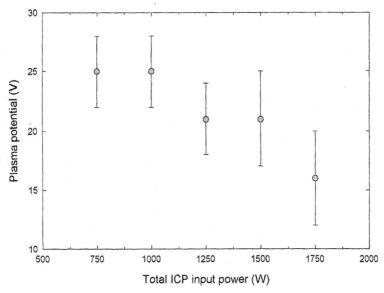

Figure 10. Plasma potential of ICP Ar/N$_2$ plasmas as a function of total rf input power.

Thus the ion energy and flux can be adjusted independently in such a deposition tool. With high-flux, low-energy ion bombardment, it was shown that densification of ceramic coatings such as TiN becomes largely independent of the deposition temperature. As an example, dense TiN coatings were deposited at ~100 K [24]. Figure 11 shows the optical reflectance of two TiN coatings deposited at substrate temperatures ~100 K. At an ion to neutral flux ratio of ~2, the coating is porous with a dark optical appearance. By increasing the ion to neutral flux ratio to ~48, the TiN coating deposited at the same substrate temperature is fully dense with the characteristic gold optical appearance.

Due to the effective separation of the ICP plasma and magnetron sputtering, a wider range of ceramic/ceramic and ceramic/metal composite coatings can be deposited by the ICP assisted hybrid PVD/CVD tool, as compared to by UBM deposition. Figure 12 shows the Ti composition in Ti-containing hydrocarbon (Ti-C:H) coatings as a function of Ti sputter gun current in ICP Ar/C$_2$H$_2$/H$_2$ plasmas. A complete compositional analysis of the Ti-C:H coatings is accomplished by combining electron probe microanalysis (EPMA) with elastic recoil detection (ERD) of hydrogen. When the hydrogen to carbon ratio of the input gas is fixed, increasing the Ti gun current systematically increases the Ti composition in the coating.

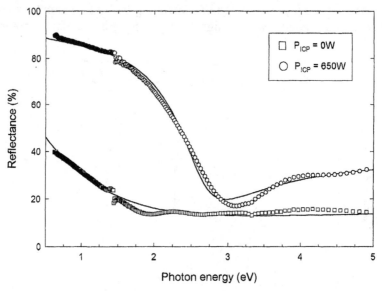

Figure 11. Near normal incidence optical reflectance of two TiN coatings deposited at ~100 K. The ratios of total ion flux to neutral flux at total rf input powers of 0 and 650 W are ~ 2 and 48, respectively.

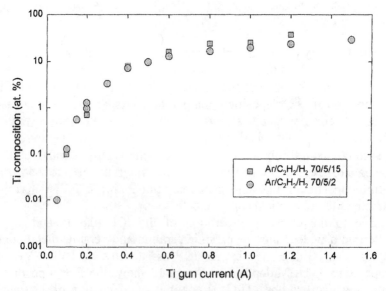

Figure 12. Ti composition in Ti-C:H coatings as a function of the Ti sputter gun current at different hydrogen to carbon ratios in the input gas.

With this hybrid PVD/CVD process, the Ti composition in Ti-C:H coatings can be varied smoothly from less than 0.1 at.% to over 40 at.%. Changing the hydrogen to carbon ratio of the input gas mixture changes the hydrogen to carbon ratio in the Ti-C:H coatings [25].

4 CERAMIC/CERAMIC NANOCOMPOSITE SURFACE COATINGS

Results on Ti-C:H coatings provide a prototypical example of ceramic/ceramic nanocomposite surface coatings. Hard metal-free amorphous hydrocarbon (*a*-C:H) coatings deposited by various plasma assisted vapor deposition techniques have been studied for over two decades [26]. More recently, metal-containing hydrocarbon (Me-C:H) coatings have been studied intensely [27]. Both a-C:H and Me-C:H coatings possess moderately high hardness [28], low coefficient of friction [29], and low wear rate [30]. They have been suggested to adhere better to metallic substrates [31]. This combination of attractive tribological properties and better adhesion makes Me-C:H coatings potentially useful for surface modification of a wide range of mechanical components. Previous plan-view TEM examination has shown that Ti-C:H coatings with Ti composition >5 at.% consist of nanocrystalline B1-TiC precipitates embedded in an a-C:H matrix [32].

Progress has been made towards understanding the microstructure and mechanical properties of Ti-C:H coatings. The structure of *a*-C:H coatings has been modeled in terms of an amorphous random network consisting of sp^2 and sp^3 bonded C atoms plus hydrogen atoms [33]. The ability of *a*-C:H to accommodate metal atoms within the random network structure has not been studied in detail in the past. Figure 13 shows Ti K-edge X-ray Absorption Near Edge Structure (XANES) spectra of 3 Ti-C:H coatings with 0.9, 2.5, and 27.9 at.% Ti, respectively. XANES spectrum of a B1-TiC standard is also shown. At Ti composition >2.5 at.%, XANES oscillations of the Ti-C:H specimens exhibit a one-to-one correspondence to those of the TiC standard. At 0.9 at.% Ti, the XANES spectrum becomes rather featureless, exhibiting none of the post-edge oscillations observed in the TiC standard and Ti-C:H specimens with Ti composition >2.5 at.%.

This lack of post-edge oscillation is consistent with the absence of Ti atoms within the first few coordination shells of Ti, which in turn suggests uniform dispersion of Ti atoms within the *a*-C:H matrix. Data shown in Figure 13 suggest that the dissolution limit of Ti in *a*-C:H is between 0.9 and 2.5 at.%. Above the dissolution limit, TiC/a-C:H nanocomposite coatings are formed. This suggestion is completely confirmed by cross-sectional high-resolution TEM [34]. A cross-sectional high-resolution TEM micrograph of a Ti-C:H coating with 17.6 at.% Ti is shown in Figure 14. Lattice fringes from B1-TiC grains, a few nm in diameter, are clearly visible. The interface between the nanocrystalline TiC precipitate and the *a*-C:H matrix appears to be atomically sharp [35].

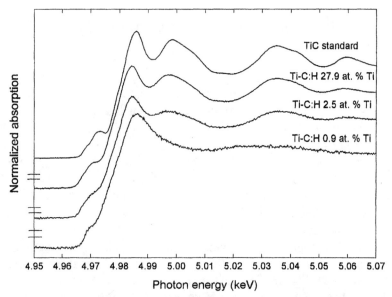

Figure 13. Ti K-edge XANES spectra of 3 Ti-C:H coatings, together with that of a B1-TiC standard. All spectra have been normalized to unit edge jump. Ti compositions of the coatings are 0.9, 2.5, and 27.9 at.%, respectively.

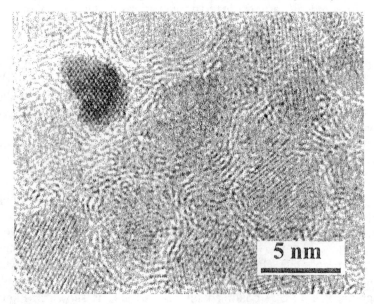

Figure 14. A cross-sectional high-resolution TEM micrograph of a Ti-C:H coating with 17.6 at. % Ti.

Since it is not possible to experimentally isolate the mechanical response of nm-sized TiC clusters embedded in the a-C:H matrix, the mechanical properties of these nanoprecipitates need to be inferred from detailed structural determination. Extended X-ray Absorption Fine Structure (EXAFS) spectroscopy has been used to define the local Ti atomic bonding environment in Ti-C:H coatings [34]. Ti K-edge absorption spectra of Ti-C:H specimens have been collected together with that of the B1-TiC standard, and the EXAFS function $\chi(k)$ extracted. At Ti compositions >2.5 at.%, EXAFS spectral features of the Ti-C:H coatings exhibit an one-to-one correspondence to those of the B1-TiC standard. These spectra can be consistently analyzed by assuming that the first Ti-C and Ti-Ti coordination shells in Ti-C:H are B1-TiC like. Bond distances and coordination numbers are found to be nearly identical to those of bulk B1-TiC [34].

Through structural characterization by high-resolution TEM, XANES and EXAFS spectroscopy, a detailed picture of the microstructure of Ti-C:H coatings has emerged. Based on the small dissolution limit of Ti atoms in an *a*-C:H matrix and the structural similarity of the TiC nanoprecipitates to bulk B1-TiC, this ternary material consisting of Ti, C, and hydrogen can be considered a pseudo-binary system of *a*-C:H and B1-TiC, in which the structure of the TiC nanoprecipitates is very much bulk-like.

In view of the small dissolution limit of Ti atoms in the *a*-C:H matrix, all Ti atoms in Ti-C:H coatings are assumed to be incorporated in the cubic B1-TiC phase. The density of the *a*-C:H matrix has been measured previously [25]. Thus the volume fraction of TiC precipitates f can be calculated from measured Ti composition. Figure 15 shows $E/(1-v^2)$ of Ti-C:H coatings as a function of f. Macromechanical bounds on bulk and shear moduli of isotropic and homogeneous two-phase composites have been given by Hashin and Shtrikman [36]. Based on the observed structural similarity, the elastic properties of TiC nanoprecipitates are taken to be the corresponding bulk values. Figure 15 shows that measured $E/(1-v^2)$ of Ti-C:H coatings are reasonably well described by the Hashin-Shtrikman lower bound, despite their nanocrystalline microstructure.

Upper and lower bounds on yield stress of two-phase composite materials have been given by Ashby in terms of the yield stresses of the precipitate and the matrix [37]. Figure 16 shows hardness H of Ti-C:H coatings as a function of f. Assuming that the measured hardness of the Ti-C:H coatings is linearly related to its yield stress with a constant proportionality, Ashby bounds on the hardness of Ti-C:H coatings can be written in terms of the hardnesses of the TiC precipitate and the *a*-C:H matrix. Based again on the observed structural similarity, the hardness of TiC nanoprecipitates is taken to be the corresponding bulk value. This assumption should be good provided that a size dependent effect on plasticity is not pronounced [38].

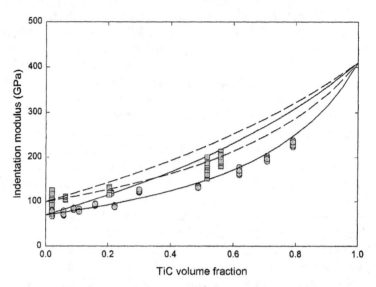

Figure 15. Indentation modulus $E/(1-v^2)$ of Ti-C:H coatings as a function of TiC volume fraction f. Circles and squares represent data on Ti-C:H coatings made in Ar/CH$_4$ and Ar/C$_2$H$_2$ plasmas, respectively. Solid and dashed lines represent Hashin-Shtrikman macromechanical upper and lower bounds for TiC – a-C:H two-phase composites. Modulus of the a-C:H matrix is found by extrapolating measured data to $f = 0$.

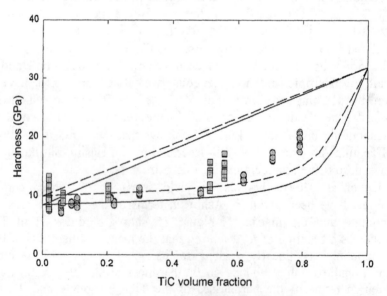

Figure 16. Hardness H of Ti-C:H coatings as a function of TiC volume fraction. Circles and squares represent data on Ti-C:H coatings made in Ar/CH$_4$ and Ar/C$_2$H$_2$ plasmas, respectively. Solid and dashed lines represent the Ashby upper and lower bounds for TiC – a-C:H two-phase composites. Hardness of the a-C:H matrix is found by extrapolating measured data to $f = 0$.

Figure 16 shows, together with the data, the Ashby upper and lower bounds on H. Figure 16 shows that measured H of Ti-C:H coatings are reasonably well described by the Ashby lower bounds. In addition, it is also found that the hydrogen composition significantly affects the elastic modulus and hardness of Ti-C:H coatings, through its influence on the mechanical properties of the a-C:H matrix [25].

Figure 17 shows the wear coefficient for a series of Ti-C:H coatings, measured by running these coatings against WC-Co balls in unlubricated ball-on-disk tests [39]. The wear coefficient increases monotonically by more than one order of magnitude as the TiC nanoprecipitate volume fraction increases.

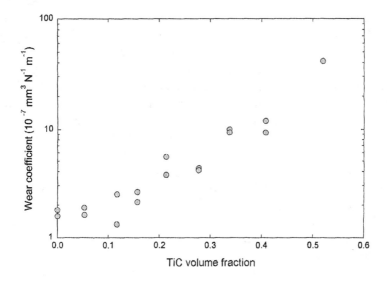

Figure 17. Ball-on-disk wear coefficient of a series of Ti-C:H coatings made by ICP assisted PVD/CVD against unlubricated WC-Co balls as a function of TiC nanoprecipitate volume fraction. Points at the same composition denote repeat measurements.

At the highest Ti composition, the Ti-C:H coating is the least wear-resistant according to the ball-on-disk test. The wear coefficient for this least wear-resistant coating, running against an unlubricated WC-Co ball, is $\sim 43 \times 10^{-6}$ mm^3/Nm, comparable to that for electroformed Ni seal surface, running against lubricated stainless steel counter faces. Furthermore, the friction coefficient of the Ti-C:H coatings running against unlubricated WC-Co balls is ~0.1. The application of suitable ceramic nanocomposite coatings, such as Ti-C:H coatings, to the surface of electroformed micro-mechanical-systems is therefore expected to greatly increase the overall tribological performance of the mechanical component.

5 SUMMARY

Considerations of durability, reliability, performance, and environmental impact of mechanical systems demand continuous improvements of traditional mechanical systems. Automotive transmissions are required to function at increasingly higher power density. Pumps in the chemical industry are required to have longer mean uptime between service and lower leakage to the environment. All mechanical systems are required to be safeguarded against catastrophic failures from loss of lubrication, change of operating environment, etc. These demands put increasingly stringent technical requirements on the mechanical components involved. Operating conditions of many traditional mechanical systems have reached a point such that traditional metallurgical improvements on the bulk material and traditional surface engineering techniques are no longer sufficient to ensure satisfactory component performance. It is also realized that in many cases, it is the thermomechanical properties of the near-surface region which dictate the performance of the overall mechanical component. We believe, based on these reasons, that the application of micro-mechanical-system techniques such as LIGA microfabrication, advanced surface engineering technologies such as high-density plasma based coating deposition tools, and advanced surface engineering materials such as ceramic nanocomposite surface coatings to engineer the next generation mechanical components such as gears, seals, and bearings is a highly promising endeavor. The following specific statements are made:

1. The application of the micro-mechanical-system concept to engineer the next generation traditional mechanical components such as seals and bearings is in its infancy. Concept validation by implementing prototype systems through careful experimentation and modeling should prove to be highly beneficial.
2. The application of advanced surface engineering technologies and materials to engineer the near-surface regions of micro-mechanical-systems is in its infancy. Results described in the present paper lead us to believe that further research and development along this direction should lead to substantial improvement of the overall system performance. Combining microfabricated component prototypes with engineered surfaces should serve to lead the transition from concept prototypes to real world applications.
3. New methods have recently been developed to apply high-density plasma technologies to synthesize a wide range of novel ceramic coatings and engineer surfaces. As described in this paper, low-pressure, high-density plasma technologies possess several intrinsic technical advantages over the current commercial coating deposition technologies. Deposition process development and research into

fundamental deposition mechanisms should help guiding this technology from laboratories to industrial production.
4. The application of ceramic nanocomposite surface coating materials to next generation mechanical components is just beginning. The basic justification for studying this interesting class of materials lies in the potential to engineer coatings for specific mechanical and tribological properties through adjusting its chemical composition and nano-scale microstructure. Detailed mechanistic understanding of the thermomechanical behaviors of nanocomposite materials is largely lacking at present and remains a challenging and exciting topic for future study. Understanding the mechanical response of the interfacial regions in nanocomposites should lead to better overall understanding of their thermomechanical properties and lead to better predictability of coating performance, life, and failure modes.

ACKNOWLEDGEMENT

WJM gratefully acknowledges project support from the College of Engineering of LSU and NSF grant DMR-9871417. KWK gratefully acknowledges project support from DARPA and Louisiana Education Quality Support Fund (LEQSF). We gratefully acknowledge the technical assistance of R. C. Tittsworth, D. M. Cao, B. Feng, J. C. Jiang, and D. Kountouris.

REFERENCES

[1] W. Cheng, H. S. Cheng, L. M. Keer, and X. Ai, ASME Trans. J. of Tribology 115, 658 (1993).
[2] R. L. Johnson and K. Schoenherr, Seal Wear, in Wear Control Handbook, edited by M. B. Peterson and W. O. Winer, ASME, New York (1980).
[3] A. O. Lebeck, Principles and Design of Mechanical Face Seals, Wiley, New York, 1991.
[4] E. W. Becker, W. Ehrfeld, P. Hagmann, A. Maner, D. Munchmeyer, Microelectronic Engineering 4, 35 (1986).
[5] A. A. Voevodin and J. S. Zabinski, J. Mater. Sci. 33, 319 (1998).
[6] J. Asmussen, in Handbook of Plasma Processing Technology, Eds. S. M. Rossnagel, J. J. Cuomo, and W. D. Westwood, Noyes Publications, Park Ridge (1989).
[7] J. Q. Zhang, Y. Setsuhara, T. Ariyasu, and S. Miyake, J. Vac. Sci. Technol. A14, 2163 (1996).
[8] J. H. Keller, J. C. Foster, and M. S. Barnes, J. Vac. Sci. Technol. A11, 2487 (1993)
[9] M. A. Lieberman and R. A. Gottscho, in Physics of Thin Films Vol. 18, edited by M. H. Francombe and J. L. Vossen, Academic Press, San Diego, 1994.
[10] M. S. Despa, K. W. Kelly, J. R. Collier, Microsystem Technol. 6, 60 (1999).
[11] K. W. Kelly, L. S. Stephens, D. Kountouris, and J. McLean, J. Microelectromechanical Systems, in progress.

12. T. S. Eyer, in Treatise on Materials Science and Technology Vol. 13, edited by D. Scott, Academic Press, New York (1979).
13. D. Kountouris, M.S. Thesis, Mechanical Engineering Department, Louisiana State University, December 1999.
14. L. E. Toth, Transition Metal Carbides and Nitrides, Academic Press, New York (1971).
15. J. Musil, S. Kadlec, V. Valvoda, R. Kuzel, and R. Cerny, Surf. Coat. Technol. 43/44, 259 (1990).
16. T. Itoh (ed.), Ion Beam Assisted Film Growth, Elsevier, Amsterdam (1989).
17. W. J. Meng and G. L. Doll, J. Appl. Phys. 79, 1788 (1996).
18. F. Adibi, I. Petrov, J. E. Greene, L. Hultman, and J. E. Sundgren, J. Appl. Phys. 73, 8580 (1993).
19. B. Window and N. Savvides, J. Vac. Sci. Technol. A4, 196 (1986).
20. W. J. Meng and T. J. Curtis, J. Electronic Mater. 26, 1297 (1997).
21. W. J. Meng, pending U.S. patent (1999).
22. M. A. Lieberman and A. J. Lichtenberg, Principles of Plasma Discharges and Materials Processing, Wiley, New York, 1994.
23. B. Chapman, Glow Discharge Processes, Wiley, New York, 1980.
24. W. J. Meng, T. J. Curtis, L.E. Rehn, and P. M. Baldo, Surf. Coat. Technol. 120/121, 206 (1999).
25. W. J. Meng, E. I. Meletis, L. E. Rehn, and P. M. Baldo, J. Appl. Phys. 87, in press (2000).
26. S. Aisenberg and F. M. Kimock, Mater. Sci. Forum 52/53, 1 (1989).
27. H. Dimigen, H. Hubsch, and R. Memming, Appl. Phys. Lett. 50, 1056 (1987).
28. R. G. Lacerda and F. C. Marques, Appl. Phys. Lett. 73, 617 (1998).
29. C. Donnet and A. Grill, Surf. Coat. Technol. 94/95, 456 (1997).
30. J. H. Arps, R. A. Page, and G. Dearnaley, Surf. Coat. Technol. 84, 579 (1996).
31. C. P. Klages and R. Memming, Mater. Sci. Forum 52/53, 609 (1989).
32. W. J. Meng, T. J. Curtis, L. E. Rehn, and P. M. Baldo, J. Appl. Phys., 83, 6076 (1998).
33. J. C. Angus and Y. Wang, in Diamond and Diamond-Like Films and Coatings, edited by R. E. Clausing, L. L. Horton, J. C. Angus, and P. Koidl, Plenum, New York, 1991.
34. W. J. Meng, R. C. Tittsworth, J. C. Jiang, B. Feng, D. M. Cao, K. Winkler, and V. Palshin, J. Appl. Phys., submitted (2000).
35. D. M. Cao, J. C. Jiang, B. Feng, and W. J. Meng, submitted to Microscopy and Microanalysis 2000.
36. Z. Hashin and S. Shtrikman, J. Mech. Phys. Solids 11, 127 (1963).
37. M. F. Ashby, Acta Metall. Mater. 41, 1313 (1993).
38. Q. Ma and D. R. Clarke, J. Mater. Res. 10, 853 (1995).
39. W. J. Meng, R. C. Tittsworth, and L. E. Rehn, submitted to Surf. Coat. Technol. 2000.

Chapter 22

NANO-TRIBOLOGY OF THIN FILM MAGNETIC RECORDING MEDIA

T. E. Karis
IBM Research Division
Almaden Research Center
San Jose, CA 95120

Abstract The magnetic recording industry is projected to continue growing into the foreseeable future. This growth is fueled by increasing data storage density through advances in channel and read/write head integration, tracking servo mechanisms, higher speed spindle motors, chemical integration, and nanotribology. The rotation rate is approaching 20,000 rpm, and the spacing between the read/write head and the disk is approaching molecular dimensions. The ceramic head rails and the disk magnetic layer are currently protected by a complex, yet robust, system comprising 5-10 nm thick carbon overcoats and a 1–2 nm thick perfluoropolyether lubricant film. The film surface energy and lubricant mobility, which control the surface diffusion, are determined by the film thickness and chemisorption of polar lubricant end groups on the carbon overcoat. Intermittent contacts between the head and the disk incrementally remove lubricant from asperities. Lubricant diffuses from the surrounding surface to restore the film thickness on the asperities.
In this paper, the principles of magnetic recording disk lubrication are reviewed and summarized in terms of a lubrication system comprising lubricant removal, reflow, and chemisorption. New test results are presented to illustrate the lubrication system components. The general principles of the magnetic recording lubrication system should also apply to lubrication of micro- and nanoscale devices.

1 INTRODUCTION

The low cost per bit of data storage, along with the reasonably high data rate, short access time, and high reliability are the most salient features of magnetic recording hard disk drives. Higher areal density enables increasing capacity. Recent trends suggest that soon the areal density will surpass 100 Gbits/in^2, the average OEM cost will be about a one cent USD per megabyte, and the data rate will exceed 200 MBytes/sec (Grochowski, 2000). Increases in areal density have been realized by advancing from thin film inductive heads to magnetoresistive and, more recently, to giant magnetoresistive read heads. Improved servo technology provides a higher track pitch, and faster channels enable higher data rates. Advanced coding schemes ameliorate signal deterioration by thermal asperities.

The rapid advancement of magnetic recording throughout the past decade has built upon the skills of manufacturing engineers who refine quality control to provide the nanoscale tolerances which are reduced in every new product. This has been accomplished with essentially the same set of materials, carbon overcoated thin film media, overcoated heads, and functional lubricant. These advances are accomplished while maintaining high reliability.

The magnetic recording head (slider) is said to be flying when it is supported over the spinning disk surface by an air bearing formed in the hydrodynamic boundary layer. Presently the physical spacing between the slider and disk (flying height) is about 20 nm, and it will be closer to 10 nm when the areal density is 100 Gbits/in^2. While the spacing between the slider and the disk is decreasing, the disk rotation rate is increasing. High end disk drives are now commonly running at 10,000 rpm, and 20,000 rpm is on the horizon. The focus of thin film magnetic recording media nano-tribology is in the gap between the head and the disk, which is approaching molecular dimensions.

In this paper, the structure and surface chemistry of the lubricant film and the carbon overcoat are described. The tribochemistry and physics during controlled tribological testing is reviewed. A general nano-lubrication system model is proposed. New tribological test results are presented to illustrate the components of the model. Areas of nano-lubrication which should be the focus of renewed industrial and academic research are identified. From this viewpoint, we see that the principles of the magnetic recording media nano-lubrication system may also be applied to micro- and nano-scale devices.

2 DISK AND SLIDER FABRICATION

The magnetic recording medium is deposited on an aluminum magnesium alloy or, more recently, glass substrate. The magnetic recording layer is typically Co alloy on a Cr underlayer, with 5 to 10 nm of amorphous carbon sputter deposited on top of the magnetic layer as a protective overcoat. After removal from the vacuum deposition system, the disks are stored, polished with a fine abrasive grit, and washed. Perfluoropolyether lubricant is deposited by dip coating from dilute solution in a fluorinated solvent (Scarati and Caporiccio, 1987). Alternatively, lubricant may be deposited within the vacuum system to control the chemisorbed fraction and to avoid the use of fluorinated solvents (Coffey *et al.*, 1994; DeKoven,

2000). After lubrication, the disks are glide burnished with a special air bearing slider to remove asperities below the specified flying height. The finished disks undergo final surface analysis and testing, and they are mounted between spacer rings on a spindle motor and assembled into the disk drive.

The slider substrate is a $TiCAl_2O_3$ ceramic wafer. Magnetic elements are sputter deposited onto the wafer, and it is sliced into thin rows (approximately 0.5×2×50 mm). One face of the row is polished by lapping and a 5–10-nm-thick carbon overcoat is deposited. The rows are photolithographically patterned with air bearing rails, and diced into sliders (~1 mm wide). The sliders are adhesively bonded to a gimbal on a stainless steel suspension (Qian *et al.*, 1999). The spring suspension provides a normal force equal to the air bearing lift force at the operating linear velocity to set the slider flying height. The head gimbal assemblies are mounted on a comb which attaches to the servo actuator, and assembled into the disk drive.

3 OVERCOATS

Strong adhesive forces are inherent between unlubricated ceramic and metal. In direct ceramic-metal contact, adhesion outweighs cohesion in the metal, and metal is transferred to the ceramic (Miyoshi, 1990). The carbon overcoats on the ceramic slider and the magnetic recording layer enable sliding between the two surfaces without cohesive failure in the magnetic layer. Addition of the carbon overcoat on the slider improves the wear durability (Wang *et al.*, 1994). Carbon overcoats also inhibit corrosion of the magnetic layer in aqueous solutions, although micro-scratches made during sliding can provide corrosion sites (Suzuki and Kennedy, 1989).

Carbon overcoats are sputtered with Ar inert gas and usually hydrogen or nitrogen to make hydrogenated CH_X or nitrogenated CN_X carbon, respectively (Anoikin *et al.*, 1998a,b). The properties and characterization of thin carbon overcoats were reviewed by Tsai and Bogy, 1987. Lower density of sputter deposited carbon is due in part to the presence of a network of voids, and decreasing the partial pressure or the rf power increases the film density (Tsai and Bogy, 1987). Carbon overcoats with a more uniform grain size distribution, more sp^3 bonding, and a more homogeneous work function distribution had the best durability (Khan *et al.*, 1988). A heterogeneous morphology for sputtered carbon was recently proposed by Kasai *et al.*, 1999. Surface roughness measurements on typical sputtered carbon were consistent with a granular structure of closely packed spheroids, 2.7 nm in diameter. The granular nature of the overcoat is consistent with the effects of oxygen, water, and alcohols on the dangling bond signal as measured by electron spin resonance (ESR) spectroscopy.

The CH_X overcoat surface contains unsaturated carbon-carbon bonds (Kasai et al., 1999), and oxides (hydroxyl, carboxyl, and carbonyl, Wang et al., 1996), while the CN_X surface contains imines and carbonitrile (Waltman et al., 1999).

Overcoats other than sputtered carbon have been considered. The durability of carbon deposited with plasma assisted chemical vapor deposition approaches that of sputtered carbon (Koishi et al., 1993). An ion beam deposited fluorinated carbon overlayer exhibited good durability (Karis et al., 1998). Recently a vapor deposited and UV polymerized cyanate ester overlayer was deposited on carbon overcoats to improve their interaction with lubricants and to provide corrosion resistance (Mate and Wu, 2000). Durability and corrosion resistance was provided by a 10 nm thick nickel oxide film and a perfluoropolyether lubricant on top of the magnetic layer (Yanagisawa et al., 1989).

4 LUBRICANTS

The lubricants used for particulate magnetic recording disks throughout the 1970's and most of the 1980's were perfluoropolyethers with nonpolar CF_3- end groups. These nonpolar perfluoropolyethers physisorb on surfaces, primarily through dispersion interaction force. The carbon overcoats of modern thin film magnetic recording media are lubricated by perfluoropolyethers with polar end groups that enhance their attachment to the surface and limit lubricant spinoff induced by centrifugal force and air shear (Merchant et al., 1990; Ruhe et al. 1994; Mate and Wilson, 2000). Lubricants with polar end groups also have better durability and static friction properties than those with nonpolar end groups (Scarati and Caporiccio, 1987; Bhushan, 1990).

The molecular structures of the commercial perfluoropolyethers referred to here are shown in Table 1. The Fomblin Z type pefluoropolyethers are random copolymers of perfluoro methylene oxide and ethylene oxide with $0.8 < m/n < 1.6$. The Z type is available with several different end groups e.g., a nonpolar Z03 with two fluoromethyl end groups, polar Zdiac with two carboxylic acid end groups, and polar Zdol with two hydroxyl end groups. The Demnum type perfluoropolyethers are homopolymers of perfluoro n-propylene oxide. The Krytox type perfluoropolyethers are homopolymers of perfluoro isopropylene oxide. The Y perfluoropolyethers are similar to Krytox with about 3% perfluoro methylene oxide randomly copolymerized into the chain, and nonpolar fluoromethyl end groups. Fomblin, Demnum, and Krytox are the registered trade name of commercial

Table 1. Perfluoropolyether lubricant molecular structures. The degree of polymerization is the sum of the alphabetic subscripts plus 2 (taking into account the end groups).

Perfluoropolyether	Molecular Structure
Zdol	$HOCH_2CF_2 (OCF_2)_m (OCF_2CF_2)_n OCF_2CH_2OH$
Zdiac	$HOOCCF_2 (OCF_2)_m (OCF_2CF_2)_n OCF_2COOH$
Z03	$CF_3 (OCF_2)_m (OCF_2CF_2)_n OCF_3$
Demnum S100	$CF_3CF_2CF_2 (OCF_2CF_2CF_2)_{xo} OCF_2CF_3$
Krytox 143AD	$CF_3CF_2CF_2 (OCF(CF_3)CF_2)_{xo} OCF_2CF_3$
Y	$CF_3CF_2CF_2 (OCF(CF_3)CF_2)_a (OCF_2)_b OCF_2CF_3$

samples produced by Ausimont S.p.A (Milan, Italy), Daikin (Osaka, Japan), and Du Pont (Wilmington, DE), respectively.

Novel lubricants have also been prepared and tested. For example, the octadecylamine salt, the amide, and the octadecanol ester with perfluorodecanoic acid, which form self assembled monolayers, were tested on carbon overcoats (Seki and Kondo, 1991). Low friction and chemisorption were observed with an oriented film of stearyl amine and a dialkyl carboxylic acid on carbon overcoats (Sano *et al.*, 1994).

Hydrocarbons have not been widely considered as lubricants for rigid magnetic recording media. Although stearic acid has been used as a lubricant in some disk drive products (Gregory *et al.*, 1988), and virtually all floppy disks are lubricated with tridecyl stearate (Gini *et al.*, 1981).

4.1 Chemisorption

Fomblin Zdol is the most widely used thin film magnetic recording disk lubricant. Zdol both physisorbs and chemisorbs on carbon overcoats. Chemisorbed (often referred to in the literature as bonded) Zdol is defined as lubricant which remains attached to the overcoat after washing with a nonpolar solvent such as perfluorohexane (Merchant *et al.*, 1990).

There are two leading mechanisms which have been proposed to account for chemisorption:

(1) proton transfer from the hydroxyl end group to dangling bonds in the overcoat (Yanagisawa, 1994; Kasai *et al.*, 1999), and

(2) hydrogen bonding between the hydroxyl end group and carboxylic acids on CH_X and imines on CN_X (Waltman *et al.*, 1999).

Chemisorption mechanism (1) is supported by the observation that Zdol chemisorption, or oxygen, quenches the ESR signal, and that the chemisorbed fraction is higher when Zdol is applied directly to the overcoat before exposure to atmospheric oxygen (DeKoven, 2000).

There are both weak and strong chemisorption sites. Table 2 shows the lubricant thickness before and after washing with nonpolar solvent perfluorohexane followed by washing with polar solvent trifluoroethanol. After washing with perfluorohexane, about 0.1 nm of more weakly chemisorbed Zdol was removed during the subsequent wash with trifluoroethanol. The weakly chemisorbed, portion of the chemisorbed Zdol is probably that which is attached by hydrogen bonding. The total amount of weakly chemisorbed Zdol was about the same for both low and high chemisorbed fraction, so that the weakly chemisorbed Zdol was 21% of the total chemisorbed lubricant at the low level of chemisorption and 12% of the total chemisorbed lubricant at the high level of chemisorption.

Table 2. The total and chemisorbed lubricant amounts on CN_x disks dip lubricated with 1 nm of Zdol 4000 from the test series shown in Fig. 1. The low chemisorbed fraction was at 65°C and 40% RH, and the high chemisorbed fraction was in ambient air at 65°C and 5% RH, both for 630 hours after lubrication. The nonpolar solvent is perfluorohexane, and the polar solvent is trifluoroethanol.

Chemisorbed Fraction	Before Wash (nm)	Nonpolar Solvent Washed (nm)	Polar Solvent Washed (nm)
Low	0.90	0.42	0.33
High	0.97	0.88	0.77

After initially depositing the Zdol film on the overcoat by dip coating from solution, the chemisorbed fraction increases with time. The chemisorption reaction is suppressed by high relative humidity (RH) (Karis, 2000). The chemisorption reaction is reversible. Completely chemisorbed Zdol returned to equilibrium with 20–30% physisorbed over several weeks time at ambient temperature and RH on a CH_x overcoat (Karis et al., 2001). Temperature and RH influence the chemisorption reaction through water adsorption on dangling bonds and surface oxides (Zhao et al., 1999).

The typical chemisorption rate of Zdol on a CN_x overcoat with time after dip coating is shown in Fig. 1. The rate of chemisorption was highest at 65°C and 5% RH. The rate decreased when the RH was decreased from 5% to 0% at 65°C, even though more surface sites should be available at the very low RH. This decrease in the chemisorption rate between low and very low RH is attributed to the decrease in Zdol mobility at very low RH as

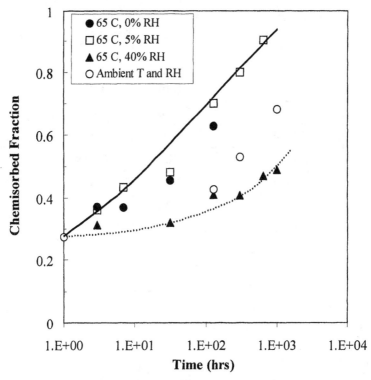

Figure 1. Chemisorption of initially 1 nm thick Zdol 4000 with O/C ratio 0.67 on CN_X overcoat as a function of time showing the effect of high and low relative humidity (RH).

observed by Min *et al.*, 1995. The chemisorption at 65°C was significantly decreased by increasing the RH to 40%. The chemisorption at ambient temperature and RH was higher than at 65°C and 40% RH. Ambient temperature was between 20 and 25°C, and the ambient RH was between 30 at 50%. These results suggest that the chemisorption of Zdol on CN_x overcoat is a strong function of both RH and temperature.

The chemisorption rate is also influenced by the Zdol polymer chain relaxation times, which govern the mobility, hence the rate at which reactive end groups diffuse to active surface sites (Oshanin *et al.*, 1998; Karis, 2000). The relaxation times for bulk perfluoropolyethers calculated from rheological measurements of the storage and loss shear modulus master curves are listed in Table 3. The relaxation time τ_c for Zdol increased by a about factor of 2 as the *O/C* ratio decreased from about 0.69 to 0.65. Assuming that the chain relaxation time on the surface scales with the bulk relaxation time, the increase in τ_c should decrease the chemisorption rate at a given time after dip coating, and increase the time to the transition from end

Table 3. The relaxation time τ_c for bulk, unfractionated, commercial perfluoropolyethers at 60°C. The number average molecular weight M_n, and the oxygen to carbon ratio O/C were measured by nmr spectroscopy The relaxation times were calculated from rheologically measured dynamic shear modulus master curves using time-temperature superposition.

Perfluoropolyether	M_n (amu)	O/C	τ_c (µs)
Zdol 4000	3,600	0.693	2.40
Zdol 4000	3,600	0.666	2.40
Zdol 4000	4,300	0.658	3.50
Zdol 4000	3,900	0.650	5.20
Z03	7,600	0.646	3.28
Demnum S100	5,000	0.333	12.50
Krytox 143AD	6,600	0.333	23.00

bead to segmental diffusion limited chemisorption. These trends are reflected in the experimental dependence of the chemisorption kinetics on Zdol chain composition in Waltman et al., 2000.

4.2 Diffusion

The disjoining pressure gradient arises from the thickness dependence of the surface energy, and it provides the driving force for surface diffusion (Tyndall et al., 1999; Karis and Tyndall, 1999). Physisorbed lubricant diffuses toward regions of higher disjoining pressure. The total surface energy of Zdol lubricated overcoats decays non-monotonically toward its bulk value with increasing Zdol film thickness due to molecular orientation induced by polar interactions with the surface (Tyndall and Waltman, 1998; Tyndall et al., 1998a, b; Karis, 2000). The typical surface energy, and the underlying film structure, as a function of Zdol thickness for low and high chemisorbed fractions are shown in Fig. 2. At the lower chemisorbed fraction, the hydroxyl end groups are exposed in some regions of film thickness, giving rise to local maxima in the polar component of the surface energy. The diffusion velocity increases rapidly with decreasing film thickness near monolayer coverage (van der Waals chain diameter ~ 0.4 nm) as the surface energy sharply increases toward that of the unlubricated carbon overcoat with decreasing film thickness. Since the disjoining pressure is the negative gradient of the surface free energy with respect to film thickness, regions of stable film thickness are those in which the surface energy decreases with increasing film thickness.

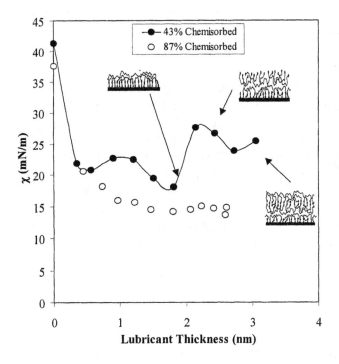

Figure 2. Total surface energy as a function of lubricant thickness and bonding for Zdol 4000 with O/C ratio 0.67 on CN_X overcoat.

The mobility of physisorbed Zdol increases with increasing relative humidity (Min *et al.*, 1995) due to an effective reduction in mobility activation energy by adsorbed water on surface sites (Karis, 2000). The mobility of physisorbed Zdol at ambient temperature and RH was lower on CN_x than on CH_x (Ma *et al.*, 1999), suggesting a higher activation energy for mobility on CN_x than on CH_x. The hydroxyl end groups of chemisorbed Zdol are strongly attached to the overcoat (*i.e.*, very high mobility activation energy) and contribute little to the polar component of the surface energy or reflow.

4.3 Degradation

Hydrocarbon oxidation starts with proton abstraction to make a carbon radical which reacts with oxygen to form a peroxy radical. Interchain proton abstraction by a peroxy radical leads to hydroxyl groups, while intrachain proton abstraction by a peroxy radical leads to chain scission. Products formed in both the interchain and intrachain pathways include more carbon radicals, and this is referred to as a chain propagation reaction (Karis *et al.*, 1999a). Perfluoropolyethers are subject to a different type of chain scission through bond dissociation induced by thermal/catalytic, mechanochemical, and triboelectric effects.

4.3.1 Thermal/Catalytic

Scission of perfluoropolyethers is catalyzed by Lewis acids (Sianesi *et al.*, 1971; Kasai, 1992a,b; Paciorek *et al.*, 1994; Kasai, 1999a,b). In catalytic scission, the Z chain dissociates at -O-CF_2-O- groups, generating a fragment with a methoxy end group and enriched in -O-CF_2CF_2-O- which produces mass peaks at 69, 119, and 135 amu in mass spectrometry of the vapor phase over the heated liquid (Kasai, 1992). The other fragment has a carbonyl fluoride end group that hydrolyzes to carboxylic acid in the presence of moisture. Perfluoropolyethers Demnum, Krytox, and Y are more stable with respect to catalytic scission because of sterically restricted accessibility to the Lewis acid surface sites. The best stability with respect to thermal/catalytic scission is for Krytox, and decreases in the order Krytox > Demnum > Z (Kasai, 1992a,b; Kasai, 1999a).

4.3.2 Thermal

In the absence of catalytic metals or oxides, the perfluoropolyethers are thermally stable up to 360–380°C (Paciorek *et al.*, 1994). Low molecular weight Z oligomers, evaporated by heating to 300°C in vacuum and subjected to ionization in a mass spectrometer produce a pattern of mass peaks distinctly different from that produced by the volatile catalytic scission fragments, with mass peaks at 47 and 66 amu (Kasai, 1992a; Kasai, 1999a).

4.3.3 Mechanochemical

Mechanochemical scission of high molecular weight polymers has long been known to occur in extensional flows. First, the chain is extended by the flow. When the opposing hydrodynamic force on the chain equals the bond dissociation energy, the chain breaks in the middle. Short chains, such as the perfluoropolyethers, are not long enough for scission by this mechanism. However, scission products were detected after milling (Karis *et al.*, 1993) or sonication (Karis *et al.*, 1994) of perfluoropolyethers in the presence of solid particles at ambient temperature.

Confinement of polymer chains at solid surfaces increases their relaxation time relative to that in the bulk fluid (Hirz *et al.*, 1992). The bulk relaxation time τ_c at 60°C measured for several different chain compositions and monomer structures are listed in Table 3. When the process time scale $\tau << \tau_c$, a viscoelastic liquid exhibits solid-like response to stress or strain.

Deformation energy is temporarily stored as increased bond strain and vibrational amplitude before it is dissipated into segmental and longer motions (viscous dissipation). During the time period between chain deformation and energy dissipation, there is a higher probability of bond dissociation or chain scission. Bond dissociation is even more likely in perfluoropolyethers such as Krytox and Y because some of the bonds in these chains are already sterically strained by the fluoromethyl side group (Pacansky *et al.*, 1991). Judging from the relaxation times in Table 3, the stability with respect to mechanochemical scission should decrease in the order Z > Demnum > Krytox, which is opposite to the ordering expected from thermal/catalytic degradation. The products of mechanochemical scission would resemble those of thermal/catalytic scission in which dissociation takes place at randomly selected perfluoromethylene oxide groups in the chain for the Z type perfluoropolyethers and at randomly selected ether linkages in the chain for D type perfluoropolyethers. The products of mechanochemical scission would be significantly different from those of thermal/catalytic scission in which dissociation takes place only at the end groups for the K and (mostly) for the Y type perfluoropolyethers.

4.3.4 Triboelectric

Much less is known about triboelectric scission as a degradation mechanism. Nonpolar Z type perfluoropolyethers were found to chemisorb onto the carbon overcoat during sliding due to scission of carbon oxygen bonds by low energy triboelectron emission (Zhao *et al.*, 1999). The scission products of Krytox produced during electron bombardment were different from those produced thermally or mechanically (Vurens *et al.*, 1992). Electron stimulated desorption of Zdol on carbon overcoats demonstrated that Zdol is dissociated by electrons with energy as low as 3 eV (Lin *et al.*, 1995). The typical open circuit electrical potential measured between the slider and the disk during a start/stop cycle, shown in Fig. 3, indirectly measures tribocharge transfer between the disk and slider. However, during normal operation the circuit between the slider and the disk is completed by a ferrofluid seal and steel ball bearings in the motor.

4.4 Lubricant Additives

Lewis acid catalysis of perfluoropolyether chain scission is inhibited by some organic compounds, *e.g.*, arylamines, benzimidazole derivatives, and selenium derivatives (Sianesi *et al.*, 1971). The only inhibitor of perfluoropolyether scission that has been widely considered for use in magnetic recording is a cyclic phosphazine. Cyclic phosphazines were developed to lubricate high-temperature turbine engines (Nader *et al.*, 1992). Zdol undergoes catalytic scission on the Lewis acid surface of the

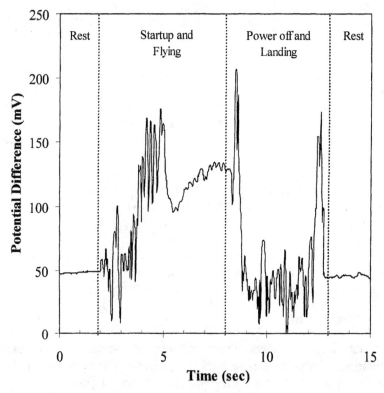

Figure 3. The open circuit potential measured across the head disk interface during one start/stop cycle. The positive terminal of an electrometer was attached to the slider, and the negative terminal was attached to the disk through a grounding tab.

ceramics at elevated temperatures in high pressure differential scanning calorimetry, while a cyclic phosphazene X-1p (partially fluorinated hexaphenoxy cyclotriphosphazene) is thermally stable under these conditions (Perrettie *et al.*, 1995). Addition of 1% by weight of X-1p to Z type perfluoropolyethers prevented catalytic scission on alumina powder by passivating the catalytic sites (Kasai, 1999a). X-1p also increased the lubricant mobility by mediating the polar interactions between the Zdol end groups and the overcoat (Mate *et al.*, 1998). Addition of X-1p to Zdol suppressed both the generation of scission fragments and desorption of oligomers (Chen at al., 1997). X-1p alone, as well as mixtures of X-1p in Zdol, provided relatively good start/stop durability (Yang *et al.*, 1994; Zhao and Bhushan 1999).

In addition to cyclic phosphazines, a number of other additives are effective at inhibiting catalytic scission of perfluoropolyethers, but these have not been evaluated for use in magnetic recording. One disadvantage of most inhibitors is their low solubility in perfluoropolyethers, making them prone to phase separation and dewetting. A perfluoropolyether substituted perfluorophosphine had improved solubility in Z and Krytox type perfluoropolyethers, and greatly improved the thermal stability in the presence of catalytic metals (Snyder et al., 1978). Several anti-wear additives known to be effective for boundary lubrication in hydrocarbon oils were modified to make them soluble in perfluoropolyethers by attaching short perfluoropolyether chains onto them. These are intended to react with active sites at the surface, inhibiting the catalytic effects of metal. Additives based on phosphate, thiophosphate, β-diketone, amide/thiol, and sulfite exhibited antiwear activity during a four-ball wear test in vacuum (Shogrin et al., 1999).

An interesting class of perfluoropolyether soluble perfluoroalkyl diphenyl ether stabilizers was developed by Gschwender and Snyder, 1994, and these stabilize perfluoropolyethers in high temperature oxidation corrosion tests. These compounds may act to stabilize perfluoropolyethers by (1) acting as free radical traps (analogous to antioxidant in hydrocarbons), (2) reacting with oxides or fluorides that catalyze degradation, and/or (3) providing a physical separation between the perfluoropolyether and the wall (Gschewnder et al., 1996).

5 TRIBOLOGY TESTING

5.1 Vacuum Tribochamber

In the vacuum tribochamber, desorbed oligomers and fragments produced during sliding (typically 0.2 m/s) in a vacuum ($<2.7 \times 10^{-6}$ Pa) are analyzed using a mass spectrometer. Early tribochamber studies with a non-carbon overcoated ceramic slider and a CH_X overcoat on the disk detected a number of mass fragments (Novotny et al., 1994). The contact temperatures were considered too low to produce significant thermal desorption or catalytic scission, and the mass spectrum obtained during sliding differed from the Z oligomer spectrum, suggesting that chains were fragmented prior to desorption from the overcoat. The chain scission was attributed to triboelectrons (Strom et al., 1993). The generation of CO_2 during sliding of ferrite sliders on nonpolar Z lubricated carbon overcoats was attributed to reaction of carbonyl fluoride with the steel walls of the tribochamber (Mori et al., 1993).

Subsequent vacuum tribochamber studies report a consistent trend of decrease in the mass peaks associated with desorption of scission fragments (69 and 119 amu) relative to those associated with desorption of Zdol oligomers (47 and 66 amu), and an improvement in the wear durability of

Zdol lubricated overcoats, when the ceramic slider is overcoated with carbon (Chen et al., 1997; Wei et al., 1998; Chen et al., 1999; Chen et al., 2000). Increasing the atomic concentration of hydrogen in the overcoat improved the friction and wear of Zdol lubricated CH_X overcoats, and decreased the fragment peaks relative to the oligomer peaks (Chen et al., 1997). The fragment and oligomer peaks, and wear durability, all increased as the Zdol thickness was increased from 0.9 to 1.2 nm with carbon overcoated sliders on CH_x overcoats (Chen et al., 1999). The oligomer peaks and wear durability decreased with increasing chemisorbed fraction (Chen et al., 2000).

Thus the vacuum tribochamber results show the following: (a) carbon overcoating the ceramic slider decreases the chain scission relative to oligomer desorption during sliding, and (b) the improved wear durability provided by increasing the Zdol lubricant thickness or decreasing the chemisorbed fraction is accompanied by increased oligomer desorption.

5.2 General Tribological Testing of the Head Disk Interface

5.2.1 Sliding Tests

Tribological tests are typically employed to study the lubrication system of magnetic recording disk overcoats near ambient pressure. In the pin on disk test, a hemispherical ceramic probe is maintained in contact with the slowly rotating disk while recording the friction. The wear track profile is measured to determine the wear rate. The unlubricated carbon overcoat was found to wear abrasively in ambient air during the pin on disk test (Karis and Novotny, 1989). There is also evidence for tribochemical effects in sliding on unlubricated carbon overcoats (Marchon et al., 1990). Start/stop tests consist of repetitively starting and stopping the disk rotation while measuring the friction force and acoustic emission during startup. Friction and wear in the start/stop test on unlubricated carbon overcoats (Raman and Howard, 1990), and also on Zdol lubricated overcoats (Raman and Tang, 1993), remained low in the presence of nitrogen but increased with time in the presence of air. The higher friction and wear in the presence of air was attributed to adhesive interaction between tribochemically produced oxides, while in nitrogen the oxides desorb and do not reform. There may also be an adhesive contribution to the friction from new dangling bonds generated by abrasion of asperities during sliding. The differences between air and nitrogen environment during sliding would then arise from an environmental effect on dangling bond reactivity.

Throughout the late 1980's and all of the 1990's there were many publications on lubricant removal, and reflow during sliding and flying, and the wear mechanisms in single disk tests on lubricated rigid disk recording media. Perfluoropolyether lubricant removal from the overcoat was observed ellipsometrically in the slider rail regions during an oscillatory test with a slider on a carbon overcoated disk. Reflow of lubricant back into the depleted region was observed following the test (Hu and Talk, 1988).

The initial studies were done with relatively thick films of nonpolar perfluoropolyethers Y or Z. In sliding tests on particulate media, the Y type lubricant was removed more rapidly than the Z type (Novonty et al., 1992), and this was attributed to primarily to mechanochemical scission due to the longer relaxation time of the Y chain structure (Karis and Jhon, 1998). The Z scission products accumulated slightly faster than Y scission products, but catalytic scission could not account for the much higher removal rate of Y as suggested in Kasai, 1999a. The mechanism of lubricant removal is common to both particulate media and carbon overcoated thin film media. During sliding, lubricant chains are broken into smaller fragments. Short fragments which have nonpolar end groups desorb from the surface, and others which contain which carboxylic acid end groups, accumulate in the test track (Novotny and Karis, 1991; Novotny and Baldwinson 1991).

Lubricant scission fragments can also accumulate on the slider rails and lead to high static friction. Accumulation of lubricant scission fragments is aggravated by organic silicone vapor contamination (Xuan et al., 1993). Certain vapor phase contaminants due to outgassing from file components can also accumulate at the slider disk interface (Jesh and Segar 1999; Segar and Jesh, 1999).

The accumulation of adsorbed fragments in test tracks was studied using high resolution surface potential measurements. Surface potential measures changes in the normal component of the surface dipole moment (or changes in the work function) while translating across the test track (Novotny et al., 1997; Karis and Novonty, 1997). Accumulation of adsorbed fragments on the disk carbon overcoat increased the surface dipole moment. In sliding, the accumulation of adsorbed fragments was nearly eliminated by carbon overcoating $TiCAl_2O_3$ sliders. The accumulation of adsorbed fragments increased dramatically with decreased flying height (inversely proportional to the fifth power of the flying height, bare ceramic slider) (Novotny et al., 1997). The sensitivity of tribochemistry to flying height suggests that accelerated tests could be used to study the removal mechanisms.

5.2.2 Low Flying Tests

Lowering the flying height increases the probability of contact between the slider and asperities on the overcoat, which is, by design, infinitesimally small at disk file operating conditions. Tribological interactions are investigated at reduced flying height by decreasing the ambient pressure, or linear velocity, or by increasing the load on the slider. The acoustic emission signal is recorded as the pressure is gradually reduced (Terada et al., 1988), or the slider load is gradually increased (Peck et al., 1995), to determine the point at which the slider begins making significant contact with asperities. As the pressure or velocity is decreased, there is a critical pressure (or velocity) at which the acoustic emission sharply increases. The critical pressure for a given linear velocity corresponds to a critical flying height at which there is a transition from stable to unstable flying (Azarian et al., 1993).

Based on an approximate uniform distribution of surface roughness and flying height in Peck et al. 1995, the probability of contact $P_c \approx \sqrt{3} z_{rms} / 4h$, where z_{rms} is the rms surface roughness and h is the average physical spacing, or the mean flying height. The flying height at which $P_c = 1$ corresponds to the critical load, pressure, or velocity at which the acoustic emission sharply increases. The functional form of the expression for P_c is more complicated for real surfaces (Hua et al., 1999), and wear is observed when $P_c \approx 1$. For example, in tests operated at low enough pressure for a $TiCAl_2O_3$ head and a SiO_2 overcoated disk to be in continuous contact, Z lubricant was gradually removed from the overcoat until there was dry contact between the slider and disk, at which point overcoat wear was observed (Terada et al., 1988).

Sub-ambient pressure test methodology was extended to reliability analysis of ultra-low flying height magnetic recording disk files. A stress factor was defined as the ratio of the rms acoustic emission voltage at sub-ambient to that at ambient pressure. A log-log plot of the measured file life vs. the stress factor was used to predict the file life at ambient pressure (Novotny, 1996).

6 NANO-LUBRICATION SYSTEM

A clear picture of magnetic recording nano-tribology emerges from the fundamental studies of the thin film magnetic media overcoat and lubricant film structure in conjunction with the tribological test results. The nano-lubrication mechanism is schematically shown in Fig. 4. The lubricant removal and reflow components of the lubrication system are summarized below, followed by new test results illustrating the effect of chemisorption on tribology and the lubrication system response at low flying height.

6.1 Lubricant Removal

Nonpolar perfluoropolyethers Z and Y in 5 nm thick films were found to undergo significant displacement during seek accessing of the sliders (Novotny and Baldwinson, 1991). These thick films of nonpolar perfluoro-

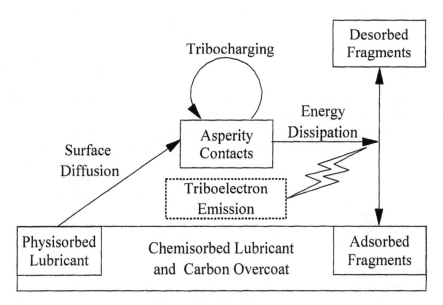

Figure 4. A block diagram of the nano-lubrication system.

polyethers are easily removed and displaced by the pressure gradients from the slider air bearing, because the dispersion force on perfluoropolyethers has decayed beyond about 1 nm from the surface (Tyndall and Waltman, 2000). The lubricant thickness on modern magnetic recording media is typically 1 nm or less, and the lubricant is Zdol. The Zdol films are under the influence of polar and dispersion forces in these thin films (Tyndall et al., 1999). The typical surface energy and film structure are shown in Fig. 2.

The pressure gradient from the slider air bearing might still contribute to lubricant removal from test tracks. The time scale for the passage of a pressure region over a point on the disk surface is estimated to be 2 µs (100 µm negative pressure pulse, linear velocity 50 m/sec). Relaxation times in confined 1 nm thick Zdol films are orders of magnitude longer than the bulk Zdol relaxation times (Hirz et al., 1992) of 2 to 5 µs in Table 3. Therefore the relaxation time of the lubricant films is much longer than the pressure impulse from the slider, and the film exhibits solid-like response on this time scale. Thus, 1 nm thick Zdol films cannot respond by moving in such a short time. However, repeated application of a transient pressure, or shear stress, impulse from the slider could produce lubricant creep out of the test track.

The lubricant can be gradually displaced radially on the whole disk surface over long periods of time by air shear and centrifugal force. Thicker lubricant films with initial thicknesses corresponding to regions of negative disjoining pressure will "phase separate" into regions of stable film thickness (Karis and Tyndall, 1999). Regions of thicker lubricant film, or droplets, are less confined so they have reduced free energy of vaporization, and are subject to spinoff and transfer to the slider.

In the lower stable region of lubricant film thickness, lubricant removal from localized regions on the disk surface is the result of microcontacts between the slider and the disk overcoat. Lubricant is removed from points of asperity contact through one of the energy dissipation mechanisms. As mentioned, the lubricant film responds as an elastic solid on the time scale of an asperity contact. Energy input as bond distortion is dissipated thermally. In general, thermal energy is partitioned between internal rotational, vibrational, and electronic energy states, and molecular translation and rotation. In a confined, molecularly thin, viscoelastic, film, a stress/strain impulse initially excites the chain vibrational modes because they have the shortest relaxation times. Vibrationally excited, or stretched, bonds have a higher probability of bond dissociation. The probability of bond dissociation is increased if the molecule is adsorbed on a catalytic Lewis acid site. Vibrational and bond distortion energy gradually transfers into segmental and progressively larger scale, slower, chain motions, dissipating energy through frictional interactions with the surroundings. Finally, if the whole chain is still intact when it begins to translate, the stored elastic energy can be dissipated into free energy of vaporization by desorption. Phonon transfer into the carbon overcoat is unlikely due to the thermal impedance mismatch (Pacansky, 2000).

Lubricant may also be removed by triboelectron emission induced bond dissociation. It is interesting to consider how tribocharging can occur between two like surfaces, since both the slider rails and the disk are overcoated with sputtered carbon. There is no difference in the average electron chemical potential between surfaces of the same material. When the slider and disk are in sliding contact, there is a high probability that a filled state in one of the surfaces will approach closely enough to an empty state in the other surface so that tunneling can take place. The slider transfers electrons that are near its surface into the track, and electrons transfer from the track to the slider. Since the slider surface is smaller than the disk surface, there are many more available non-equilibrium electrons in the larger surface area of the disks for transfer into empty states of the slider overcoat, the net electron flow is from the disk to the slider. The sign and magnitude of the charge transfer is determined by the densities of states and the probability of transfer between states of different energy (Lowell and Truscott, 1986). The sign of the open circuit potential difference between the slider and disk shown in Fig. 3 is consistent with triboelectron transfer from the disk to the slider.

The tribocharge should increase with decreasing flying height due to the higher frequency of contacts, thus giving rise to a higher electrostatic potential difference between the slider and disk. From Fig. 3, the electric field between the slider and disk could be as high as 10^6 V/m. (The dielectric strength of air is 3×10^6 V/m.) Presumably, the charged slider could emit low energy electrons into the electric field, and these electrons could induce bond dissociation in perfluoropolyether lubricant molecules as suggested by Zhao et al., 1999. The spatial distribution of removal, and adsorbed degradation products, resulting from low energy electron emission by the slider should be different from that induced by mechanical contact. The electron emission would probably occur more or less uniformly throughout the region of slider accessing, while contact-induced thermal/mechanochemical scission would be concentrated on higher points of the surface topography or asperities with which the slider makes physical contact.

6.2 Lubricant Reflow

The feature of the nanolubrication system which enables it to perform so well is the surface diffusion of physisorbed lubricant from the surroundings to replace lubricant which has been removed by scission and desorption. The surface diffusion is driven by the disjoining pressure gradient, and the diffusion flow rate is proportional to film thickness. The volumetric flow rate per unit length of spreading front is given by $j_0 = mz(x)\{\partial\Pi[z(x)]/\partial x\}$ (Karis and Tyndall, 1999) where m is a mobility (Ruckenstein and Rajora, 1983), and $z(x)$ is the lubricant film thickness at position x along the surface. The disjoining pressure $\Pi[z] = -\partial\gamma/\partial z$, where $\gamma(z)$ is the total surface energy at lubricant thickness z. A typical plot of the total surface energy vs. lubricant thickness for Zdol is shown in Fig. 2.

The chemisorbed fraction of Zdol increases with time, as shown in Fig. 1. Chemisorbed Zdol is immobilized, effectively reducing the reflow rate according to $j = (1-\phi)j_0$, where ϕ is the chemisorbed fraction. Thus if the removal rate $R = kP_c$, where k is a removal rate constant, the lubricant will become depleted from asperity tips when $R > j$. This inequality is satisfied by lowering the flying height h or by increasing the chemisorbed fraction ϕ.

6.3 Lubrication System Response Tests

6.3.1 Effect of Chemisorption on Tribology

Start/stop tests were done on disks with different levels of chemisorption (Karis et al., 2001). The maximum friction coefficient during startup during the first 5,000 cycles on carbon overcoats lubricated with 1 nm of Zdol 4000 at low, intermediate, and high levels of chemisorption is shown in Fig. 5. The friction force during startup on these disks after 20,000 cycles is shown in Fig. 6. The average number of start/stop cycles before the friction coefficient during startup exceeded 2.2 upon three consecutive cycles (failure criterion), was > 82,000 > 59,000 and 2,700 cycles for the low, intermediate, high levels of chemisorption, respectively. (Greater sign > indicates that some of the tests in the group of four at each level of chemisorption were truncated prior to failure.)

The friction coefficient is approximately given by $\mu = \mu_0 /[1-(2W/rp)\cot\theta]$, where μ_0 is the friction coefficient when $(2W/rp)\cot\theta = 0$, W is the work of adhesion, r is the average junction radius, p is the compressive strength of the overcoat, and θ is the average surface roughness angle (Rabinowicz, 1965). The cycle time and acceleration are fixed throughout the start/stop tests, so the same increment of lubricant is removed in each cycle. The decrease in the reflow rate with increasing chemisorbed fraction eventually increases the friction coefficient during startup. The friction coefficient increases because the surface energy (hence the work of adhesion) increases sharply as the film thickness decreases below 0.5 nm, as shown in Fig. 2. Therefore, friction was higher following repeated start/stop cycling at the higher levels of chemisorption as shown in Figs. 5 and 6. The large fluctuations in Fig. 6c are due to stick-slip behavior that occurs in the presence of a high friction coefficient.

6.3.2 Low Flying Height

The friction coefficient and cycles to failure provide limited insight into the nano-tribochemistry. A low-velocity test was adopted to investigate the tribochemistry during the early stages of wear. Testing at low velocity avoids the possibility of lubricant removal by centrifugal force and air shear, so that lubricant removed in the low-velocity test is due only to asperity contacts. A typical plot of the friction and acoustic emission signals as a function of linear velocity used to measure the critical velocity (where $P_c = 1$) is shown in Fig. 7. The linear velocity is decreased from 4 m/sec by incrementally decreasing the disk rotation rate while measuring the friction and acoustic emission. The critical velocity is the velocity at which both the friction and acoustic emission sharply increase, near 1.5 m/sec in this case. In the low-velocity test, the disks are typically rotated at 400–500 rpm, with 20 Hz random accessing in a test region between 27.5 and 33.5 mm radius. This is near the critical velocity of about 1.4 to 1.6 m/sec. The disks are periodically removed from the testers and the lubricant thickness is measured across the radius by Fourier Transform Infrared Spectroscopy (FTIR, 3 mm spot size) in reflection. The ellipsometric angle Δ is measured from 20 to 40 mm radius with an imaging ellipsometer in 250-μm steps using a 20× objective lens (resolution 1 μm). The angle of incidence is 50°,

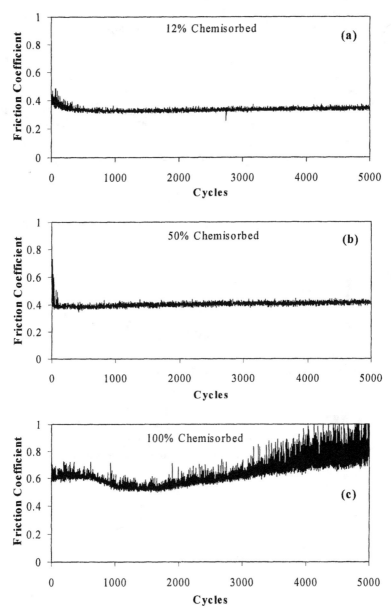

Figure 5. Maximum friction coefficient during startup while start/stop cycling. This shows the effect of chemisorbed fraction (a) 12% chemisorbed, (b) 50% chemisorbed, and (c) 100% chemisorbed. Zdol 4000 with *O/C* ratio 0.65 on CH_x overcoat at 65°C and 5% RH, 500 msec stiction window, acceleration 2.33 m/s^2, maximum rotation rate 7,200 rpm, 18 mm radius, 1×2 mm carbon overcoated negative pressure air bearing slider, 45 mN normal load, laser textured landing zone, AlMg substrate.

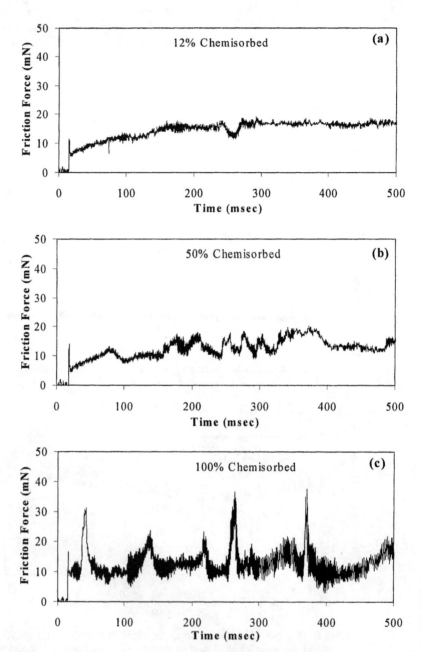

Figure 6. Friction force during startup after 20,000 start/stop cycles. This shows the effect of chemisorbed fraction (a) 12% chemisorbed, (b) 50% chemisorbed, and (c) 100% chemisorbed, during continuation of the tests shown in Fig. 5.

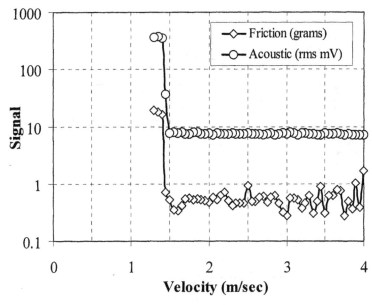

Figure 7. A typical spin down curve used to measure the critical velocity for the low-velocity test.

and the laser wavelength is 532 nm. The beam is incident along the sliding direction, and the ellipsometric null is determined in a region 29×44 µm, with the long side of the region in the sliding direction. After measurement, the disks are replaced in the testers and ovens and the test is restarted for another interval of time (24 to 240 hours).

The low-velocity test was used to characterize lubricant removal and adsorbed fragments. Typical lubricant thickness and ellipsometric profiles from the low-velocity test are shown in Fig. 8. Initially the lubricant thickness was relatively uniform across the radius. After 1160 hours a significant amount of the lubricant had been removed from the tested region, as shown in Fig. 8a. The decrease in lubricant thickness extends slightly outside of the tested region because the surrounding lubricant was reflowing into the tested region. Small amounts of lubricant detected on the slider by secondary ion mass spectrometry could not account for the much larger amount of lubricant that was removed from the tested region. The lubricant was removed from the tested region by desorption following chain scission. The ellipsometric profile across the tested region is shown in Fig. 8b. The profile baseline was leveled to remove linear changes across the radius due to the radial ramp in the carbon overcoat thickness, and offset by Δ_0 to zero the untested regions. The high frequency oscillations on the Δ profile are from angular positioning error in the optical train of the ellipsometer. In this range of film thickness, Δ decreases linearly with increasing film thickness.

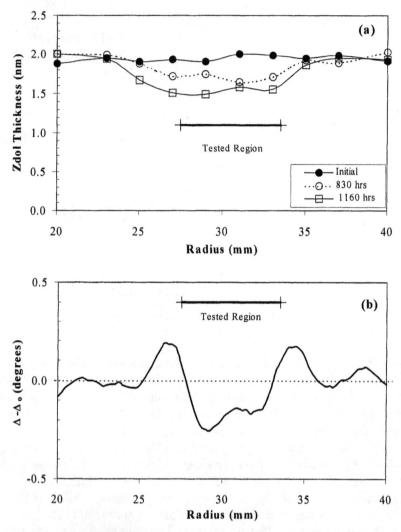

Figure 8. (a) lubricant thickness measured by FTIR, and (b) ellipsometric angle across the tested region after 1160 hours of the low-velocity test. Zdol 4000 with O/C ratio 0.67, on CH_x overcoat, chemisorbed fraction 65%, ambient temperature and humidity environment, rotation rate 500 rpm, smooth data zone, AlMg substrate.

As expected, Δ increased near the outer edges of the tested region. However, Δ decreased within the interior of the tested region. This pattern of change in Δ always accompanied lubricant removal prior to detectable wear of the carbon overcoat. The decrease in Δ within the tested region was unchanged by rotating the sample so that the beam was incident across the sliding direction, so that it was not due to polymer chain orientation in the sliding direction (Gao et al., 1998). An ellipsometric image near the center of a similarly tested region is shown in Fig. 9 (glass instead of AlMg substrate). The image was recorded near the ellipsometric null, and the splotchy pattern of gray scale contrast in the image is due to spatial variations in the polarization of rays within the gaussian beam. (The fringe pattern is due to reflections within the optical path of the ellipsometer.) The decrease in Δ within the tested region, and the splotchy pattern in the image are due to birefringence from oriented carboxylic acid end groups on the adsorbed fragments. The accumulation of dipoles oriented normal to the overcoat in sliding and low-flying test tracks was previously detected by high resolution surface potential measurements (Novotny et al., 1997; Karis and Novotny, 1997).

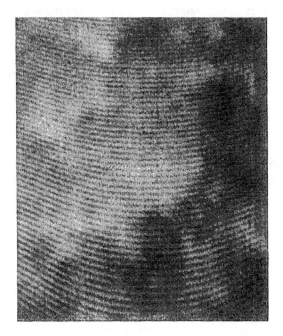

Figure 9. Ellipsometric image (220 μm wide) in the tested region showing splotchiness from birefringence of adsorbed fragments with caboxylic acid end groups. The fringe pattern is due to reflections within the optical elements of the ellipsometer.

The elemental composition and chemical environment of surface layers can be measured using X-ray photoelectron spectroscopy (XPS). The XPS peak areas for carbon, fluorine, and oxygen measured in steps across the region of accessing are shown in Fig. 10. These peak areas are qualitative measurements of atomic concentration because they are not corrected for scan time or capture cross section and kinetic energy. The change in the carbon peak area in Fig. 10a is close to the background noise level. The decreased lubricant film thickness in the region of accessing detected by FTIR in Fig. 8a also shows up as a decrease in the fluorine peak area in Fig. 10b. The increase in the oxygen peak area within the region of accessing shown in Fig. 10c is consistent with oxidation of the carbon overcoat or carboxylic acid groups on adsorbed fragments.

The atomic concentration of C, F, and O in each chemical environment were quantitatively compared with one another inside and outside the tested region to look for any changes in the chemical composition. Quantitative atomic concentrations were calculated from the XPS peak areas, and chemical environments were assigned by peak fitting software. The oxygen to carbon ratio in the perfluoropolyether was the same both inside and outside of the tested region, so that the polymer chain composition of the adsorbed fragments is the same as that of the chains prior to scission. The carbonyl carbon increased relative to the fluorocarbon in the tested region.

The carbonyl absorption of the carboxylic acid is also detected in the FTIR spectrum. Reflection FTIR spectra measured outside and inside of the tested region are shown in Fig. 11a and b, respectively. The height of the large perfluoromethylene absorption band near 1280 cm^{-1} is used to measure the lubricant thickness (Gao et al., 1998). The broad carbonyl band near 1680 cm^{-1} was identified by measuring the spectrum for the sodium salt of Zdiac shown in Fig. 11c (Kasai, 2000). There is a similar absorption band near 1680 cm^{-1} inside the tested region, as shown in Fig. 11b This absorption band is absent outside the tested region, as shown in Fig. 11a.

Similar results were reported by Anders et al., 1998, who used X-ray photo emission electron microscopy (PEEM) to look for carbonyl formation in vacuum tribochamber test tracks by. Carbonyl groups were detected in test tracks on Zdol lubricated overcoats but were absent from test tracks on unlubricated carbon overcoats. Therefore the carbonyl groups detected in the low-velocity test described above are carboxylic acid end groups on adsorbed scission fragments in the tested region.

6.4 Summary of the Lubrication System

In the presence of significant asperity contacts, lubricant is gradually removed from the overcoat by chain scission into smaller fragments. Most of the fragments desorb from the overcoat. Some fragments with carboxylic acid end groups remain adsorbed on the overcoat. The polymer chain of the fragments was indistinguishable from that of the original lubricant. The lubricant removal rate increases with increasing chemisorbed fraction and decreasing relative humidity. Lubricant removed from asperities is replaced by surface diffusion from surrounding regions. Increasing the chemisorbed fraction, or decreasing the relative humidity, decreases the reflow rate. The friction coefficient eventually increases with continued asperity contacts due

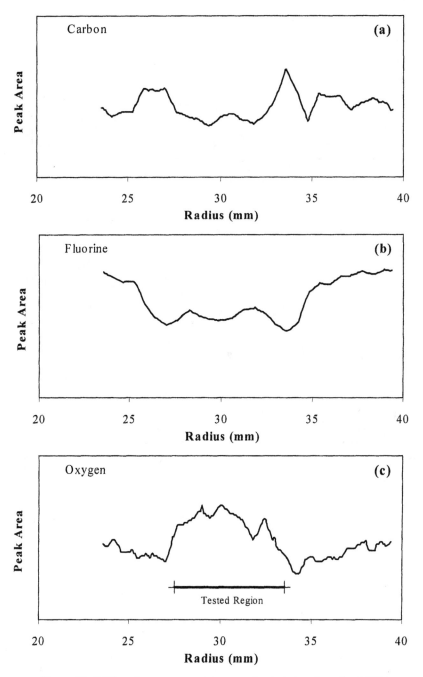

Figure 10. XPS peak area measured across the tested region after 1160 hours of the low-velocity test (a) carbon peak, (b) fluorine peak, and (c) oxygen peak. Same test as shown in Fig. 8.

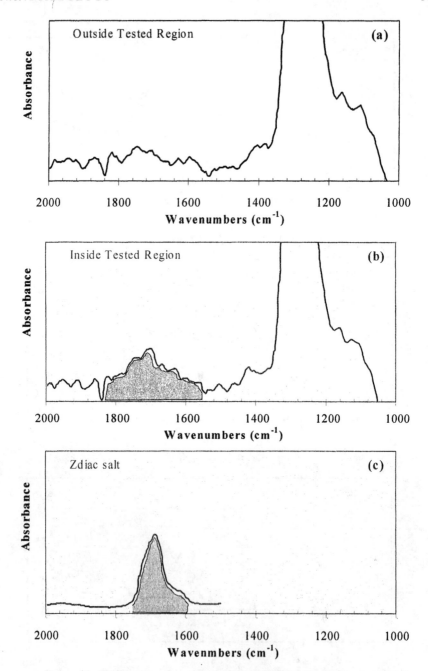

Figure 11. FTIR spectrum measured (a) outside and (b) inside the tested region after 1160 hours of the low-velocity test on the same sample shown in Fig. 8. (c) shows the spectrum of Zdiac sodium salt for comparison. The crosshatched region shows the carboxylic acid absorbance.

to the increase in the surface energy (work of adhesion) with decreasing lubricant thickness. Once the physisorbed (mobile) lubricant is sufficiently depleted from regions near asperity contacts, the friction has significantly increased, and sustained asperity contacts begin to wear the carbon overcoat.

7 RESEARCH RECOMMENDATIONS

The nano-tribology of thin film magnetic recording is a complex and multifaceted process. Following several decades of research, the components of the lubrication system are now reasonably well-known. Present day disk drives are highly reliable because there is a low probability of slider disk contact. The robust lubrication system acts to quickly replenish lubricant that is removed by a rare intermittent contact between asperities.

The nanolubrication system, Fig. 4, is extendible to future magnetic recording interfaces in which there may be a higher probability of contact. The principles learned from magnetic recording are applicable to lubrication of micro- and nano-scale devices. For magnetic recording, durable thin overcoats (<5 nm thick) with precisely controlled surface chemistry are needed. Further research should focus on the lubricant reflow and removal. The interactions between the lubricant and the overcoat must provide adequate reflow. Lubricant mobility is mediated by the polar and dispersive interaction forces between the physisorbed lubricant and the overcoat. The chemisorbed lubricant fraction must be controlled within limits at technologically encountered temperatures and relative humidities.

One of the greatest challenges, and the phenomenon that has received the most widespread attention from the tribological community, is the lubricant removal mechanism. Improved durability in the presence of concentrated asperity contacts may be achieved by decreasing the incremental rate of lubricant removal in these contacts. Lubricant removal mechanisms are classified either as energy dissipation or triboelectron emission. Energy dissipation removal includes thermal desorption of intact lubricant molecules and chain scission through bond dissociation during transfer of elastic energy. Dissociation is enhanced by the presence of catalytic surfaces. Triboelectron emission can add an electron to a lubricant molecule, which leads to bond dissociation.

Research aimed at extending the nanolubrication system should address all of the removal mechanisms. Improved lubricant molecular structures will have higher bond dissociation energies. Self assembling nano-film structures will efficiently dissipate stored elastic energy. Additives and special end groups will be used control the equilibrium chemisorption with the overcoat, and to block catalytic activity. The triboelectric cycle of contact charging and triboelectron emission must be interrupted. Slider overcoats can be modified to reduce the number of vacancies, and disk overcoats can be modified to hold fewer occupied states. This may be achieved by balancing the energy and spatial distribution of available and occupied states between the slider and disk overcoats. Developing the nano-lubrication system of the future relies on a fruitful collaboration of multidisciplinary research in vacuum science, coating technology, polymer physics, surface chemistry, chemical engineering, thin film rheology, and nano-tribology.

ACKNOWLEDGMENT

The author is grateful to G.W. Tyndall for his insights into the surface energy and chemisorption. All of the disks for this study were graciously provided by R.J. Waltman. Thanks are due to P.H. Kasai for discussions on catalysis and for the IR spectrum of Zdiac salt. Thanks are also due to J. Pacanksy for his thoughts on bond dissociation and energy dissipation. The author appreciates technical discussions with B. Marchon, and the technical support of J.L. Miller, A.M. Spool, D.C. Miller, and H.R. Wendt. The author thanks O. Melroy and J.R. Lyerla for their encouragement and support throughout this work.

REFERENCES

Anders, S., Stammler, T., Bhatia, C.S., Stohr, J., Fong., W., Chen, C-Y., and Bogy, D.B., "Study of Hard Disk and Slider Surfaces Using X-Ray Photoemission Electron Microscopy and Near Edge X-Ray Absorption Fine Structure Spectroscopy," Mat. Res. Soc. Symp. Proc., 571,415-420(1998).

Anoikin, E.V., Yang, M.M., Sullivan, M.T., Chao, J.L., and Ager, J.W. III, "Effects of Substrate Cooling in Hard Magnetic Disk Sputtering Process on Protective Overcoat and Magnetic Layer Properties," Acta Mater., 46(11)3787-3791(1998a).

Anoikin, E.V., Yang, M.M., Chao, J.L., Elings, J.R., and Brown, D.W., "Nanoscale Scratch Resistance of Ultrathin Protective Overcoats on Hard Magnetic Disks," J. Vac. Sci. Technol., 16(3)1741-1744(1998b).

Azarian, M.H., Bauer, C.L., O'Connor, T.M., and Jhon, M.S., "Head-Disk Interaction in Gas-Lubricated Slider Bearings," Wear, 168,49-57(1993).

Bhushan, B., "Magnetic Media Tribology: State of the Art and Future Challenges," Wear, 136,169-197(1990).

Chen, C-Y., Bogy, D.B., and Bhatia, C.S., "Tribochemisty of Various Lubricants on Chx Coated Disks Using DLC Coated and Uncoated AL2O3/TiC Sliders in UHV," Computer Mechanics Laboratory, Department of Mechanical Engineering, University of California, Berkeley,CA, Technical Report No. 97-003(1997).

Chen, C-Y., Fong, W., Bogy, D.B., and Bhatia, C.S., "Lubricant Thickness Effect on Tribological Performance of ZDOL Lubricated Disks with Hydrogenated Carbon Overcoats," Trib. Lett., 7,1-10(1999).

Chen, C-Y., Bogy, D.B., and Bhatia, C.S., "Effect of Lubricant Bonding Fractions at the Head-Disk Interface," Trib. Lett., submitted (2000).

Coffey, K.R., Raman, V., Staud, N., and Pocker, D.J., "Vapor Lubrication of Thin Film Disks," IEEE Trans. Magn., 30(6)4146-4148(1994).
DeKoven, B., Intevac Corporation, Santa Clara, CA (2000).
Gao, C., Vo., T., Weiss, J., "Molecular Orientation of Polymer Lubricant Films: Its Tribological Consequence," ASME J. Trib., 120,369-378(1998).
Gini, D., Larson, T.L., Merten, R.A., and Simonetti, A., "Magnetic Media Having Tridecyl Stearate Lubricant," US patent 4,303,738 (1981).
Gregory, T.A., and Keller, C.G., Kennedy, B.E., Murray, B.A., and Rothschild, W.J., "Method and Apparatus for Lubricating a Magnetic Disk Continuously in a Recording File," US patent 4,789,913 (1988).
Grochowski, E., Almaden Research Center, www.storage.ibm.com (2000).
Gschwender, L.J., and Snyder, C.E., "Stability Additive for Perfluoropolyalkylethers," US 5,302,760(1994).
Gschwender, L.J., Snyder, C.E. Jr., Oleksiuk, M., and Koehler, M., "Computational Chemistry of Soluble Additives for Perfluoropolyalkylether Liquid Lubricants," Trib. Trans., 39(2)368-373(1996).
Hirz, S.J., Homola, A.M., Hadziioannou, G., and Frank, C.W., "Effect of Substrate on Shearing Properties of Ultrathin Polymer Films," Langmuir, 8,328-333(1992).
Hu, Y., and Talke, F.E., "A Study of Lubricant Loss in the Rail Region of a Magnetic Recording Slider Using Ellipsometry," STLE SP-25, 43-48(1988).
Hua, W., Liu, B., Sheng, G., and Zhu, Y., "Probability Model for Slider-Disk Interaction," J. Info. Storage Proc. Syst., 1(3)273-280(1999).
Jesh, M.S., and Segar, P.R., "The Effect of Vapor Phase Chemical on Head/Disk Interface Tribology," Trib. Trans., 42(2)310-316(1999).
Karis, T.E., and Novotny, V.J., "Pin-on-Disk Tribology of Thin Film Magnetic Recording Disks," J. Appl. Phys., 66(6)2706-2711(1989).
Karis, T.E., Novotny, V.J., and Johnson, R.D., "Mechanical Scission of Perfluoropolyethers," J. Appl. Polym. Sci., 50,1357-1368(1993).
Karis, T.E., and Novotny, V.J., Johnson, R.D., and Jhon, M.S., "Ultrasonic Scission of Perfluoropolyethers," J. Magn. Soc. Japan, vol. 18, No. S1,509-519(1994).
Karis, T.E., Novotny, V.J., "Surface Potential of Thin Perfluoropolyether Films on Carbon," Appl. Phys. Lett., 71(1)52-54(1997).
Karis, T.E., Tyndall, G.W., Crowder, M.S., "Tribology of a Solid Fluorocarbon Film on Magnetic Recording Media," IEEE Trans. Magn., 34(4)1747-1749(1998).
Karis, T.E, and Jhon, M.S., "The Relationship Between PFPE Molecular Rheology and Tribology," Trib. Lett., 5(4)283-286(1998).
Karis, T.E., and Tyndall, G.W., "Calculation of Spreading Profiles for Molecularly Thin Films from Surface Energy Gradients," J. Non Newt. Fluid Mech., 82,287-302(1999).
Karis, T.E., Miller, J.L., Hunziker, H.E., de Vries, M.S., Hopper, D.A., and Nagaraj, H.S., "Oxidation Chemistry of a Pentaerythritol Tetraester Oil," Trib. Trans., 42(3),431-442(1999a).
Karis, T.E., Tyndall, G.W., and Waltman, R.J., "Lubricant Bonding Effects on Thin Film Disk Tribology," Trib. Trans., 44(2)249-255(2001).
Karis, T.E., "Water Adsorption on Thin Film Magnetic Recording Media," J. Coll. Int. Sci., 225(1)196-203(2000).
Kasai, P.H., "Degradation of Perfluoropolyethers Catalyzed by Lewis Acids," Adv. Info. Storage Syst., 4,291-314(1992a).
Kasai, P.H., "Perfluoropolyethers: Intramolecular Disproportionation," Macromolecules, 25,6791-6799(1992b).
Kasai, P.H., Wass, A., and Yen, B.K., "Carbon Overcoat: Structure and Bonding of Z-DOL," J. Info. Storage Proc. Syst., 1(3)245-258(1999).

Kasai, P.H., "Degradation of Perfluoropoly(ethers) and Role of X-1P Additives in Disk Files," J. Info. Storage Proc. Syst., 1(1)23-31(1999a).

Kasai, P.H., "Perfluoropoly(ethers): Intramolecular Disproportionation II," J. Info. Storage Proc. Syst., 1(3)233-243(1999b)

Kasai, P.H., IBM Almaden Research Center, private communication (2000).

Khan, M.R., Heiman, N., Fisher, R.D., Smith, S., Smallen, M., Hughes, G.F., Veirs, K., Marchon, B., Ogletree, D.F., and Salmeron, M, "Carbon Overcoat and the Process Dependence on its Microstructure and Wear Characteristics," IEEE Trans. Magn., 24(6)2647-2649(1988).

Koishi, R., Yamamoto, T., and Shinohara, M. "Chemical Structure and Durability of a Plasma-Polymerized Protective Layer for Thin-Film Magnetic Disks," Trib. Trans., 36(1)49-54(1993).

Lin, J-L., Bhatia, C.S., Yates, J.T.Jr., "Thermal and Electron-Stimulated Chemistry of Fomblin-Zdol Lubricant on a Magnetic Disk," J. Vac. Sci. Technol. A, 13(2)163-168(1995).

Lowell, J., and Truscott, W.S., "Triboelectrification of Identical Insulators: I. An Experimental Investigation," J. Phys. D: Appl. Phys. 19,1273-1280(1986) and Lowell, J., and Truscott, W.S., "Triboelectrification of Identical Insulators: II. Theory and Further Experiments," J. Phys. D: Appl. Phys. 19,1281-1298(1986).

Ma, X., Giu, J., Grannen, K., Smoliar, L., Marchon, B., Jhon, M.S., and Bauer, C.L., "Spreading of PFPE Lubricants on Carbon Surfaces: Effect of Hydrogen and Nitrogen Content," Trib. Lett., 6, 9-14(1999).

Marchon, B., Heiman, N., and Khan, M.R., "Evidence for Tribochemical Wear on Amorphous Carbon Thin Films," IEEE Trans. Magn., 26(1)168-170(1990).

Mate, C.M., Kasai, P.H., Tyndall, G.W., Lee, C.H., Raman, V., Pocker, D.J., and Waltman, R.J., "Investigation of Phosphazene Additive for Magnetic Recording Lubrication," IEEE Trans. Magn. 34(4)1744-1746(1998).

Mate, C.M., and Wu, J., "Nanotribology of Polymer Surfaces for Disk Drive Applications," in Microstructure and Microtribology of Polymer Surfaces, eds V.V. Tsukruk, K.J. Wahl, ACS Symposium Series 741,405-417(2000).

Mate, C.M., and Wilson, R.S., "Grand Unified Theory of Fluorocarbon Polymer Mobility on Disk Surfaces," Chapter 7, pp. 83-95, ACS Symposium Series 787, *Fluorinated Surfaces, Coatings, and Films*; editors: D.G. Castner and D.W. Grainger, American Chemical Society, Washington, D.C. (2001).

Merchant, K., Mee., P., Smallen, M., and Smith, S., "Lubricant Bonding and Orientation on Carbon Coated Media," IEEE Trans. Magn., 26(5)2688-2690(1990).

Min, B.G., Choi, J.W., Brown, H.R., Yoon, D.Y., O'Connor, T.M., and Jhon, M.S., "Spreading Characteristics of Thin Liquid Films of Perfluoropolyalkylethers on Solid Surfaces. Effects of Chain-End Functionality and Humidity," Trib. Lett. 1, 225-232(1995).

Miyoshi, K., "Fundamental Considerations in Adhesion, Friction, and Wear," Wear, 141,35-44(1990).

Mori, S., Onodera,N., and Itoh, M., "Tribochemical Reactions of Very Thin Layers on Magnetic Recording Disks," Wear, 168,85-90(1993).

Nader, B.S., Kar, K.K., Morgan, T.A., Pawloski, C.E., and Dilling, W.L., "Development and Tribological Properties of New Cyclotriphosphazene High Temperature Lubricants for Aircraft Gas Turbine Engines," Trib. Trans., 35(1)37-44(1992).

Novotny, V.J., and Karis, T.E., "Sensitive Tribological Studies on Magnetic Recording Systems," Adv. Info. Storage Syst., 2,137-152(1991).

Novotny, V.J., and Baldwinson, M.A., "Lubricant Dynamics in Sliding and Flying," J. Appl. Phys., 70(10)5647-5652(1991).

Novotny, V.J., Karis, T.E., and Johnson, N.W., "Lubricant Removal, Degradation, and Recovery on Particulate Magnetic Recording Media," J. Trib., Trans. ASME, 114(1)61-67(1992).

Novotny, V.J., Pan, X., and Bhatia, C.S., "Tribochemistry at Lubricated Interfaces," J. Vac. Sci. and Technol. A, 12(5)2879-2886(1994).

Novotny, V.J., "Mechanical Integration of High Recording Density Drives," IEEE Trans. Magn., 32(3)1826-1831(1996).

Novotny, V.J., Karis, T.E., Whitefield, R.J., "Surface Potential and Magnetic Recording Media Tribology," Tribol. Trans., 40(1)69-74(1997).

Oshanin, G., Nechaev, S., Cazabat, A.M., and Moreau, M., "Kinetics of Anchoring of Polymer Chains on Substrates with Chemically Active Sites," Phys. Rev. E., 58(5) 6134-6144 (1998).

Pacansky, J., Miller, M., Hatton, W., Liu, B., and Scheiner, A., "Study of the Structure of Perfluoro Ethers: Structural Models for Poly(perfluoroformaldehyde), Poly(perfluorethylene oxide), and Poly(perfluoropropylene oxide)," J. Am. Chem. Soc., 113,329-343(1991).

Pacansky, J., private communication, IBM Almaden Research Center (2000).

Paciorek, K.J.L., and Kratzer, R.H., "Stability of Perfluoralkylethers," J. Fluorine Chem., 67,169-175(1994).

Peck, P.R., Kono, R.N., Jhon, M.S., and Karis, T.E., "Acoustic Emission Analysis During High Velocity Accelerated Wear Test," Adv. Info Storage Syst., 6,121-135(1995).

Perettie, D.J., Johnson, W.D., Morgan, T.A., Kar, K.K., Potter, G.E., DeKoven, B.M., Chao, J., Lee, Y.C., Gao, C., and Russak, M., "Cyclic Phosphazenes as Advanced Lubricants for Thin Film Magnetic Media," ISPS-Vol. 1, Advances in Information Storage and Processing Systems, eds. G.G. Adams, B. Bhushan, D. Miu, and J. Wickert, Book no. H01016,117-123(1995).

Qian, W., Childers, W.W., Prime, R.B., Karis, T.E., and Jhon, M.S., "Crown Sensitivity of the Magnetic Recording Head Gimbal Assembly Bonded by a Viscoelastic Adhesive," Mech. Time Dep. Mater., 2,371-387(1999).

Rabinowicz, E., Friction and Wear of Materials, John Wiley & Sons, Inc., New York, p. 66, 1965.

Raman, V., and Howard, J.K., "Mechanical Properties and Wear Behavior of Sputtered Carbon Overcoats," Proc. Of Japan International Trib. Conf., Nagoya, 1-6(1990).

Raman, V., and Tang, W.T., "Environment Dependent Stiction and Durability in Hydrogenated Carbon Overcoated Thin Film Disks," IEEE Trans. Magn., 29(6)3933-3935(1993).

Ruckenstein, E., and Rajora, P., "On the No-Slip Boundary Condition of Hydrodynamics," J. Coll. Int. Sci., 96(2)488-491(1983).

Ruhe, J., Blackman, G., Novotny, V., Clarke, T., Street, G.B., and Kuan, S., "Terminal Attachment of Perfluorinated Polymers to Solid Surfaces," J. Appl. Polym. Sci., 53, 825-836 (1994).

Sano, K., Murayama, H., and Yokoyama, F., "Lubricant Bonding via a Hydrogen Bond Network," IEEE Trans. Magn., 30(6)4140-4142(1994).

Scarati, A.M., and Caporiccio, G., "Frictional Behaviour and Wear Resistance of Rigid Disks Lubricated with Neutral and Functional Perfluoropolyethers," IEEE Trans. Magn., MAG-23(1)106-108(1987).

Segar, P.R., and Jesh, M.S., "The Effect of Pressure Sensitive Adhesive Outgassing on Head/Disk Interface Tribology," J. Info. Stor. Proc. Syst.,1(2)125-133(1999).

Seki, A., and Kondo, H., "FT-IR Reflection Absorption Spectra of Novel Lubricant Layers on Magnetic Thin Film Media," J. Magn. Soc. Japan, 15, Supplement No. S2, 745-749(1991).

Shogrin, B.A., Jones, W.R. Jr., Herrera-Fierro, P., Lin, T-Y., and Kawa, H., "Evaluation of Boundary-Enhancement Additives for Perfluoropolyethers," Trib. Trans., 42,747-754(1999).

Sianesi, D., Zamboni, V, Fontanelli, R., and Binaghi, M., "Perfluoropolyethers: Their Physical Properties and Behaviour at High and Low Temperatures," Wear, 18,85-100(1971).

Snyder, C.E. Jr., Tamborski, C., Gopal., H., and Svisco, C., "Synthesis and Development of Improved High-Temperature Additives for Polyperfluoroalkylether Lubricants and Hydraulic Fluid," Lub. Eng., 35(8)451-456(1978).

Strom, B.D., Bogy, D.B., Walmsley, R.G., Brandt, J., and Bhatia, C.S. "Gaseous Wear Products from Perfluoropolyether Lubricant Films," Wear, 168,31-36(1993).

Suzuki, S., and Kennedy, F.E. Jr., "Incipient Damage on Thin Film Magnetic Disks and its Detection with a Corrosion Test," in Tribology and Mechanics of Magnetic Storage Systems, Vol. VI, STLE SP-26, 111-119(1989).

Terada, A., Ohtani, Y., Kmachi, Y., and Yoshimura, F., "Wear Properties of Lubricated Medium Surface Under High Velocity Head Sliding," in Tribology and Mechanics of Magnetic Storage Systems, Vol. V, STLE SP-25, 69-73(1988).

Tsai, H-C., and Bogy, D.B., "Critical Review Characterization of Diamondlike Carbon Films and their Application as Overcoats on Thin-Film Media for Magnetic Recording," J. Vac. Sci. Technol. A, 5(6)3287-3312(1987).

Tyndall, G.W., and Waltman, R.J., "Structure of Molecularly-Thin Perfluorpolyether Films on Amorphous Carbon Surfaces," Mat. Res. Cos. Symp. Proc., 517,403-414(1998).

Tyndall, G.W., Waltman, R.J., and Pocker, D.J., "Concerning the Interactions Between Zdol Perfluoropolyether Lubricant and an Amorphous-Nitrogenated Carbon Surface," Langmuir, 14,7527-7536(1998a).

Tyndall, G.W., Leezenberg, P.B., Waltman, R.J., and Castenada, J., "Interfacial Interactions of Perfluoropolyether Lubricants with Magnetic Recording Media," Trib.Lett., 4,103-108(1998b).

Tyndall, G.W., Karis, T.E., Jhon, M.S., "Spreading Profiles of Molecularly Thin Perfluoropolyether Films," Trib. Trans., 42(3)463-470(1999).

Tyndall, G.W., and Waltman, R.J., "Thermodynamics of Confined Perfluoropolyether Films on Amorphous Carbon Surfaces Determined from the Time-Dependent Evaporation Kinetics," J. Phys. Chem. B, 104,7085-7095(2000).

Vurens, G., Zehringer, R., and Saperstein, D., "The Decomposition Mechanism of Perfluoropolyether Lubricants during Wear," in Surface Science Investigations in Tribology: Experimental Approaches, ed's Y-W. Chung, A.M. Homola, and G.B. Street, ACS Symposium Series 485, 1992, pp.169-180.

Waltman, R.J., Tyndall, G.W., and Pacansky, J., "Computer-Modeling Study of the Interactions of Zdol with Amorphous Carbon Surfaces," Langmuir, 15,6470-6483(1999).

Waltman, R.J., Tyndall, G.W., Pacansky, J., and Berry, R.J., "Impact of Polymer Structure and Confinement on the Kinetics of Zdol 4000 Bonding to Amorphous-Hydrogenated Carbon," Trib. Lett., 7(2,3)91-102(1999).

Wang, G., Yeh, T-A., Sivertsen, J.M., Judy, J.H., and Chen, G-L., "Tribological and Recording Performance of Carbon-Coated Thin Film Head Sliders on Unlubricated and Lubricated Thin Film Media," IEEE Trans. Magn., 30(6)4125-4128(1994).

Wang, R-H., White, R.L., Meeks, S.W., Min, B.G., Kellock, A., Homola, A. and Yoon, D., "The Interaction of Perfluoro-Polyether Lubricant with Hydrogenated Carbon," IEEE Trans. Magn., 32(5)3777-3779(1996).

Wei, J., Fong, W., Bogy, D.B., and Bhatia, C.S., "The Decomposition Mechanisms of a Perfluoropolyether at the Head/Disk Interface of Hard Disk Drives," Trib. Lett., 5,203-209(1998).

Xuan, J, Chen, G-L., and Chao, J., "Organic Buildup on Slider Leading Edge Tapers and its Effect on Wet Stiction," IEEE Trans. Magn., 29(6)3948-3950(1993).

Yanagisawa, M., "Adsorption of Perfluoropolyethers on Carbon Surfaces," in Tribology and Mechanics of Magnetic Storage Systems, Vol. IX, STLE SP-36, 25-32(1994).

Yanagisawa, M., Tsukamoto, Y., and Goto, F., "Wear Mechanism on Thermal Oxidation Film for Plated Magnetic Disks," in Tribology and Mechanics of Magnetic Storage Systems, Vol. VI, STLE SP-26, 153-159(1989).

Yang, M., Talke, F.E., Perettie, D.J., Morgan, T.A., and Kar, K.K., "Environmental Effects of Phosphazene Lubricated Computer Hard Disks," IEEE Trans. Magn., 30(6)4143-4145(1994).

Zhao, Z., Bhushan, B. and Kajdas, C. "Effects of Thermal Treatment and Sliding on Chemical Bonding of PFPE Lubricant Films with DLC Surfaces," J. Info. Storage Proc. Syst., 1(3)259-264(1999).

Zhao, Z., and Bhushan, B., "Tribological Performance of PFPE and X1-P Lubricants at Head-Disk Interface. Part I Experimental Results," Trib. Lett., 6,129-139(1999).

Chapter 23

NANOLUBRICATION: CONCEPT AND DESIGN

Stephen M. Hsu
National Institute of Standards and Technology
Gaithersburg, MD 20899

Abstract The advent of microelectromechanical system (MEMS) devices, sensors, actuators, microsystems, and nanotechnology have called to attention the effect of friction on moving parts in nano/micro devices. To take full advantage of the opportunity to sense, compute, and actuate in real time, fast-moving parts are often necessary or desirable. As the scales of the components shrink, adhesion, stiction, friction, and wear become a significant technological barrier for the successful deployment of durable devices. Most current devices in production avoid such contacts.
Surface forces that normally are dwarfed by mechanical loading, as component scale moves from macro to micro to nano, dominate the nature of the surface contacts. Therefore nanolubrication needs to take into account different factors than conventional lubrication concepts. This paper compares traditional lubrication concepts and those necessary for nanolubrication and proposes various nanometer scale thick lubricating film designs as a means to control the surface properties of surfaces at nano/micro scales.
Many of the concepts derive their origin from studies and observations from the magnetic hard disk technology where a monolayer of lubricant protects the system and has proven to be robust and safe. Examples from magnetic hard disks will be used to illustrate some of the concepts.

1 INTRODUCTION

Lubrication is one of the oldest technologies in human history. It dates back to the use of animal fats and water in the days of Egyptian Pharaohs in building the pyramids [1]. Historically, lubrication is more an art than science. In the 19th century, amidst the industrial revolution, the need to consistently lubricate machine components sparked much-needed research. Fluid mechanics and Reynolds' Equations were developed for bearing design. As machinery advanced, from steam engines to jet fighters, lubrication became an interdisciplinary science involving physics, chemistry, materials, fluid mechanics, and contact mechanics. Monolayer of fatty acids was examined as early as 1947 to study the fundamentals of lubrication [2]. In recent years, the discovery of buckyballs or fullerenes ignites the imagination of many people. Yet one of the first proposed applications for the new materials was for lubrication as friction reducing nano-bearings in microsystems. Even though these materials did not succeed in becoming the new age wonder-lubricants, it illustrates the position lubrication holds in new technologies.

Lubrication research thrives when there is a demand for it. In the past, these demands arose when new technologies pose new challenges, *e.g.*, space stations, adiabatic diesel engines, and ultra-high storage density in magnetic hard disks. These new technologies pose demands outside of knowledge base such as high temperatures, radiation, and nanometer scale precision. Demands like these often force practitioners and researchers to go back to first principles and try to create new means to meet these demands. Sometimes one wonders why accumulated knowledge in lubrication does not provide answers for these new application needs? The answer may lie in the fact that we do not have a thorough understanding of how lubrication works and its relationship with material science. Our knowledge base tends to rely on past solutions so as we move from one material system into new materials, gaps in our knowledge emerge. As we learn from one material system to the next, our knowledge improves. The majority of our understanding, however, is based on our experience in lubricating steel bearings and it depends on the concept of sacrificial wear of the lubricating films to protect the steel surfaces. When it comes to other materials such as titanium, alumina, and new composite coatings, our knowledge base is very limited.

Recent developments in micro-electromechanical devices (MEMS), microsystems, nanotechnology, nano-electromechanical devices (NEMS) in communication, microelectronics, aerospace, biomedicine, and micro-analytical tools such as lab-on-a-chip devices pose brand new challenges for lubrication at the micro and nanoscale. The requirements for stiction, friction control at such small scale on unfamiliar materials again tax our lubrication knowledge to the limit. Many of these micro devices are made of silicon (single crystalline or polycrystalline) using conventional semiconductor manufacturing technology [3]. Other materials being used are silicon nitride, silicon carbide, nickel, and diamond-like carbon. These materials are often deposited on silicon substrates in the form of nanometer thick films to enhance mechanical strength, wear resistance, and corrosion resistance. Under repeated contacts, without effective lubrication, the components have very short lives.

Lubrication of such devices requires new concepts and revolutionary thinking. The spacing between surfaces in such devices is often separated by only nanometers. Applications in these new technologies desire high-speed relative motion under light load, billions of duty cycles, and environmental stability. At nanometer scale, the surface area to volume ratio for a typical component is very high, surface forces become the dominant forces governing the contact behavior. Therefore, adhesion, stiction, and friction are the critical technological issues to resolve for the present and beyond. Additionally, the mechanical strength of components (silicon beams) is necessarily weak because of the small dimensions of the components. Coefficient of friction, while very useful at macro-scale (a normalized number for comparison different materials) is not very useful at nanoscale. The magnitude of the lateral loading determines whether the component will break or not, not the coefficient of friction.

Nano-lubrication therefore can be defined as the arts and sciences necessary to control adhesion, stiction, friction, and wear of surfaces coming into contacts at the micro/nano-scale.

2 TRADITIONAL LUBRICATION PRINCIPLES

Lubrication, traditionally, has been based on two main principles: generation of fluid pressure to separate the surfaces to avoid contacts; easily sheared (sacrificial) chemical films formed on the surface to redistribute the stresses and sacrificially worn off to protect the surface. Fluids are used to generate hydrostatic and hydrodynamic pressures to support the load. Under high load and/or low speed, boundary chemical films from chemical additives are used to generate protective thin films against the inevitable asperity contacts. When the surfaces come into contact, many of the asperities undergo elastic deformation. This condition is generally referred to as elastohydrodynamic lubrication (EHL). The EHL theories are well developed. They describe and predict the surface temperatures, fluid film thickness, and the hydrodynamic pressures. Further increase in the contact pressure beyond the EHL conditions causes the asperities to deform plastically and the thickness of the fluid film to decrease. When the average fluid film thickness falls below the average surface roughness, the asperity contacts between the surfaces become the dominant load supporting mechanism. This condition is referred to as the boundary lubrication regime (BL). Under such conditions, the temperatures at the asperity tips (flash temperatures) have been shown to be high enough under most lubrication conditions to cause chemical reactions to take place between the lubricant and the solid surface forming a chemical film. By design, this film is easily sheared and thereby protects the surface. The chemical film can be organic, inorganic, and mixture of inorganic/organic in nature. The exact nature of the film and the chemical kinetic processes associated with its formation and destruction are system and environmental dependent and are not clearly understood. A detailed discussion on the subject can be found in (4).

Analyses performed on these films using ESCA, Auger, SIMS, EDAX, and XPS suggest a mixture of organic polymeric materials, oxides, wear particles, and inorganic reaction products from phosphorus- and sulfur-containing additives (FeO, Fe_2O_3, Fe_3O_4, $FePO_4$, FeS, $Fe_xO_yS_z$, *etc.*). The lubricant oxidizes forming surface-active carboxylic acids (or the surface-active additives added to the lubricant). The polar molecules then react with the iron surface forming an iron-organo compound, which polymerizes through condensation reactions to form high molecular weight products. This high molecular weight product (3000–5000) is critical in providing lubrication for typical bearing surfaces (0.05–0.1 µm Ra) [4]. When the base oil or the high molecular weight products are absent, pure antiwear additives in solvents do not provide effective lubrication. When the molecular weight reaches 100,000, it becomes insoluble in the liquid lubricant phase, producing what is generally referred to as "sludge." If the reaction rate with the surface is too rapid, chemical corrosion would result and wear would increase. An insufficient reaction rate, of course, does not form a film quickly enough to protect the surface. When chemical antiwear additives (phosphorus, sulfur, nitrogen, and chlorine containing compounds) are used, inorganic "glassy" films are formed (structures such as $FePO_4$ and FeP_4 have been identified). Oftentimes, the metallo-organic polymers work synergistically with the inorganic reaction products forming a reinforced film with stronger mechanical properties. The thickness of the films is typically between 100 to 300 nm depending on the contact pressures.

The fact that a film exists in the contact does not necessarily mean the surfaces are automatically protected. The adhesive and cohesive strength of the film, density of the film, and thickness of the film all contribute to the effectiveness of the film. The uncertainty associated with our knowledge base lies in our inability to predict reactivity and the resultant film property for a given material system operating under a specific contact condition.

Therefore, the basic principle of traditional lubrication is either to avoid contacts using fluid pressures or when inevitable, use chemistry to generate a sacrificial film to protect the surfaces. The characteristics of a good lubricating boundary film are: good adhesion; good cohesion; thick enough in relation to the surface roughness and wear mode to effect protection for the contact conditions. The film also functions to redistribute the stresses at the interface; provide a sacrificial easily shearable layer; increase the real area of contact by physically smoothing out the relative roughness thereby lowering the contact pressure.

3 NANOLUBRICATION & ITS REQUIREMENTS

At nanoscale, the load is not likely to be supported by the macro-scale squeezed-film mechanism due to the lack of controlled contact geometry and insufficient number of molecules for such films to form. The nano-sacrificial film mechanism is also not very likely since it is difficult to continuously supply molecules to replenish films and the difficulty in disposing the degradation products. Therefore, a new principle of lubrication is needed.

If the lubricant supply is limited, *i.e.*, only one or two layers of molecules are available for the duration of the component life, the effectiveness of the film in resisting shear stresses will be essential. The ability of the film to withstand repeated contacts over a long period of time is also critical. Lubricating films at nanoscale need to have two essential qualities: ability to withstand shear stresses without being ruptured, ability to withstand repeated contacts for a long duration. The lubricating films will have to be tenacious (strong cohesion), self-repairing (surface mobility), and stable (strong adhesive bonding to surface).

Early controlled experiments by Bowden and Tabor demonstrated that a monolayer of fatty acids on glass surfaces reduced friction [2]. However, these monolayer films did not last long after repeated sliding. When some of the molecules were removed from the surface by the mechanical rubbing, the film failed.

Usefulness of solid lubricating films such as MoS_2, graphite, Teflon should be examined. Micro- and nano-scale devices need to operate in a clean environment. Any inorganic degradation products such MoOx or carbon particles potentially could cause severe abrasive wear. If continuous exits and swiping actions for the degradation products are available, such as roller bearings, these films could be effective. The disposal of the unwanted products could pose a problem for wide spread applications.

This leads to an all organic film. If an organic film can perform effectively, the issue is how to keep the film intact under impact and if damaged, how to repair it or to resupply molecules for the film to repair itself. When the lubricant molecules are removed from a location either by shear or oxidation/evaporation, molecules from other location can move in to cover the surface, this can be defined as self-repairing. Under typical high speed contacts, asperity temperatures are high. So molecules will have to withstand thermal decomposition and oxidation. The vapor pressure and volatility, therefore will have to extremely low. Resistance to oxidation and decomposition will need to be high.

In magnetic head-disk lubrication practice, the lubricant used is perfluoropolyether (PFPE). The lubricant layer thickness is between one to two nanometers. PFPE is highly oxidation resistant, low volatility (10^{-6}

Torr), and tend to adhere to the surface through hydrogen bonding to withstand the high centrifugal force from disk rotation.

To summarize, lubrication at nanoscale (control of adhesion, stiction, friction) requires lubricant molecules which are non-volatile, oxidation and thermal decomposition resistant, good adhesion and cohesion, and self-repairing or self-regenerating.

4 NANOLUBRICATING FILM PROPERTIES

Effective nanolubricating films should have strong adhesive strength and cohesive strength to resist shear stresses. Organization of such films, whether single molecular species or mixed species, has profound influence on film performance. The most important property of the film is its bonding characteristics with the surface (adhesive strength of the film).

4.1 Adhesion and Bonding

Molecules attach themselves to a surface via either physical or chemical adsorption. The bonding force depends on the nature of the bonds, number of bonds per molecule (in high molecular weight molecules), and the molecular orientation and packing density per unit area. On an ideal surface (atomically flat and uniform surface energy), molecular attachment to the surface can be uniform and evenly distributed. On engineering surfaces (atomically rough surfaces with non-uniform surface energy), molecules tend to preferentially bond to the surface defect sites (high energy sites such as steps, twins, dislocations, lattice defects, *etc.*). Even though the initial film may be uniform, subsequent surface reorganization and reconstruction will distribute the molecules into islands and clusters. In such a case, measurement of the precise adhesive strength of the monolayer will be difficult. Thus, the nature of the surface and its surface energy are important considerations in adhesion and bonding of molecules on a surface.

Since silicon is a semiconducting material, electrical double-layer force can easily give rise to electro-static charges. This introduces two effects: electrochemistry and additional surface forces to contend with. At micro-nano-scales, surface forces are often the dominant forces that control the adhesion and bonding. Surface charge (and surface charge density) therefore, when ignored can produce unanticipated movements and adhesion. The charges at the surface also influence interfacial chemistry and adsorption of molecules. The silicon surface can be passivated during processing with hydrogen or nitrogen. But when the surface is disrupted under rubbing conditions, the electro-static charge will develop again. In the presence of adsorbed ionic species, electrochemistry can occur giving rise to reaction products that otherwise may not take place.

The classical technique to measure the bonding strength of molecules on surfaces is by measuring the heat of adsorption and desorption [5]. This technique works well for ideal surfaces where uniform coverage can be reasonably assured. Translating the heat of adsorption data to shear strength, however is tenuous at best. AFM has been used to measure the lateral resistance to motion of some self-assembled molecular layers on ideal surfaces [6]. Results tend to vary with molecules and surfaces and the conditions the films are deposited. Inter-laboratory comparison of such measurements is currently not possible due to the lack of standards and calibration artifacts.

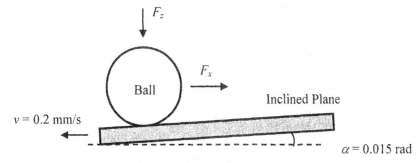

Figure 1. Schematic diagram of the ball-on-inclined plane test.

A ball-on-inclined plane apparatus has recently been developed in our laboratory to address this issue. The test technique measures the shear rupture strength of monolayer films on atomically rough surfaces. Experimental details are described in reference [7]. Figure 1 shows the schematic set up of the technique. A ball of various diameters (3 to 15 mm) slides on a highly polished flat plane (mica, silicon, diamond like carbon surface on magnetic hard disks). The load is controlled by the geometric interference of the two surfaces and the forces in the x, y, and z directions are continuously recorded by a force transducer mounted on top of the ball. The angle of incline of the flat plane controls the initial loading. The geometric interference loading by the two smooth surfaces provides a well-controlled loading mechanism. The elastic deformation of the surfaces also provides a continuously conformal contact under which the molecular film will act to resist the increasing shear actions. The large area of contact between the ball and the flat captures a large number of molecules under relatively low contact pressure as compared to the high contact pressure of an AFM tip. The load increases linearly until the monolayer cannot sustain the shear stresses and solid to solid contact occurs. This is accompanied by a sudden increase in lateral force, F_x. Normal force measured at this point can be translated into contact pressure by measuring the spot with AFM. The resulting contact pressure indicates the maximum film strength. Figure 2 shows typical data on model compounds deposited on a single silicon surface at one nanometer film thickness. The data suggest that the adhesive strength follows somewhat the polarity of the functional groups and the packing density. This is reasonable. Experiments using AFM lateral force measurements did not distinguish various molecular structures (7).

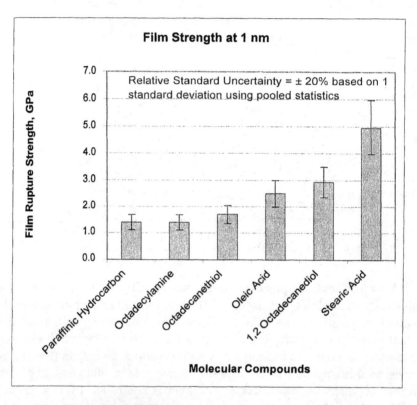

Figure 2. Typical results of film rupture strength of model compounds.

The measurement of adhesive strength of monolayer on surfaces, sometimes, is also complicated by unintended chemical reactions (hydration and tribochemistry).

4.2 Materials and Tribochemistry

Chemical reactions between the molecules and the surface under rubbing conditions depend on the nature of surface bonds and the chemical potential between the molecule and the surface structure. The surface chemistry of silica is well known [8]. Since a silicon metal surface is usually covered by a layer of oxides, the surface chemistry of silicon tends to be governed by the silica chemistry. Silica is known to react with and sometimes dissolve in water forming silica gel and silicic acids via the alkoxide route [9]. Polymerization of silica to form gels is well known and the reactions are controlled by pH of the environment. Under rubbing conditions, tribochemistry takes place and water reacts with silicon surfaces to form SiH and SiO. Subsequent alkoxide reactions lead to high molecular weight products, which provide lubricating property under certain speed and load conditions [9-11].

The chemical reactivity and mechanical properties of silicon are also influenced by crystalline orientations and the phases present at the surface. When the surface is mechanically disrupted or thermally excited, dangling bonds are generated providing high-energy active sites for adsorption and interaction.

Silicon nitride and silicon carbide surface layers can be formed on silicon via gas phase sintering in vacuum. They can provide a wear resistant surface layer several nm thick on surfaces subjected to rubbing in a device (rotating hubs, sliding interfaces). These surfaces are covered with an oxide layer, similar to silicon under static conditions. However, under dynamic rubbing conditions, they behave differently. Silicon nitride is much more reactive than silicon carbide. Detailed discussions on the chemistries are described in references [11-15].

Nickel and diamond-like-carbon are the other two materials often used in microsystems. Nickel historically is not known to be a good tribomaterial due to the combination of a hard oxide layer and a soft substrate. Nickel is chemically reactive and forms organometallic compounds under rubbing conditions. Diamond-like-carbon is a family of materials and its properties heavily depend on the processing conditions and the resulting microstructure. Chemically, it is relatively inert and reacts via dangling bonds under tribological conditions [16].

Water is a very polar molecule and the hydration force of water is very strong. It can preferentially adsorb on surfaces and alter the bonding characteristics of molecules with that particular surface. Under rubbing conditions, water will also compete with other molecules to react with the surface forming unintended bonds, interfering with the film design.

4.3 Wear Resistance and Self-repairing Property

The wear resistance of a monomolecular film can be related directly to the bonding strength and the cohesive strength of the monolayer [7]. The classical Langmuir-Blogett films (fatty acids, alkyl thiols, siloxanes) typically behave like a solid. Solid films will deform under contact stresses, produce defects (Gauche defects), and eventually fracture which lead to film failure. The durability of such solid-behavior films under repeated contacts is typically short.

If the spacing between molecules is larger than individual molecular size or volume, then the film behaves like a liquid or a solid-liquid mixture. The durability of the film will increase due to the ease of mobility of the molecules to accommodate stresses but the magnitude of shear stress it can withstand will decrease. Molecule size (molecular weight) also influences the shear strength of the film when the size of the molecule is larger than the asperity contact radius. If the molecular size is sufficient large and several asperities are in contact with the same molecule, shear resistance will increase due to the large energy necessary to rupture the molecule under shear motion.

How long will such a layer last? The durability of monolayer can be related the ability of self-repairing property which can be defined as the ability of the molecules to reorganize themselves into the original state after being mechanically disrupted by the contact. In order to have a self-repairing property, the ability of the molecules to self-assemble or reorganize on the surface will be essential, but the requirement for self-assembly would almost dictate that molecules are free to move about on the surface. This mobility implies that the molecules cannot be chemically bonded to the surface, hence low bonding strength. This diametrically opposite requirement of strong adhesive strength and self-repairability poses an interesting challenge. To meet this dual requirement, one way is to use a mixed molecular system where one species will bond to the surface and another specie will be allowed to move freely on the surface. Thus, a carefully designed molecular assembly consisting of different molecules arranged at a monolayer level with each molecule having a specific function or characteristics will give the best combination of tribological properties. This idea leads to the concept of molecular engineering of a monolayer film specifically for lubrication, stiction, friction control, or wear protection.

5 THIN FILM ORGANIZATION (SINGLE MOLECULE OR SAME FUNCTIONAL GROUP)

A monolayer film can be deposited on a flat surface using a single molecular species. There have been many surface science studies on molecular interactions, surface forces, and molecular organization using Langmuir-Blogett (LB) technique to deposit surfactant molecules (amphiphiles) on mica, glass, gold, or silicon [17]. The technique utilizes the ability of the surfactant molecules to pack closely at an air-liquid interface with an ordered monolayer structure. When a surface is immersed, a monolayer coating is adsorbed onto the surface. Over the years, many molecules have been studied, such as fatty acids, silanes, thiols, phospholipids, and polymeric films. Cadmium arachidate forms a very high quality monolayer and it has been used as a reference film for calibration purposes. Most of these films behave like a solid, the molecules are tightly packed. When they are under contact stress, the film fractures and break down. Hence the durability of the films is relatively short. Langmuir-Blogett (LB) films can also be deposited as liquid and mixed solid-liquid state on well-defined surfaces. However, non-polar molecules or high molecular weight polymers are difficult to deposit uniformly on a surface.

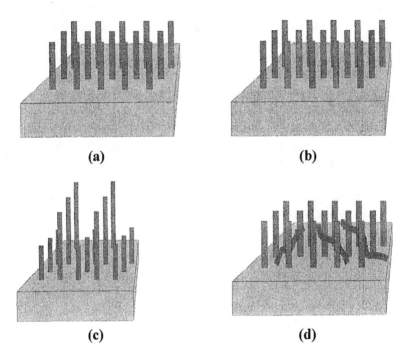

Figure 3. Schematic drawings for different molecular designs.

What are the conceptual designs possible to control friction more effectively using a monolayer? Figure 3 provides some illustrations. Figure 3a shows the schematic representation of molecular arrangement of polar molecules distributed uniformly on the surface. In reality, surface reconstruction often occurs and some molecules will aggregate into different crystalline structures. Figure 3b shows two rows of molecules with one row of molecules chemically bonded to the surface and one row of molecules physisorbed on the surface. This way the film can have both mobility and adhesive strength. Figure 3c shows a molecular film consisted of molecules with the same functional group but with different chain lengths. Since the adsorption is governed by the functional groups, the lower half the film will be closely packed behaving like a solid. The molecules with longer chain length will have part of the molecules sticking above the solid layer. Since they have much larger molecular volume to move about, they will behave like a liquid. Thus, a film with both liquid and solid properties can be constructed. The liquid layer provides damping and low friction and the solid layer provides load supporting strength. Figure 3d is a variation of the same concept. In this case, a molecular species is first adsorbed on the surface below the monolayer coverage. They are made to bond chemically with the surface either by heat or by UV irradiation. The surface is rinsed with solvent to remove the unreacted species and deposited again with a long chain polymeric molecule similar in structure to the bonded ones with or without the same functional groups. The similarity in molecular structure sometimes is necessary to avoid solubility/compatibility at the monolayer level. This kind of molecular organization, therefore provides a way to achieve high adhesive strength without sacrificing self-repairability.

In practice, most single molecular monolayer are used to control the surface state, *e.g.*, hydrophobicity, surface charge passivation on relatively ideal surfaces. On engineering surfaces, the roughness and non-uniform surface energy will require much more than simple monolayer coverage. Three dimensional stacking of different molecules are required to completely cover the surface and provide a thermodynamically stable surface film for long durable coating.

6 MIXED MOLECULAR FILMS

To extend these concepts further, different molecules assembled together will provide greater opportunities for more effective films. This leads to the concept of mixed molecular film design. One could sprinkle molecule A which would form a strong bond immediately with the surface at half the surface coverage. Molecule B can be introduced to cover the remainder of the surface area and B possesses different properties such as self-assembly, cross-linking, oxidation resistance, etc. The resulting film then has desired properties for different applications.

When molecules of different structures, chain length, and functional groups are put on a surface at a monolayer level, molecular compatibility and solubility between molecules are unexplored territories. Molecular dynamic studies may yield some insights on how to define two molecules of different structures interact with each other on a surface. At the simplest level, the two molecules will either attract or repulse one another. If they attract, how close they can come together or even merge into one entity? If they repulse, what will be the distance between them? On a broader scale, would molecules A aggregate with each other and exclude B forming islands of A and B?

In terms of imparting special properties to the film by adding a small amount of molecules of particular functions and design, *i.e.*, an additive on the molecular scale, how will this work at the monolayer level? The traditional concept of additives in lubricants is that the additives are surface active, therefore a small amount will alter the surface chemistry therefore achieve the objectives to control friction and wear. At a monolayer level, this obviously is not the case. Concentration of one species necessary to affect the outcome needs to be defined as the percentage of surface covered in relation to the contact area. If the contact area is large, 10% coverage may be sufficient. If the contact area is very small, 50% coverage may be necessary for the particular molecular species to have an impact on the outcome of the contact.

The ability of the additive molecules to stay at one place is also important. In order to maintain the effective concentration (area coverage ratio), the mobility of the additive molecules to move about should be restricted. This suggests chemically bonded molecules. With active functional groups such as acids or amines, the danger of these molecules to aggregate on the surface after initial deposition is real. Surface reconstruction and migration of molecules have been observed in catalysis [18]. Under rubbing conditions, the mixing action of moving surfaces also tends to promote self-aggregation once the molecule is dislodged from the surface by shear stresses.

What kind of functionality do we want in an additive? Our recent data [19] suggest that we need antioxidants and antiwear functions. Particular additives will depend on the surfaces to be applied to and the intended applications.

7 DURABILITY MEASUREMENT TECHNIQUES

We have shown that the ball-on-inclined plane apparatus under one pass condition can provide a reasonable estimate of the intrinsic molecular effect on shear strength which is a combination of adhesion and cohesion of the film. We have discussed extensively of how to organize different molecules into long lasting monolayer films. To test such concepts, we need a test to measure the durability of monolayers as deposited on the surface. The deposition techniques will be discussed later on.

We had previously developed a simple ball-on-inclined-plane test to simulate the CSS test [19]. In this case, a ruby ball of 3 mm diameter was used to control the contact stresses. The linear speed was 2 m/s. A sudden friction rise was used to monitor the failure point. The number of cycles to failure was used to rank different chemistries.

This measurement method was used to measure the durability of thin films (1nm–2.5 nm). The results are shown in Table 1.

Table 1. Comparison of one pass experiments and durability

	One Pass Failure Load (diamond)	Cycle to Failure
Silicon	2.4±0.1 g (6.9 GPa[1])	instant
$C_{18}OH$	2.7±0.1 g (7.2 GPa)	1000
Stearic acid	2.9±0.1 g (7.4 GPa)	5500
$C_{15}F_{31}COOH$	3.2±0.1 g (7.6 GPa)	7500
$C_{18}NH_2$	3.2±0.1 g (7.6 GPa)	>>35,000

[1] Initial Hertzian contact pressure based on nominal diameter of apparent contacts

The results in Table 1 illustrate the fact that while one pass experiments are valuable to measure the molecular structural effects, durability of the film is much more complex and depends on many other parameters. One of the parameters is the surface mobility of the molecules or the spreading coefficient of the molecules on that particular surface.

8 DEPOSITION OF MOLECULES ON SURFACES

The traditional Langmuir-Bloggett technique only works well in a system that is well-defined, i.e., purity of molecule, hydrophobic functional head group, and well-defined surfaces. When it comes to depositing complex mixtures of molecular weights, mixed functional groups, dip coating is better suited. Vapor phase deposition of molecules in vacuum is also feasible if the vapor pressure of the molecule to be deposited is low enough. Spin coating can also be done. All these techniques have problems in depositing mixed molecules especially in controlling the film thickness and surface concentrations. We have tried a successive dip-coating technique with annealing to induce bonding of the deposited molecules before the second dip-coating of another molecule. Vapor phase deposition was also used. The deposition area needs a lot of research to define the parameters necessary to control thickness and surface concentrations.

The success of depositing different molecules on a surface at a monolayer level with different molecular weight, functional groups, and molecular structures depends largely on how the molecules pack together (geometric compatibility, orientations, size, shape). If the molecules are not anchored through chemical bonding, then surface reorganization often takes place forming various islands of molecular aggregations of various thicknesses.

9 CHEMICAL STRUCTURAL INFLUENCE ON DURABILITY

While complex molecular assembly can be achieved by using sequential vapor phase deposition of selected molecules, uniformity of molecular distribution on the surface cannot be guaranteed. Preliminary results using simple molecules dip coated on smooth surfaces can be used to demonstrate the chemical structural effects to validate the test method and concepts. Table 2 lists some of the chemical compounds tested and Fig. 4 shows the durability results of various chemical structures on supersmooth magnetic

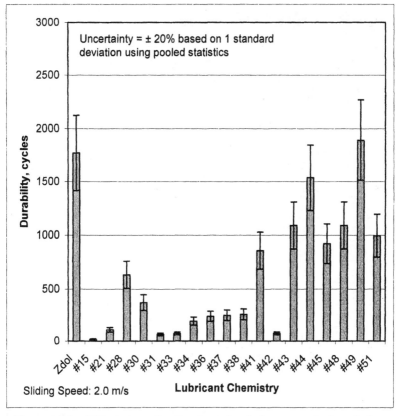

Figure 4. durability test results of various chemical compounds deposited on hard disks using dip coating technique at a nominal film thickness of 1 nm as measured by FTIR/XPS.

Table 2 Chemical compounds used in durability tests

ID	Chemical Compound
#4	1-OH Fluorobenzyl chloro ester
#15	Fluorobenzyl chloro ester
#16	Benzyl chloro ester
#17	1-H Fluorobenzyl chloro ester
#18	Perfluoropolyether (Ausimont Z-tetrol)
#19	Perfluoropolyethjer (Ausimont Z-dol 2000)
#20	Octabutoxy naphthalocyanine
#21	Phthalocyanine
#22	Tetrafluorophenyl porphine
#23	Halocarbon Oil 4.2
#24	Halocarbon Oil 27
#28	Phalothcyanine
#30	Mono-Hydroxy terminated ethylene/butylene polymer
#31	Di-Hydroxy terminated ethylene/butylene polymer
#33	Fluorinated surfactant (3M FC-1265)
#34	Fluorinated surfactant (3M FC-135)
#36	Polystyrene Polymer
#37	1% Functionalized Polystyrene Polymer
#38	10% Functionalized Polystyrene Polymer
#39	Fluoroaliphatic Polymeric Esters
#41	#21 by LB deposition
#42	Co Phthalocyanine
#43	Cu Phthalocyanine
#44	Cu tetra-tert-butyl Phthalocyanine
#45	Di-tert-butyl Dibenzo-18-crown-6 by LB deposition
#48	Fluorinated surfactant (3M FC-740)
#49	Fluorinated surfactant (3M FC-431)
#51	Fluorinated surfactant (3M FC-740)

hard disks (7.5 nm CHx, isotropic roughness of 3–5 angstrom). Some of these molecules have been tested in constant-start-and-stop simulation sequence tests by cooperating industrial partners and the ranking appears to be correct. Generally, the presence of metal ions seems to increase the durability cycles and too short a molecular weight tends to give short durability. Surface mobility, *i.e.*, the ability of the molecule to move back to replenish lubricant depleted from the contact points, is also important. At this time, we do not understand enough to explain the detail mechanism of durability. Availability of such a test set up will provide future opportunity to elucidate the basic working mechanisms.

10 NANOLUBRICATION AS PRACTICED IN HARD DISK TECHNOLOGY

Nanolubrication can be defined as the means to achieve control of adhesion, friction, stiction, and wear at a scale where properties at nanoscale control the outcome. Magnetic hard disks operate in this regime. Since the technology has been successful, it will be instructive to examine some of the concepts proposed in this paper in the context of magnetic hard disk lubrication.

In magnetic hard disks lubrication, the hard disk is protected by a monolayer of perfluoropolyether (PFPE). The commonly used PFPE has two -OH functional groups at the ends of the molecules. The film is deposited by dipcoating and the thickness is controlled at 1.0 ± 0.1 nm. The magnetic head is an air bearing design and flies over the surface at a height of 10–15 nm. Current design calls for the head to act as an air bearing flying over the disk reading and recording magnetic signals. The head lands and starts on the disk each time the computer is shut down. Therefore one of the simulation test methods is so-called constant start/stop (CSS) test which is usually run for 20,000 cycles. Failure of the system can be caused by many factors: stiction, wear, and high frictional resistance. As the gap between the head and the disk continues to shrink (20 to 5 nm) to increase the areal density, constant start and stop arrangement become too severe. Ramp load and unload (the head is parked on a stationary ramp either outside or inside the disk and move across the disk without stopping) technology is being developed. Occasional high speed (10 to 40 m/s) impacts of the flying head with the disk requires even more effective surface films to prevent surface damage.

There are many studies investigating the mechanisms of why such a system works [20-25]. The lubricant has been fractionated using supercritical solvent extraction into 31 fractions of different molecular weights. The range of molecular weight for this PFPE is 800 to 12,000 MW with the average at 2000 to 4000. There is an optimum molecular weight that appears to work much better [26]. Low molecular weight and the very high molecular weight fractions do not have long durability.

At such a molecular weight range, the lubricating film does not exist in solid form as defined by the Langmuir-Blogett convention. Surface mobility of the molecules is an important parameter that governs the lubricant effectiveness [25]. One hundred percent bonded lubricant layers at the same thickness has been shown to perform poorly. This is in line with our principle described earlier. When the molecules are bonded, there was no self-repairability hence limited durability.

The PFPE lubricant was discovered in the 1960s to be useful in the magnetic hard disk application. PFPE was originally developed for fire-resistant hydraulic fluid application in the US Air Force. It is remarkable, in fact, that PFPE appears to be such an ideal lubricant for the magnetic hard disk technology. It has the right vapor pressure (10^{-6} torr), stickiness (resists spin-off at high disk rotational speeds), and the oxidative stability.

The hard disk surface has been textured at the micron/nanometer level to control the contact area hence contact stresses. Bumps microns wide and 50 nm wide have been created in the landing zone to control the contact area between the head and the disk to minimize stiction. It also provides a reservoir for additional lubricant molecules when needed. The surface texture is accomplished by mechanical polishing and energy beam surface reconstruction by ion beams, lasers, and x-rays. These have proven to be very useful in assisting the lubrication of hard disks.

11 CONCLUSIONS

Nanolubrication concepts have been examined systematically and certain principles have been proposed in order to provide effective surface protection. Monolayer films need to be shear resistant, self-repairing, and long lasting in order to protect microsystems and devices. Lessons learned from magnetic hard disk technology provide useful insights to validate some of the proposed concepts. Monomolecular films can be designed to consist of different molecules to impart different degrees of protection on various surfaces commonly used in microsystems and magnetic hard disks. Validity of some of the concepts remains to be proven by careful experiments in the future. The basic principle of nanolubrication, however, is important in order for the field to advance.

REFERENCES

[1] "History of Tribology," Duncan Dowson, Longman Group Limited, London, 1979.
[2] Bowden, E. P., Tabor, D., *The Friction and Lubrication of Solids*, Oxford University Press, New York, 1950.
[3] "Microelectromechanical Systems" NMAB-483, issued by the committee on Advanced Materials and Fabrication, Methods for MEMS, National Academy Press, Washington,

D.C., 1997.
[4] Hsu, S. M., Gates, R. S.,"Boundary lubrication and boundary lubricating films," in *Modern Tribology Handbook*, Bharat Bhushan, ed., CRC Press, New York, 2000, pp. 455-492.
[5] K. Paserba, N. Shukla, A.J. Gellman, J. Gui, B. Marchon, ""Bonding of Ethers and Alcohols to a-CN_x Films"" Langmuir 15(5), (1999), 1709-1715
[6] "Comparative Atomic Force Microscopy Study of the Chain Length Dependence of Frictional properties o fAlkanethiols on Gold and Alkylsilanes on Mica," A. Lio, D. H. Charych, M. Salmeron, J. Phys. Chem. B. 1997, 101, 3800-3805
[7] "Scaling Issues in the Measurement of Monolayer Films," S. M. Hsu, P. M. McGuiggan, J. Zhang, Y. Wang, F. Yin, Y. P. Yeh, and R. S. Gates, in 'Micro- Nano-Tribology' NATO ASI series, edited by Bharat Bhushan, Kluwer Academic Press, to be published in June, 2001.
[8] Iler, R. K., *The chemistry of silica*, John Wiley & Sons, New York, 1979.
[9] "Tribochemical Reaction of Oxygen and Water on Silicon Surfaces," K. Mizuhara and S. M. Hsu, in *'Wear Particles,'* Edited by D. Dowson et al., Elsevier Science Publishers B. V., 323-328, 1992.
[10] "Tribochemical Reaction of Oxygen and Water on Silicon Surfaces," K. Mizuhara and S. M. Hsu, in *'Wear Particles,'* Edited by D. Dowson et al., Elsevier Science Publishers B. V., 323-328, 1992.
[11] "Effect of Selected Chemical Compounds on the Lubrication of Si_3N_4," R. G. Gates and S. M. Hsu, *Tribology Trans.*, **34**, 3, 417 (1991).
[12] "Silicon Nitride Boundary Lubrication: Effect of Phosphorus-containing Organic Compounds," R. S. Gates and S. M. Hsu, *Tribology Trans.*, **39**, 795-802 (1996).
[13] "Silicon Nitride Boundary Lubrication: Effects of Oxygenates," R. S. Gates and S. M. Hsu, *Tribology Trans.*, **38**, 3, 607-617, (1995).
[14] "Silicon Nitride Boundary Lubrication: Lubrication Mechanism of Alcohols," R. S. Gates and S. M. Hsu, *Tribology Trans.*, **38**, 3, 645-653 (1995).
[15] "Effect of Selected Chemical Compounds on the Lubrication of Silicon Carbide," D. Deckman, C. I. Chen, S. M. Hsu, *Tribology Trans.*, **42**, 3, (1999).
[16] "Carbon Overcoat: Structure and bonding of Z-Dol," P. H. Kasai, A. Wass, B. K. Yen, *J. Info. Storage Proc. Syst.*, 1, 245-258, (1999).

[17] "Ultrathin Organic films" Abraham Ulman, Academic Press, Inc., 1991.
[18] "Morphology of Graphite-supported iorn-manganese catalyst particles - Formation of hollow spheres during oxidation," A. A. Chen, M. A Vannice, J. Phillips, *J. Catal*, **116**, (2), 568-585 (1989).
[19] "An Accelerated wear test method to Evaluate Lubricant thin films on Magnetic Hard disks," X. H. Zhang, R. S. Gates, S. Anders, S. M. Hsu, submitted to *Trib. Lett.*.

[20] "Surface Potential of thin perfluoropolyether films on carbon," V. J. Novotny, T. E. Karis, *Appl. Phys. Lett.*, **71**, 1, 52-54 (1997).
[21] "Tribochemistry at the Head/disk Interface," C. S. Bhatia, W. Fong, C. Y. Chen, J. Wei, D. B. Bogy, S. Anders, T. Stammler, J. Storhr, *IEEE Trans. on Magnetics*, **35**, 2, 910-915 (1999).
[22] "Degradation of perfluoropolyethers catalyzed by aluminum oxide," P. H. Kasai, W. T. Tang, P. Wheeler, *Applied Surface Science*, **51**, 201-211 (1991).
[23] "Lubricant Distribution on Hard Disk Surfaces: effect of humidity and terminal group reactivity," Q. Dai, G. Vurens, M. Luna, M. Salmeron, *Langmuir*, **13**, 4401-4406 (1997).
[24] "Impact of polymer structure and confinement on the kinetics of Zdol 4000 bonding to amorphous-hydrogenated carbon," R. J. Waltman, G. W. Tyndall, J. Pacansky, R. J.

Berry, *Tribology Lett.*, **7**, 91-102 (1999).
[25] "Complex terraced spreading of perfluropolyakylether films on carbon surfaces," X. Ma, J. Gui, L. Smoliar, K. Grannen, B. Marchon, C. L. Bauer, M. S. Jhon, *Physical Review E*, **59**, 1, 722-727 (1999).
[26] US Patent number 5562965, Oct. 1996.

Chapter 24

MOLECULARLY ASSEMBLED INTERFACES FOR NANOMACHINES

Vladimir V. Tsukruk
Materials Science and Engineering Department
Iowa State University
Ames, IA 50011

Abstract State-of-the-art development of the field of ultrathin organic and polymeric molecular coatings is briefly reviewed from the prospective of long-term applications for microelectromechanical systems, microfluidic devices, and microanalytical instrumentation. A possible evolution of the field of protective/lubricate coatings towards intelligent (responsive, sensing, self-repairing, and self-reinforcing) molecular interfacial assemblies acting as an adaptive nanointerface between core elements of nanodevices and molecular environment is discussed.

1 INTRODUCTION

The field of organic and polymeric coatings is traditionally focused on the fabrication of relatively thick and robust films for surface protection of large-scale parts and devices. Classical examples are automotive paints and lubrication layers [1,2]. In this kind of applications, surface coatings act as a passive (although, very complicated) protective film that, among many other functions, can inhibit electrochemical reactions, prevent direct physical contact of metal surfaces, distribute mechanical stresses, and dissipate mechanical/thermal/light energy.

Modern developments in the field of microscale devices and their prospective applications ignite new activities in the coating fabrication and study that go far beyond their traditional role [3-5]. Current developments in this field include such topics as the fabrication of sub-micrometer thick lubricant protection of high-density magnetic discs and microelectronic packaging of electronic circuits [3]. New organic and polymeric coatings shield electromagnetic field in microelectronic packaging, selectively reflect light in antireflective coatings, change hydrophobic/hydrophilic surface balance in antifogging layers, and reduce friction/adhesion on micromotor surfaces [3-8]. These recent developments explore new ways of how active, "built-in", molecular functions of molecular materials can be exploited for the functional molecular coating fabrication. For example, selective adsorption or scattering of specific wavelength by molecules/nanoparticles can be used as an active element of antireflective layers [9]. Polar/non-polar balance and molecular flexibility of long chain molecules with reactive terminal groups can be used in the fabrication of robust grafted lubrication layers [8]. Current and future developments in device nanomitiarization and the creation of molecular-based machines and mechanisms will require very different kinds of protective/interactive interfaces.

In this essay, we focus on one particular development, as we see it, namely, in the field of molecularly assembled interfaces for nanotribological applications. These nanointerfacial assemblies can be designed to serve as a sophisticated buffer that controls molecular-scale mechanical and energetic interactions between nanoscale-sized mating and interacting objects.

Current developments that are directly related to the topic discussed here include molecular protective coatings for microelectromechanical systems (MEMS), where micrometer-sized parts, fabricated from solids like silicon, mechanically interact with each other in the course of their sliding, vibration, rotation, and stop-go motion [3-5]. Another example is microfluidic systems (MFS), where microfluid is a subject of constrained flow within microchannels that create significant shearing stresses at solid-liquid interfaces [10]. Finally, we need to mention that we only focus on organic and polymer-based interfacial films, as applied to nanodevices and put aside other possible developments, such as hard coatings or new microfabrication routines, which can be instrumental for some nanoscale applications.

2 CURRENT DEVELOPMENTS AND PROSPECTIVES

Several recent developments, which are directly related to nanotribological applications discussed here, are brought to a life by the severe problems related to operational reliability of MEMS devices such as micromotors and comb-like drivers [3-5]. These microdevices, fabricated from traditional electronic material, silicon, display high stiction that prevents free, non-destructive motion of microparts. In addition, their short operational lifetime is caused by high surface energy and brittleness of silicon oxide surface layer with well-developed nanograin surface texture [6, 7] (Fig. 1).

High surface energy of such surfaces is due to the high concentration of silanol (Si-OH) groups on a silicon oxide surface that results in extremely high hydrophilicity (low contact angle, close to 0° for clean surfaces). This, in turn, leads to complete surface wetting under normal air conditions and very strong capillary forces being developed between micrometer-sized parts. On the other hand, high brittleness of silicon and silicon oxides contributes to the fast deterioration of micromachined silicon surfaces being the subject of high local shear stresses. The microcracks propagate via intergrain grooves with high local friction, as can be seen from lateral/friction force microscopy images (Fig. 1). Several ways to improve microtribological properties via the chemical modification of such devices were proposed and tested to date.

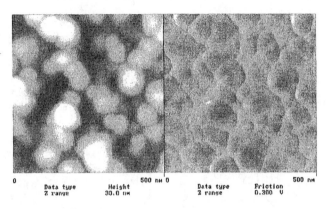

Figure 1. Scanning probe microcopy images of the micromachined silicon surface of a microelectromechanical device that demonstrate nanograiny surface topography on topographical images (left) and uneven distribution of friction forces on lateral force image (right).

First approach includes the chemical treatment of the silicon oxide surface with hydrofluoric acid, which increases hydrophobicity of the treated surface by reducing the number of silanol surface groups and reducing oxide layer thickness [11]. After this treatment, the contact angle increases to 60-70°, this makes the surfaces partially wet and significantly reduces capillary interaction. Another approach explores self-assembling monolayers (SAMs) to modify surface properties [6,7,12]. Alkylsilane and thiol-based organic molecules with non-polar terminal groups can be used to fabricate SAMs on silicon oxide and gold surfaces respectively, and make these surfaces completely hydrophobic. Various versions of composite molecular layers with interesting nanotribological properties can be fabricated by polymer grafting to functionalized SAMs (see several examples in Fig. 2) [12].

Dense monomolecular layer chemically attached to the appropriate solid surface with functional terminal groups can serve as reactive interface for such build-ups (Figure 2).

As the result of this fabrication, surface properties can be controlled in a wide range from completely hydrophobic (contact angle of 100–110°) with low capillary forces to partially/completely hydrophilic with high chemical reactivity [8, 12]. Interfacial shear strength of such modified surfaces decreases significantly due to the presence of thin molecular layer with low shear strength. As the result, the greatly reduced friction coefficient is usually observed. This approach, in fact, resembles classical boundary lubrication model, with the only difference being the firm chemical attachment of the organic monolayer to solid surface [2].

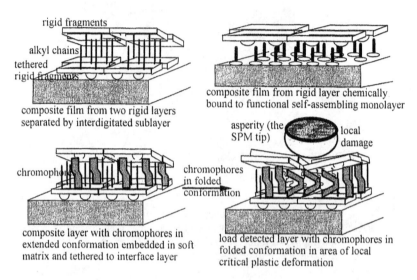

Figure 2. Several examples of nanocomposite molecular layers for surface modification.

Finally, the chemical attachment (grafting) of the polymer layers onto the functionalized silicon surface is explored in order to fabricate wear-resistant and superelastic molecular coatings [13]. The latest development displays that the tethered nanocomposite molecular coating can improve surface stability and enhance the shearing surface response (Fig. 3).

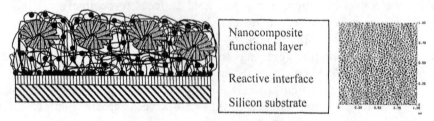

Figure 3. Nanocomposite/reinforced elastomeric layer chemically grafted to the silicon substrate via reactive interfacial SAM and bearing surface functional groups for further modification [13]: sketch on molecular structure (left) and SPM image of 2D network of reinforcing nanodomains (<10-nm across).

Controlled binding to the surface with the presence of rigid polymer fragments or grafted elastic phase reinforced by the glassy nanodomains can significantly enhance wear resistance of such nanocomposite coatings in comparison with one-component low-molecular mass organic SAM layer [13].

These approaches have been introduced very recently and initially showed promising results, enhancing wear resistance and reducing surface friction [6-8, 12, 13]. However, further development is required to improve molecular coating performance for micro- and especially nano-scale devices and explore sophisticated molecular designs. Here, we discuss the future needs and prospective developments that are still untapped today but will be critical, in our opinion, for the prospective implementation of operationable and reliable nanomachines.

3 TRENDS

Here, we will speculate on the question of what kind of surface coatings and why should be developed for future nanomachines. We will overview current understanding of some critical issues in the nanotechnology development and what role will be devoted to the molecular coatings. We will briefly discuss fundamental understanding and possible impact of these developments on nanotechnology. We need to mention that the current discussion is too brief to be comprehensive and is based on the author's own experience in the field of organized molecular coatings as well as on a critical review of literature data available.

3.1 Fundamental Problems and Approaches

We are not going to discuss about what kind of nanomachinery will be developed within next 10 years, or predict a new revolution in this field. Instead, for further discussion, we define two possible types of these devices, which can be created by using two very different approaches. The first type of nanoscale machinery can be developed by a gradual downsizing of current lithographical microfabrication technique [3]. The current trends and possible development of new light sources (like X-rays) will, probably, result in routine nanofabrication of nanodevices with spatial dimensions less than 20 nm. This is much smaller than current 100–1000 nm scale and will represent a critical step to a nanoscale operation. On this scale, the contact area of interaction between mating parts as well as sizes of surface asperities will be in the range from 1 to 5 nm. This truly atomic scale includes only several dozen interacting atoms and molecular groups. This number marks the limit of the applicability of the continuum mechanics, as we know, and the transition to molecular mechanics laws [5].

Obviously, the basic principles of the design of molecular coatings for such nanoscale parts should be reconsidered. Instead of relaying on laterally homogeneous organic layers with the thickness on a sub-microscopic scale, one should design molecular coatings with cluster/nanodomain lateral dimensions of several tens of molecules and the layer thickness not exceeding several nanometers. These molecular assemblies should still possess interfacial properties required for stable interfacial interactions and keep their integrity under substantial shear and normal stresses.

On the other hand, for nano-sized devices, the local radius of curvature of various areas of solid surfaces becomes so small that a planar molecular packing of symmetrical molecules in organic layer is not going to satisfy the conditions for dense and organized molecular surface ordering (Figure 4). Indeed, for molecular coatings of 2–5 nm thick on a surface feature with the radius of curvature of 5–10 nm, the mismatch between the surface area per molecule on solid substrate and layer surface can easily reach 100% (Fig. 4).

Obviously, none of the current symmetrical molecular designs of protective/lubricant layers can satisfy these severe constrains. Other molecular structures should be thought of to match these extreme curved surfaces. One of the possible candidates is the recently introduced class of tree-like, dendrimer molecules, which can grow from one center and follow local surface curvature on a nanometer scale (Fig. 5) [14]. These molecules and macromolecules can be considered be promising candidates for supramolecular assemblies of complex shape and high mobility.

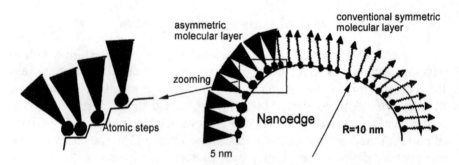

Figure 4. An example of the "nanoedge" (right) coated with conventional symmetric molecular layer (right side) and asymmetric molecular layer (left side) and "zooming" out crowded molecular assembly attached to "curved" surface wit atomic steps (left).

Another aspect of the molecular coating design is brought in by atomistic, stepped structures of nano-beams/edges/channels at this spatial scale with inhomogeneous chemical/atomic structure across the atomic planes (Fig. 4). No longer can solid asperities and parts be considered as having smooth homogeneous surface. This feature should play a crucial role in the mechanism of the molecular coating formation and greatly enhance their surface stability. The molecular behavior of such "stepped" assemblies is absolutely unknown. In this aspect, the direct molecular dynamics simulation of the molecular interactions on a spatial scale comparable to actual atomistic situation can bring invaluable insight [16].

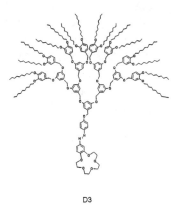

Figure 5. The chemical scheme shows an example of a monodendron molecule (generation 3) that bears major features of asymmetrical molecular design required for self-organization of ordered and dense molecular assemblies on highly curved surfaces of nanodevices such as presented in Figure 4. (See Ref. [15] for dendrimer synthesis).

Different stress distribution arises for molecular coatings tethered to nano-sized surfaces due to the type of the deformational contact of molecular assemblies during interfacial interactions (Fig. 4). Unlike current molecular coatings, where the sliding behavior is determined by the shearing deformation of predominantly vertically oriented molecules, for highly curved coatings, splay deformational mode should play a dominant role in their nanomechanical behavior. Thus, the whole pattern of interfacial interactions of nanoscale surfaces will be changed and new, unknown, interfacial nanomechanical behavior can emerge.

Second scenario of nanomachine fabrication relies on supramolecular synthesis of complex functional molecules capable of specific functions [17]. Supramolecular interactions allow them to be organized or self-organized in the ordered array of molecular mechanisms capable of doing a specific function [8]. First examples of building such molecular mechanisms with scanning probe microscopy (SPM) assistance are demonstrated [18]. The role of supramolecular chemistry in the design of complex, stable, and functional molecules is crucial. However, even for these molecular mechanisms, some sort of interface with environment (such as atomistic lattice of supporting substrate) should be thought of. The ways to deal with such problems as the configurational/conformational behavior of molecular mechanisms constrained to specific atomistic interfaces should be explored. However, atomistic incompatibility of organic molecular mechanisms and inorganic solid substrates is obvious.

The ultimate solution for this dilemma could be complete abandoning solid-state platform for nanomachinery. In this scenario, "play field" should be transferred to mobile "soft matter" environment analogous to human cell structures. In the framework of this approach, molecular mechanisms with specific functions should be kept separately within/on, *e.g.*, artificial membrane-like layers, which provide support and energy/signal exchange with environment. In this development, molecular coatings can be transformed into complicated multifunctional interfaces analogous to molecular membranes of cell structures. Apparently, the biomimetic approach will play a dominant role in the design and development of these systems. The fundamental question of interactions of synthetic molecular mechanisms with the environment via adaptive molecular interfaces will be a major task to be addressed.

3.2 Technological Developments: Fabrication, Measurement and Manipulation

From the technological point of view, the major question in the field of the development of molecular assembled coatings for nanomachinery is how these molecular coatings will be fabricated and how their interfacial behavior will be controlled. Current technologies for molecular coating fabrication and studies include Langmuir-Blodgett monolayer deposition and dip- and spin-coatings [5]. Although very versatile for model studies, these techniques are not useful for future nanoscale applications. The most probable candidates for the nanotechnological development is molecular organization through self-assembly that is known in different versions such as chemical self-assembly for SAM fabrication or electrostatic self-assembly for layer-by-layer deposition [19, 20].

We can speculate that specific conditions for the local adsorption of the specially designed molecules would require for further refining of these technologies. Exploiting additional weak interactions to organize molecular assemblies or external field assistance in aligning molecular ordering can be explored. Very complicated supramolecular structures can, in fact, be build. Two examples include 3D ordered nanodomain networks from associated dendrimers [21] or tubular superstructures from lipid complexes with helical surface patterns, spontaneously formed from self-assembled multilayer polyelectrolyte-nanoparticules complexes (Fig. 6) [22].

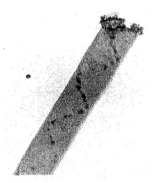

Figure 6. Transmission electron microscopic image of the lipid tubule (600 nm in diameter) covered with electrostatically assembled polymer-nanoparticles multilayers and silicon oxide nanoparticles in a patterned manner (see Yu. Lvov et al. [22] for detail discussion).

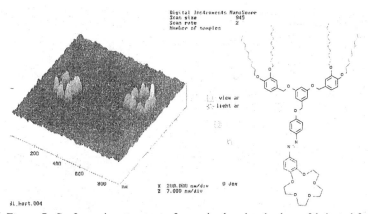

Figure 7. Surface microstructure of organized molecular layer fabricated from dendrimer photosensitive molecules (right) with embedded transmembrane proteins [23] (Tsukruk et al., unpublished).

Another example of an asymmetric molecular layer deposited on silicon single crystal and composed from photosensitive dendrimer molecules with embedded transmembrane proteins is presented in Fig. 7.

Such complex molecular assemblies as presented in the figure above, may serve several purposes simultaneously: reduce adhesion and friction, bring hydrophobicity to the silicon surface, provide selective molecular permeability through protein pore, and allow controlling mechanisms for the variation of surface properties and layer permeability through photoisomerization reaction of photosensitive molecules. In addition, asymmetrical shape of dendrimer molecules should assist in adaptation of this assembly to the highly curved nanostructures (Figs. 4, 5). These and other initial tests demonstrate that by exploiting the combination of self-assembling principles and sophisticated organic chemistry, one can nanofabricate very complicated superstructures and not just planar uniform layers or microscale-patterned planar surfaces. In the end, the nature relies only on multiple/competitive weak interactions to build, keep functioning, and repair very complex and mobile nano- and microscopic structures of very high complexity.

An additional challenge in monitoring molecular assembly behavior and measuring their properties down on a nanometer scale should be addressed for a successful implementation of this strategy. The natural candidate for such measuring and controlling tools is SPM technique introduced at the end of 80^s [24]. A combination of the latest SPM modes with other tools such as optical tweezers and near-field microscopy can constitute a very effective nanomanipulation set. However, the level of sophistication should be brought to a much higher level in comparison with today's development. Much higher force sensitivity, multi-array probing capability, and truly molecular resolution should be achieved. Recent developments show the use of functionalized nanotubes as atomistic probes and demonstrate perspectives of this technique (Figure 8) [25].

Finally, the effective mechanism of self-repair in these molecular assemblies must be incorporated. For interacting areas of several tens of molecules across, even one molecular defect can completely disturb normal functioning. The mechanism of finding and replacing such molecular defects should be thought of. Wide operational limits of molecularly

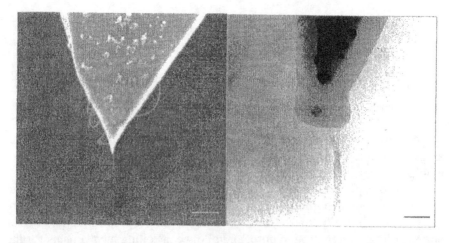

Figure 8. Nanotubes grown on the end of the SPM tip as a truly nanoscale tool for molecular and atomic manipulations. Left: SEM image, a scale bar is 500 nm; right: TEM image, a scale bar is 20 nm. J. H. Hafner et al. [25]. Reproduced from Ref. 25 with permission.

assembled interfaces should be assured for stable and reliable functioning under fluctuating environmental conditions. Various bioinspired schemes of molecular replication and multiple duplications can be explored.

3.3 Applications and Impact

Here, we will speculate on the impact that resulted from the development of molecular assembled coatings for nanotribological applications. We should emphasize that does not matter how great of a progress in nanomachinery fabrication will be achieved; this nanotechnology cannot work in vacuum (except, probably, some space-related projects and computer simulations). In real-life situation, application of these technologies must be accompanied by the development of sophisticated nanoscale interfaces to accommodate massive arrays of nanodevices, protect them from environment, and provide adequate exchange of energy between these devices and environment. Therefore, we see several major activities in the field of molecularly assembled coatings for task-oriented developments.

First, new molecular protective coatings will be developed for nanodevices and molecular mechanicsms, which will integrate great sophistication and versatility of naturally occurred supramolecular assemblies. Such molecular coatings will have highly mobile and adoptive molecular organization that can change their molecular ordering at a complex nanointerface in response on external stimuli and adopt their properties to provide the highest survivability of the nanodevice protected. Such protective molecular assemblies will provide required structural stiffness and integrity in case of "soft matter"-based molecular mechanisms and compensate for constrained atomistic configurations of nanodevices. These assemblies will serve as an adaptive buffer between internal molecular motions and mobile environment providing reduced interfacial stresses and appropriate conformational mobility.

Second, in addition to their mostly protective functions, new generation of molecularly assembled coatings will serve as an active interface between internal organization of nanodevices and external stimuli/disturbance. These interfaces should be able to support control delivery of energy in/out nanodevices in the form of excessive heat dissipation and thermal equilibration with environment. Controlled permeability is required for functioning of chemical "nanofactory" and selective removal of "molecular waste." Timely transfer of external signal to nanomachinery and molecular feedback can be provided by this interface.

Third, the long-term development of such molecular interfacial assemblies should result in, and cannot be successful without, much deeper understanding of fundamental principles of local molecular/supramolecular organization at nanostructured surfaces/interfaces, role of weak multidirectional interactions in self-organization of nanoscale assemblies, and their adaptive/responsive behavior. New molecular and atomistic designs of organic molecules and macromolecules should be searched to make them compatible with nanomachinery. As we mentioned before, highly asymmetric molecular structures with carefully tailored intra/inter molecular/interfacial interactions should be designed for assembling at nanointerfaces.

Finally, we think that on the path towards such nanoassembled interfaces, we are going to get closer to self-consistent nanodevices resembling the efficient cell-like structures. And, inevitably, we should learn, adopt, and implement basic principles of self-organization of most sophisticated nanomachines, namely, a human cell. Only integration of these principles with the logistic of deterministic artificial nanodevices may lead to successful exploration of the nanoworld. And, in this world, the molecularly assembled interfaces will be considered as an integral part of nanodevices and not a post-added function.

ACKNOWLEDGMENT

The author's work in the field of nanotribology and surface molecular engineering is supported by The National Science Foundation, CMS9996445 Grant, AFOSR F49620-98-1-0480 Contract, NATO OUTR.LG 973008, and ACS-PRF 33867-AC7. The author thanks members of his research group, who performed studies mentioned in this review and prepared figures: M. Melbs (MEMS, Figure 1), Dr. I. Luzinov and D. Julthingpiput (grafted layers, Figure 3), Dr. A. Sidorenko and C. Houphouet-Boigny (dendrimer layers, Figures 5 and 7).

We are grateful for collaboration with many colleagues who provided us with materials and microdevices to work with. MEMS (Figure 1) were provided by Dr. M. Dugger (Sandia National Lab), dendrimer compounds (Figures 5 and 7) were provided by Prof. D. McGrath (University of Arizona), and transmembrane proteins were cordially supplied by Prof. H. Bayley (Texas A&M University). Figure 6 was courtesy provided by Prof. Yu. Lvov (Louisiana Technical University) and Figure 8 from Ref. 25 was courtesy provided by C. Lieber and J. H. Hafner (Harvard University).

REFERENCES

1. L. Lin, G. S. Blackman, R. R. Matheson, in: *Microstructure and Microtribology of Polymer Surfaces*, Eds. V. V. Tsukruk, K. Wahl, ACS Symposium Series, v. 741, 2000, p. 428.

2. *Fundamentals of Friction,* Eds. E. Singer, H. Pollack, Kluwer Acad. Press, 1992.
3. Muller, R. S., in: *Micro/Nanotribology and Its Applications,* B. Bhushan, Ed., Kluwer Press, 1997, p. 579. *Tribology Issues and Opportunities in MEMS,* Ed. B. Bhushan, Kluwer Academic Press, 1998.
4. M. T. Dugger, D. C. Senft, G. C. Nelson, in: *Microstructure and Microtribology of Polymer Surfaces,* Eds. V. V. Tsukruk, K. Wahl, ACS Symposium Series, v. 741, 2000, p. 428.
5. *Micro/Nanotribology and Its Applications,* B. Bhushan, Ed., Kluwer Press, 1997.
6. K. Komvopoulos, *Wear, 200,* 305, 1996
7. V. V. Tsukruk, T. Nguyen, M. Lemieux, J. Hazel, W. H. Weber, V. V. Shevchenko, N. Klimenko, E. Sheludko, in: *Tribology Issues and Opportunities in MEMS,* Ed. B. Bhushan, Kluwer Academic Press, 1998, p. 608.
8. V. V. Tsukruk, *Progress in Polymer Science, 22,* 247, 1997
9. *Organic Thin Films,* Ed. C. W. Frank, ACS Symposium Series, v. 695, 1998.
10. G. Blankenstein, U. D. Larsen, J. Branebjerg, *SPIE Proceedings, 2998,* 2982, 1997.
11. M. R. Houston, R. T. Howe, R. Maboudian, *J. Appl. Phys., 81,* 3474, 1997.
12. V. N. Bliznyuk, M. P. Everson, V. V. Tsukruk, *J. Tribology, 120,* 489, 1998; V. V. Tsukruk, V. N. Bliznyuk, J. Hazel, D. Visser, M. P. Everson *Langmuir, 12,* 4840, 1996.
13. I. Luzinov, D. Julthongpiput, H. Malz, J. Pionteck, V. V. Tsukruk, *Macromolecules, 33,* 1043, 2000; I. Luzinov, D. Julthongpiput, V. Gorbunov, V. V. Tsukruk, *Tribology Intern.,* submitted
14. D. A. Tomalia, *Adv. Materials, 6,* 529, 1994. J. M. Frechet, *Science, 263,* 1711, 1994. J. F. Jansen, E. M. de Brabander-van den Berg, E. W. Meijer, *Science, 266,* 1226, 1994. V. V. Tsukruk, *Advanced Matl., 10,* 253, 1998
15. M. Hashemzadeh, D. V. McGrath, *Polym. Prepr. 39(2)* 338, 1998.
16. Drexler, K.E., *Nanosystems: molecular machinery, manufacturing, and computation,* Wiley&Sons, 1992.
17. T. Cagin, J. Che, Y. Qi, Y. Zhou, E. Demiralp, G. Gao, W. A. Goddard III, *Journal of Nanoparticle Research, 1,* 51, 1999; T. Cagin, A. Jaramillo-Botero, G. Gao, W. A. Goddard, III, *Nanotechnology, 9 (3),* 143, 1998; J. Gao, W. D. Luedtke, U. Landman, *J. Chem. Phys., 106,* 4309, 1997.
18. D.M. Eigler, E.K. Schweizer, *Nature, 344,* 524, 1990; M. T. Cuberes, J. K. Gimzewski, R. R. Schlittler *Applied Physics Letters, 69,* 3016, 1996.
19. C. D. Bain, G. M. Whitesides, *J. Am. Chem. Soc., 111,* 7164, 1989. Y. Xia, G. M. Whitesides, *Angew. Chem., 37,* 550, 1998.
20. G. Decher, Yu. Lvov, J. Schmitt, *Thin Solid Films, 244,* 772, 1994. G. Decher, *Science, 277,* 1232, 1997.
21. V. Percec, C.-H. Ahn, T. U. Bera, G. Ungar, D. J. Yeardley, *Chem. Eur. J., 5,* 1070, 1999.
22. Yu. Lvov, R. Price, A. Singh, J. Selinger, M. Spector, J. Schnur, *Langmuir,* 2000, in press
23. L. Song, M. R. Hobaugh, C. Shustak, S. Cheley, H. Hayley, J. E. Gouaux, *Science, 274,* 1859, 1996.
24. G. Binnig, C.F. Quate, Ch. Gerber, *Phys. Rev. Lett. 12,* 930, 1986.
25. J. H. Hafner, C. Li Cheung, C. M. Lieber, *J. Amer. Chem. Soc., 121,* 9750, 1999.

Chapter 25

STM-QCM STUDIES OF VAPOR PHASE LUBRICANTS

B. Borovsky, M. Abdelmaksoud and J. Krim
Physics Department
North Carolina State University
Raleigh, NC 27695

Abstract Vapor phase lubricants have been studied for well over 40 years, generally within the context of macroscopic system performance. Nonetheless, they may well prove to be of critical importance to the tribological performance of MEMS devices, since the vapor phase may ultimately prove to be an effective, and perhaps exclusive, means to deliver and/or replenish lubricants. With the intent of developing a realistic laboratory test set-up for actual MEMS contacts, we have combined a Scanning Tunneling Microscope (STM) with a Quartz Crystal Microbalance (QCM). The STM-QCM allows unique and detailed investigations of the simple nanomechanical system formed by a contacting tip and surface. Both STM images of the contact and the response of the QCM are monitored throughout the course of the measurements, which are performed in realistic sliding conditions of over 1 m/s. We report here on both (vapor phase) lubricated and non-lubricated contact.

1 INTRODUCTION

Studies of the atomic-scale origins of friction and adhesion have undergone rapid progress in recent years with the development of new experimental and computational techniques for measuring and simulating tribological phenomena at atomic length and time scales.[1] Employing established technologies, such as ultra-high vacuum, for the preparation of crystalline samples, nanotribologists have been gathering information in situations where the nature of the contacting surfaces is determined in advance of the measurement. They have collectively measured friction forces per unit true contact area which span twelve orders of magnitude,[1] with no wear or damage occurring at the sliding interface. Faster computers have in turn allowed large scale molecular dynamics (MD) simulations of condensed systems to be performed for physically significant time periods, enabling numerical results to be increasingly comparable to experiment.[2]

If the precise nature of the contacting asperities between macroscopic objects in sliding contact could be determined (such studies do in fact represent an area of high research activity within the tribological community), then the results of nanotribological studies could begin to be directly implemented into mainstream tribological considerations. Meanwhile, the results are most applicable to friction at the interface between liquid and solid materials,[3,4] where the complicating factors associated with asperity contacts are less of an issue, and to the MEMS community,[5] where machine components with astoundingly small dimensions are rapidly approaching the length scales routinely probed by the nanotribological community. Indeed, solid surface nanocontacts abound among MEMS devices, and a myriad of device complications and failures are associated with their friction, adhesive and wear characteristics.[6] Because each of these contact areas is small, perhaps a few tens of atoms in extent,[7] both the topology and the mechanics of the contacting asperities must be investigated at the atomic scale in order to optimize device performance.

Current issues of importance to this area include: (1) Development of realistic laboratory test set-ups which are both well-controlled and relevant to operating machinery, (2) Understanding the chemical and tribochemical reactions which occur in a sliding contact, and the energy dissipation mechanisms associated with such a contact, (3) Merging and coordinating information gained on the atomic-scale with that observed at the macroscopic scale, (4) Characterization of the microstructural and mechanical properties of the contact regions between the sliding materials, and (5) Development of realistic interaction potentials for computer simulations of materials of interest to tribological applications.

We focus here on points (1), (2), and (3), and describe herein our efforts to study the adhesion, friction and tribochemistry of atomic-scale contacts in a controlled environment by means of a Scanning Tunneling Microscope (STM) which is operated in conjunction with a Quartz Crystal Microbalance (QCM). A schematic of the experimental set-up is shown in Fig. 1. The STM allows a single asperity contact to be formed, and allows the buried contact to be imaged in both stationary and sliding conditions. The microbalance, whose surface is oscillating in transverse shear motion at speeds near 1 m/s, can be employed to measure the uptake rate and frictional properties of adsorbed lubricant species. The STM-QCM combination [8] allows access to a range of contact pressures and sliding speeds which are comparable to those encountered by actual MEMS devices.[9] Traditional instruments such as the Atomic Force Microscope (AFM) and the Surface Forces Apparatus (SFA) fail to access either the required sliding speeds or contact pressures, respectively. Furthermore, the use of STM avoids the problem of jump-to-contact associated with AFM, allowing the applied normal force to be varied continuously in a controlled fashion.

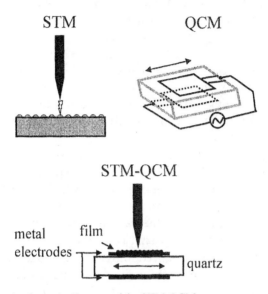

Figure 1. A schematic diagram of the STM-QCM apparatus.

In what follows, we first describe how a QCM acting alone can provide important information about lubricant layers adsorbed from the vapor phase. Next, we discuss our results obtained with the combined STM-QCM in three situations: (1) measurements of the amplitude of a vibrating QCM electrode with the STM tip in tunneling, but not physical, contact with the QCM, (2) unlubricated metal-metal contact of the STM tip with the QCM electrode, and (3) lubricated contact using a vapor phase lubricant which is known to reduce wear in macroscopic applications. In the case of unlubricated contact, we observe significant wear and evidence of increasing sliding friction with normal load. Application of molecularly thin quantities of lubricant to the same contact dramatically alters both the STM and QCM responses in a manner which is highly suggestive of the lubricant's known friction and wear reducing properties at the macroscopic scale. Our measurements also reveal a potential tribochemical reaction which is highly localized at the point of contact and is associated with the realistic rubbing conditions provided by the STM-QCM apparatus.

2 QCM STUDIES OF VAPOR PHASE LUBRICANTS

The concept of lubricating high temperature bearing surfaces with organic vapors has existed for at least forty years, with substantial efforts beginning in the 1980's and continuing on to the present day.[10] Vapor phase lubrication occurs via three distinct forms: (a) organic films which are intentionally reacted with a surface to form a solid lubricating film, (b) vapors which condense to form a lubricating liquid film on the surface of interest, and (c) light weight hydrocarbon vapors deposited onto hot catalytic nickel surfaces. Vapor phase lubricants are advantageous for use at high temperature, (meaning that either the ambient temperature of the entire system is elevated, or the local surface ``flash point'' temperature due to frictional heating is elevated) and in situations where the vapor can be used as a reservoir for replenishment of areas where the lubricant has been depleted in the course of the bearing lifetime.

Several organic vapors have been identified which exhibit desirable tribological properties,[11] but a detailed and fundamental understanding of their surface chemical reactions in the tribological processes of interest is far from complete. In particular, an understanding of the specific surface reactions which occur, and how they are affected by tribological conditions such as temperature and/ or lubricant concentration level in the carrier gas remains inadequate. Modern nanotribological techniques can be brought to bear on these issues by examining in detail the properties of a known (macroscopic-scale) vapor phase lubricant. The knowledge gained (if not the lubricant itself) is extremely likely to be applicable to NEMS/MEMS operations as well. Indeed, the vapor phase may ultimately prove to be the most effective, if not only, means to deliver and/or replenish a lubricant in the case of a submicron scale device.[12]

The vapor phase lubricant chosen for our study consists of a blend of tertiary-butyl phenyl phosphate (TBPP) molecules,[13] whose atomic constituents are carbon, hydrogen, oxygen and phosphorus. It demonstrates high quality performance at elevated temperatures [10] and exhibits oxidation inhibiting characteristics as well as a number of other desirable tribological properties, such as the reduction of wear. While the precise mechanisms for its beneficial properties are uncertain, it is believed that after reacting with the surface, the phosphate contained in the original lubricant molecule acts as a binder for graphitic carbon, which in turn may be the actual lubricant.

The QCM is particularly well-adapted for measurements of uptake rates of vapor phase lubricants.[14] It has been used for decades for microweighing purposes, and was adapted for friction measurements in 1986-88 by Widom and Krim.[15] Specifically, a QCM consists of a single crystal of quartz that oscillates in transverse shear motion with a quality factor near 10^5. The driving force (supplied by a Pierce oscillator circuit) has constant magnitude and is periodic with frequency $f = 4-10$ MHz, the series resonant frequency of the oscillator. Two metal electrodes, which serve as the substrates upon which adsorption occurs, are deposited in vacuum conditions onto each major face of the crystal.

Film adsorption onto the microbalance produces shifts in both the frequency and amplitude of vibration, which are simultaneously recorded as a function of pressure. Amplitude shifts are due to frictional shear forces exerted on the surface electrode by the adsorbed film (or alternatively by a three dimensional vapor or fluid phase). Our microbalance in its present arrangement can detect shear forces in excess of 2.5×10^{-7} N.[15]

Figure 2 depicts a typical QCM response for the uptake of TBPP lubricant on an iron surface at room temperature. The uptake is relatively slow, corresponding to a total of 4 monolayers in approximately one hour's exposure to 234 mTorr of the vapor. At the pressures we have studied, the rate is independent of TBPP gas pressure. Moreover, the uptake rate returns to its $t = 0$ value when the sample is left for a day with no exposure to further TBPP, even though the TBPP is not observed to desorb from the surface. These observations are consistent with slow diffusion of the TBPP or fragments thereof into the iron substrate, consistent with previous studies.[10] The fact that amplitude shifts are in fact observed indicate that the film is either slipping on account of the oscillatory motion of the microbalance, or else exhibiting an internal molecular motion within the TBPP molecules themselves. (A solid film adsorbed on a surface with no slippage would produce no shift in the amplitude of vibration of the microbalance).

The frequency shifts observed during the uptake of TBPP in Fig. 2 are negative, simply due to the added mass of the film to the oscillating system. In Section 3, we discuss the need for more sophisticated modeling of the STM-QCM geometry to account for the frequency shifts observed with this apparatus, which may be either positive or negative.

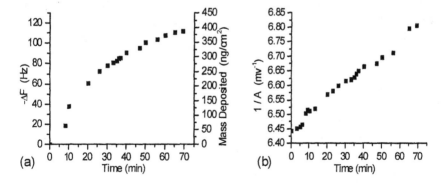

Figure 2. Uptake of TBPP on iron: The frequency drop indicates mass adsorption. Shifts in amplitude are related to adsorbate motion, either slippage or internal molecular motion of the adsorbed species.

3 STM IMAGING OF A MOVING QCM ELECTRODE

A remarkable consequence of the STM-QCM combination is that the amplitude of vibration at the surface of the QCM may be directly measured using the STM images.[16] Such measurements are an important prelude to our studies of friction with this apparatus, allowing accurate determination of the maximum sliding speed of the tip-surface interface. We note that the QCM frequency response to applied forces plays no role in this measurement. In fact, we observe no frequency shift upon engaging the tip into tunneling contact, to within the resolution of ±0.1 Hz out of 5 MHz. This indicates that there is no observable normal load on the crystal when the tip is in tunneling contact, provided that both the tip and electrode surfaces are bare metals in high vacuum ($\sim 10^{-8}$ Torr in our experiments). As will be discussed in Section 5, the presence of even a thin adsorbed layer on the tip and electrode surfaces results in a QCM frequency shift during tunneling contact, due to the normal load required to squeeze the two metal surfaces sufficiently close together for tunneling.

The ability to image a vibrating surface with an STM, while unexpected, may be attributed to the three widely separated time scales involved, and the exponential dependence of the tunneling current on tip-surface separation. The characteristic frequencies of the scanner (Hz), the feedback loop (kHz), and the QCM (MHz) are each separated by three orders of magnitude. Conventional STM operation relies on the feedback loop being much faster than the scanner (in constant-current mode). For STM-QCM, the fact that the vibrations of the QCM are in turn much faster than the feedback loop causes the tip to be held at a separation from the surface which on average (over many surface oscillation cycles) gives the desired tunneling current (typically 1 nA). Qualitatively, the closest approach of the surface to the tip during each cycle is weighted most heavily in this average, due to the exponential dependence of the tunneling current on separation, so the tip is held at an altitude sufficient to avoid direct contact of the tip and surface. This allows the STM tip to image the vibrating surface without crashing into it, as is evident from a lack of damage to the surface.

The amplitude measurements proceed as follows: We employ an AT-cut quartz crystal for our QCM, which oscillates in transverse shear mode whereby displacements are in the plane of the surface. The STM tip is held in tunneling contact (in constant current mode) with the QCM electrode (see Fig. 1). Simultaneous operation of STM and QCM is accomplished using the electrical circuit detailed in Ref. [16]. The amplitude of vibration is measured by observing the lateral smearing of features along the shear direction as the drive voltage of the QCM is increased.

Figure 3 displays a pair of STM images of the QCM surface. The surface is that of a copper electrode deposited *in situ* in high vacuum. The same 500 × 500 nm region is shown while the QCM is stationary (a) and vibrating (b). In image (b), the crystal is shown oscillating at its fundamental frequency of 5 MHz, with a quality factor of 54,000, and a peak drive voltage of 0.93 V applied across its electrodes. The horizontal elongation of features apparent in this image depicts the lateral extent of the surface vibrations. By comparison with image (a), the amplitude was determined to be 75 nm. A full exploration of the dependence of vibrational amplitude on drive voltage and quality factor is presented in Ref. [16]. In the present case, the maximum speed of the surface is $v = A\omega = 2.3$ m/s, a direct verification that the STM-QCM apparatus achieves realistic interfacial sliding speeds.

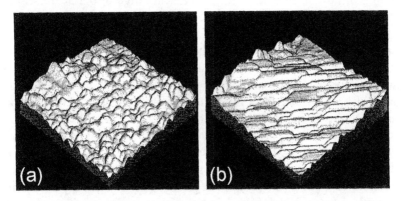

Figure 3. A pair of 500 × 500 nm STM images of the surface of a QCM. The full vertical scale is 40 nm. The same region on the QCM is shown both stationary (a) and oscillating (b). The apparent horizontal elongation of features in (b) indicates the lateral extent of surface vibrations. By comparison with (a), the amplitude of vibration was measured to be 75 nm. This results in a maximum sliding speed of 2.3 m/s at the interface of tip and surface.

4 UNLUBRICATED CONTACT

In the previous section, tunneling contact was established between the STM tip and QCM electrode, whose bare metal surfaces were maintained in high vacuum. There was no QCM response or any significant wear associated with this type of contact. In this section, we establish unlubricated physical contact between the tip and surface, and describe the resulting wear (shown in STM images) and associated QCM response. In Section 4, we compare these results to the case of lubricated contact between the same tip and surface, using TBPP as the lubricant species. In this way, we study the action of a (macroscopic) lubricant on a known interface using the combined capabilities of the STM-QCM.

For our system, the hallmark of even light physical contact between tip and surface is the presence of deformation and wear (a situation familiar to many STM operators, and often referred to as crashing the tip!). Figure 4 shows a case of carefully controlled wear: the scratching or machining of a platinum surface (again deposited in situ in high vacuum) using a tungsten tip. The region shown is 1700 × 1500 nm. The central mound in image (a) is roughly 50 nm high. Image (b) shows the result of making two horizontal scratches in the surface by scanning the tip across the mound at a velocity too fast for the feedback loop to move the tip vertically enough to prevent it from crashing into the mound (5 µm/s). We estimate that the tip was pressed 20 to 40 nm into the mound. The full vertical range of our scanner piezo is 800 nm, so only a small fraction of the available normal force was employed to produce these scratches. Similarly, various holes and depressions may be produced by holding the tip fixed, or by scanning, while pressing the tip into the surface. We observe that machining of this surface with the tungsten tip is always accompanied by a *negative* frequency shift of the QCM as the tip is forced into the surface. The frequency shift is increasingly negative as the normal load is increased. In the next section, we observe that after lubricating this contact, the frequency shift is increasingly *positive* as the normal load is increased, and we present a plot of the frequency shifts versus normal load for these different types of contact (see Fig. 5). The observation of both positive and negative QCM frequency shifts is not without precedent when considering different regimes of coupling for a point contact under shear stress.[17] In our geometry, an increasing normal load almost certainly results in an increase in sliding friction at the interface.

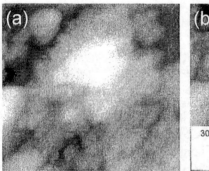 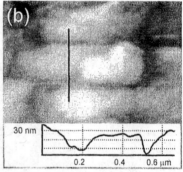

Figure 4. A pair of 1700 × 1500 nm images showing the machining of a platinum QCM electrode by a tungsten tip. The central mound in image (a) was scratched horizontally in two places by scanning the tip quickly across the mound. The final result is shown in (b). The inset in (b) shows a line section across the two scratches. This unlubricated metal-metal wear is accompanied by a negative QCM frequency shift.

We find that this is associated with an increase in the absolute value of the frequency shift.

Clearly, careful modeling is needed to account for the various QCM frequency responses observed here. In general, positive frequency shifts are associated with increased stiffness of the system and negative shifts are associated with added mass or increased damping. In previous experiments where small spheres or wires were pressed into QCM's operated in air, positive frequency shifts were exclusively observed.[18,19] These results were attributed to plausible mechanisms which would serve to increase the stiffness of the system in the geometry of probe and surface. Our results with STM-QCM in high vacuum conditions, exhibiting both positive and negative frequency shifts, show that more complete modeling is necessary to describing the competing effects of increased mass, stiffness, and/or damping in the coupling between tip and surface. For quantitative predictions, such modeling will likely need to incorporate a variety of factors, including the tip and surface materials, the presence or absence of intervening adsorbed layers, tip length, contact area, and normal load. For the time being, we focus on comparative studies where the exact same surface and tip are used before and after introducing an intervening lubricant layer. For instance, in the next section we present our results for a lubricated contact, employing the same tungsten tip and platinum surface as were used in our experiments with the unlubricated contact.

5 LUBRICATED CONTACT

We now study the case of lubricated contact between the tip and surface, employing the (macroscopic) lubricant and anti-wear additive TBPP, which was discussed in Section 2. This case is particularly relevant to the possibility of vapor phase lubrication of MEMS devices. Having characterized the unlubricated contact of our tungsten tip and platinum surface, as discussed in the previous section, we exposed the same system to TBPP vapor. The QCM registered an uptake of approximately 1 monolayer. The tip is expected to have adsorbed a comparable amount. The electrode and tip surfaces were at room temperature during deposition.

The response of the QCM to the normal force of the tip undergoes a dramatic change as the lubricant layer is introduced. Figure 5 shows the frequency shift versus normal load for both lubricated and unlubricated contact between our tungsten tip and platinum surface. We find that upon adsorption of TBPP, the QCM responds to the normal force of the tip with a positive frequency shift rather than negative, as observed for unlubricated contact. Another contrast is that the frequency shift remains increasing and positive up to the maximum normal load for lubricated contact, whereas the shift remains negative and decreasing over the range investigated for unlubricated contact. This striking change in the response of the QCM may well be a nanometer-scale signature of the friction and wear reducing properties of TBPP known from macroscopic observations.

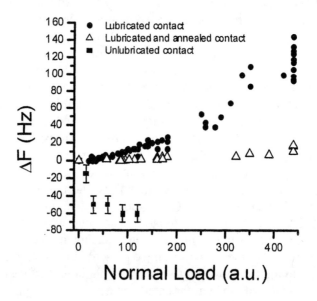

Figure 5. A plot of the QCM frequency shift versus normal load of the STM tip for lubricated, lubricated and annealed, and unlubricated contact. (The arbitrary units for normal load are obtained from the Z voltage of the scanner piezo. The full range represents a deflection of 800 nm in the vertical direction in the absence of a contacting surface). The same tungsten tip and platinum surface are used in each case. The lubricated contact involves a molecularly thin layer of TBPP, a known lubricant on the macroscopic scale, on the tip and surface. The direction of frequency shift for lubricated contact is opposite that observed for unlubricated, metal-metal contact. This striking contrast may be a nanometer scale signature of the film's lubricating properties. The magnitude of the shift is smallest by far for the lubricated and annealed contact, suggesting that sliding friction is most reduced in this case.

Additional insight into the properties of the lubricated interface may be obtained with the STM. Figure 6 shows a 350 × 350 nm STM scan of the surface immediately after exposure to the lubricant. We find that in order to obtain an image, it is necessary to run the QCM while scanning. No stable image is obtained when the QCM is off. We understand this result by considering the following: When the QCM is off, the tip-surface junction is quite insulating. The "Z" voltage, which controls the vertical position of the tip, is unstable while a constant tunneling current "I" is maintained through feedback. The so-called I-Z dependence is as follows: A change in voltage corresponding to a deflection of 100 to 300 nm into the surface is required to raise the tunneling current by 10 nA! Most likely, increases in the normal force and contact area are responsible for the increased current, not an actual change in tip-surface separation of such a large magnitude. (In contrast, for a clean metal-vacuum-metal tunneling junction, a change in Z voltage corresponding to a deflection of less than 1 Å changes the tunneling current by 10 nA). When the QCM is turned on and the tip engaged at 10 nA, the junction becomes more conductive over time (a few seconds or minutes) until an I-Z dependence close to that of tunneling through vacuum is achieved. This applies only to the region in close proximity to the tip. Changes in the Z voltage and QCM frequency indicate a decreasing normal load as the junction becomes more conductive. Under these conditions a recognizable image of the surface may be obtained. However, the junction returns to insulating as soon as the QCM is shut off, or if the tip is moved away from the region or raised up from the surface.

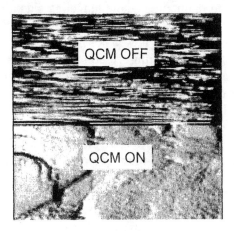

Figure 6. A 350 nm × 350 nm image of the TBPP-lubricated platinum surface at room temperature. The rubbing action of the vibrating QCM electrode against the STM tip is required to produce a stable image. The rubbing appears to clear away the lubricant, producing localized holes in which the surface may be imaged. These holes fill-in with additional lubricant once the QCM is turned off or the tip is moved away from the region, indicating that the TBPP film is mobile at room temperature.

Our interpretation of these observations is that the rubbing action of the vibrating QCM electrode against the STM tip effectively brushes the insulating lubricant layer aside to allow imaging of the underlying surface, similar to a windshield wiper except that the vibration of the QCM surface is critical in the process. Holes or clear areas produced in this way are not permanent, however, and readily fill in with additional lubricant once the QCM is turned off or the tip is moved to a different position. We therefore conclude that there is a high degree of mobility associated with the TBPP layer at room temperature. This corresponds to the results of adsorption studies with QCM alone discussed in Section 2, in which the observed change in amplitude of the QCM during adsorption of TBPP indicated either wholesale slippage of the layer or intermolecular mobility. The ability of a lubricating film to replenish its depleted areas is indeed a known requirement for good lubrication.

Thus far, our studies of the lubricated interface have been performed exclusively at room temperature. TBPP is known to undergo a chemical reaction above 100°C which renders the film more conductive.[20] In view of this, we have investigated the lubricated interface after annealing for several hours near 100°C. (The entire vacuum system was raised to elevated temperature). Indeed, we find that the TBPP-lubricated platinum surface is much more readily imaged after annealing. The tip-surface junction is conductive enough to maintain a stable tunneling signal with or without the QCM in operation. Figure 5 shows that the magnitude of the positive QCM frequency shift over the full range of normal load is significantly reduced. This suggests a reduction in sliding friction at the interface. Moreover, STM images of the annealed surface help identify a more subtle rubbing-induced response, which may provide evidence for a tribochemical reaction.

Figure 7 displays a pair of images of the lubricated surface after annealing. Image (a) shows an non-rubbed 70 × 70 nm region, for which the QCM has not yet been operated with the tip engaged. The image has a fringed appearance. (Due to the presence of the lubricant film, the appearance of the surface in these constant-current images is not to be interpreted literally as the topography of the surface. The imaging mechanism is unknown, and electronic effects are clearly present). Image (b) shows the same region after the QCM was operated for a few minutes with the STM tip held in tunneling contact at 2 nA. This short rubbing experiment produced two distinct regions in the vicinity of the tip. The majority of image (b) has a uniform appearance, but a region in the upper left remains which bears the fringed appearance of image (a). Interestingly, the two regions are accompanied by different QCM frequency responses. In both (a) and (b), the fringed region registers a small positive frequency shift when the tip is briefly engaged in tunneling (at 2 nA). The uniform region in (b) registers no frequency shift. In general, we find that if the tip is held in tunneling contact at a fringed region while the QCM is operated, a small positive frequency shift appears and decays over a period of a minute or so, leaving a uniform region with no frequency shift for the same tunneling conditions.

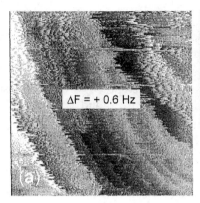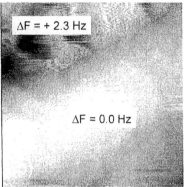

Figure 7. A pair of 70 nm × 70 nm images of the TBPP-lubricated platinum surface after annealing for several hours near 100 °C. QCM frequency shifts upon establishing tunneling contact at different places are shown in boxes. Image (a) shows a non-rubbed region. Image (b) shows the same region after a rubbing period of a few minutes, during which the QCM was vibrating with the tip engaged in tunneling contact. Two regions may be distinguished: non-rubbed regions have a fringed appearance and are accompanied by a small positive frequency shift when the tip is first engaged in tunneling. Rubbed regions have a more uniform appearance and exhibit no frequency shift when tunneling. The electrical properties of the tunneling junction suggest that the lubricant film is not simply worn away by rubbing. These results suggest a possible tribochemical reaction associated with the interfacial sliding conditions achieved by the STM-QCM.

This rubbing-induced effect is both localized to the region of rubbing and permanent. The uniform regions do not revert to fringed, and fringed regions are only converted to uniform by rubbing the tip against the vibrating QCM. The ability to distinguish rubbed and non-rubbed regions on the nanometer scale by both appearance and corresponding frequency response is intriguing, since improvements in the protective properties of lubricant films after interfacial sliding are well-known, yet poorly understood, on the macroscopic scale.[21] While the mechanism responsible for the change in frequency response observed here is unknown, we find that the tunneling junction never regains the conductivity (I-Z dependence) of a clean metal-vacuum-metal junction. This suggests that the film is not simply worn away by rubbing. Together these results provide evidence for a potential tribochemical reaction triggered by the interfacial sliding conditions attained with the STM-QCM.

6 CONCLUSION

In conclusion, we have combined a Scanning Tunneling Microscope (STM) with a Quartz Crystal Microbalance (QCM) with the intent of developing a realistic laboratory test set-up for MEMS contacts. The STM-QCM allows unique and detailed investigations of the simple nanomechanical system formed by a contacting tip and surface. We have discussed our results obtained with STM-QCM in three situations: (1) measurements of the amplitude of a vibrating QCM electrode with the STM tip in tunneling contact with the QCM, (2) unlubricated metal-metal contact of the STM tip with the QCM electrode, and (3) lubricated contact (before and after annealing) using a vapor phase lubricant which is known to reduce wear in macroscopic applications. In the case of unlubricated contact, we observe significant wear and evidence of increasing sliding friction with normal load. Application of molecularly thin quantities of lubricant to the same contact dramatically alters both the STM and QCM responses in a manner which is highly suggestive of the lubricant's known friction and wear reducing properties at the macroscopic scale. Our measurements also reveal a potential tribochemical reaction which is highly localized at the point of contact and is associated with the realistic rubbing conditions provided by the STM-QCM apparatus.

ACKNOWLEDGMENTS

This work has been supported by NSF grants DMR9896280 and CMS9634374, and AFOSR grant F49620-98-1-0201.

REFERENCES

1. Krim, J., "Friction at the atomic scale," *Scientific American*, **275**, 74-80 (1996).
2. Cieplak, M., Smith, E.D., and Robbins, M.O. , "Molecular origins of friction: The force on adsorbed layers", *Science*, **265**, 1209-1212 (1994); Robbins, M.O. and Krim, J., "Energy Dissipation in Interfacial Friction," *MRS Bulletin*, 23, 23 (1998).
3. Duncan-Hewitt, W.C. and Thompson, M., "Four-layer theory for the acoustic shear wave sensor in liquids incorporating interfacial slip and liquid structure," *Anal. Chem.*, **64**, 94-105 (1992).
4. Yang, M , Thompson, M., and Duncan-Hewitt, W.C., "Interfacial properties and the response of the thickness-shear-mode acoustic wave sensor in liquids," *Langmuir*, **9**, 802-811 (1993).
5. Maboudian, R. and Howe, R.T., "Stiction reduction processes for surface micromachines," *Tribology Letters*, **3**, 215-222 (1997).
6. Maboudian, R. and Howe, R.T., "Critical Review: Adhesion in Surface Micro-Mechanical Structures," *J. Vac. Sci. and Tech.*, **15**, 1-20 (1997).

7. Zavracky, P.M., Majumder, S., and McGruer, N.E., "Micromechanical switches fabricated using nickel surface micromachining," *J. Electromech. Sys.*, **6**, 3 (1997).
8. Krim, J., Dayo, A., and Daly, C., "Combined scanning tunneling microscope and quartz microbalance study of molecularly thin water layers", *Atomic Force Microscopy/Scanning Tunneling Microscopy*, S.H. Cohen, ed., (Plenum Press, New York, 1994), pp. 211-215.
9. Dugger, M.T., this issue.
10. Forster, N.H. and Trivedi, H.K., "Rolling contact testing of vapor phase lubricants - part I: material evaluation," *Trib. Trans.*, **40**, 421-428 (1997).
11. Faut, O.D. and Wheeler, D.R., "On the mechanism of lubrication by Tricresylphosphate (TCP)-The coefficient of friction as a function of temperature for TCP on M-50 steal," *A.S.L.E. Trans.*, **26**, 344-350 (1983).
12. Martin, J., this issue.
13. Marino, M.P. and Placek, D.G., *CRC Handbook of Lubrication and Tribology Vol. III*, R.R. Booser, ed., (CRC Press, Boca Raton, 1994), pp. 269-286.
14. Lu, C. and Czanderna, A., *Applications of Piezoelectric Quartz Crystal Microbalances*, (Elsevier, Amsterdam, 1984).
15. Krim J. and Widom, A., "Damping of a crystal oscillator by an adsorbed monolayer and its relation to interfacial viscosity" *Phys. Rev.*, **38B**, 12184-12189 (1983); Widom, A. and Krim, J., "Q factors of quartz oscillator modes as a probe of submonolayer-film dynamics," *Phys. Rev.*, **34B**, 1403-1404 (1986).
16. Borovsky, B., Mason, B.M., and Krim, J., "Scanning Tunneling Microscope Measurements of the Amplitude of Vibration of a Quartz Crystal Oscillator," submitted to *J. Appl. Phys.*
17. Dybwad, G.L., "A sensitive new method for the determination of adhesive bonding between a particle and a substrate," *J. Appl. Phys.*, **58**, 2789-2791 (1985).
18. Laschitsch, A. and Johannsmann, D., "High frequency tribological investigations on quartz resonator surfaces," *J. Appl. Phys.*, **85**, 3759-3765 (1999).
19. Martin, B.A. and Hager, H.E., "Velocity profile on quartz crystals oscillating in liquids," *J. Appl. Phys.*, **65**, 2630-2635 (1989).
20. Gellman, A., private communication.
21. Ludema, K., *Friction, Wear, Lubrication – A Textbook in Tribology*, (CRC Press, New York, 1996), p. 160.

Chapter 26

MICRO-HARDNESS AND MICRO-WEAR MEASUREMENTS ON MAGNETIC HEADS

V. Prabhakaran, F. E. Talke and J. T. Wyrobek*
Center for Magnetic Recording Research
University of California
San Diego, La Jolla, CA 92093

* Hysitron Inc.
5251 W. 73rd St., Suite A
Minneapolis, MN 55439

Abstract   Wear of carbon coated sliders during contact start stop testing was measured by monitoring changes in the depths of ion milled "trenches" in the air-bearing surface. Wear was determined as a function of sliding distance and glide height of the media. In addition, nano-hardness measurements were used to characterize the carbon overcoats.

1 INTRODUCTION

DLC (Diamond like Carbon) is a class of amorphous form of carbon with high hardness and good frictional properties. Amorphous carbon films have some advantages over crystalline films because they can be deposited by chemical vapor deposition at relatively low temperatures and they have lower surface roughness than large grain size crystalline diamond. Diamond like carbon films have been used as protective overcoats on magnetic recording heads to reduce friction [1,2], wear and corrosion of the magnetic poles. Bogy et al. [2] reported significant improvement in the durability of the head-disk interface after coating sliders with DLC.

In proximity recording, wear of the slider and the disk are important factors limiting the reliability of the head-disk interface. Hence measurement of the wear rate of the carbon overcoat on the head is required to evaluate slider designs and to make the right choices of carbon overcoat and disk parameters.

Most methods of measuring wear of the head have been indirect methods. Varanasi et al. [1] studied the wear of DLC coated sliders using Raman spectroscopy. However, changes in the structure of the DLC during wear and fluorescence of the underlying Al_2O_3 limit the applicability of

Raman spectroscopy as a tool for quantitative wear measurements. Bogy et al. [2] inferred wear at the head-disk interface from increases in the touch down velocity of the slider. Gupta et al. [3] studied micromechanical properties of overcoats using nano-indentations and micro-scratch measurements. The effect of carbon film hardness on wear resistance has been studied using micro-scratch tests [3,4] and pin on disk tests [5]. DLC overcoats with higher hardness are generally considered more durable.

However, the DLC overcoat on heads is becoming progressively thinner in order to decrease the magnetic spacing. As a consequence, the measurement of hardness and the elastic modulus of DLC overcoats without substrate effects has become increasingly difficult. In order to avoid ambiguities resulting from substrate effects (hardness measurements of DLC), structural changes in the DLC and calibration (Raman measurements of wear of DLC) a direct, quantitative measurement of head wear in CSS tests and sweep tests is desirable.

In this paper we measure wear of DLC overcoats at different locations on a slider during CSS testing using a direct method described by Hsia et al. [6]. In this method, "trenches" are created at various locations on the air-bearing surface by focussed ion beam milling. The change in the depth of the trenches after sliding tests gives a measure of the wear in the region adjacent to the "trench." We also evaluate the hardness of the DLC overcoats using a commercially available nano-indenter (Hysitron Inc).

2 EXPERIMENTAL PROCEDURE

2.1 Properties of the DLC Overcoat

The DLC was deposited at 200 V, 150 mA and contained approximately 34% H. Raman spectroscopy indicated that this DLC had an I_D/I_G ratio of 0.75. The same DLC overcoat was deposited on Al_2TO_3-TiC coupons for hardness measurements. Figure 1 shows the air-bearing of the DLC coated sliders used in contact start stop (CSS) wear tests. The sliders were coated with a 2 nm under-layer of Si_3N_4 and then a 8-nm layer of DLC.

2.2 Wear During Contact Start Stop Tests

The initial nominal depth of the "trenches" created by FIB milling was 20 nm with a width of 2 μm and a length of 20 μm. The position of the "trenches" in the air-bearing surface (denoted T1, T2, P1 and P2) is shown schematically in Fig. 2a. The position of the "trenches" in the pole tip region of the slider is shown in Fig. 2b. An atomic force microscope (AFM) was used to measure the depth of the trenches.

Micro-Hardness and Micro-Wear Measurements on Magnetic Heads 379

The difference in "trench" depth before and after CSS testing is the average wear depth for the region near the "trench."

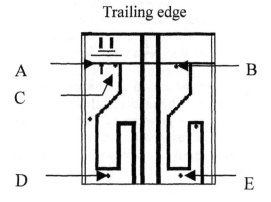

Figure 1 Air-bearing design of sliders used.

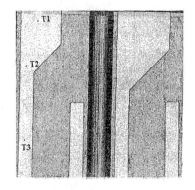

Figure 2. (a) "Trenches" T1, T2 and T3 in the air-bearing surface and (b) trenches near the poles of the slider.

2.3 Disks Used in CSS Tests

All disks used in CSS tests were characterized by roughness and glide avalanche height. The glide avalanche height was defined as the flying height of the slider below which the acoustic emission from a glide head shows a very large increase. CSS tests were first conducted using lubricated laser zone textured disks with a bump height of 12 nm and a glide avalanche height of 15 nm. In an attempt to accelerate wear, CSS tests against unlubricated mechanically textured disks (RMS roughness 5 nm, glide avalanche height 26 nm) were performed. Constant speed drag tests were also performed. Drag tests resulted in the generation of excessive debris. Debris on the air-bearing surface cause errors in the measurement of the depths of the "trenches." Hence the drag testing method of evaluating wear durability was discarded in this study.

2.4 Flying Height of Sliders

A commercially available dynamic flying height tester (Phase Metrics) was used to measure the flying characteristics of the sliders. The points of measurement are denoted as 'A', 'B', 'C', 'D' and 'E' as shown in Fig. 1. Figure 3 shows the flying height of the sliders at point 'A'. At 'A', the outer trailing edge, the sliders had a flying height of 21–23 nm at 9.4 m/s.

Since the glide avalanche height of the mechanically textured disks used in CSS testing was 26 nm, there was "interference" between the slider and the disk at maximum speed during CSS.

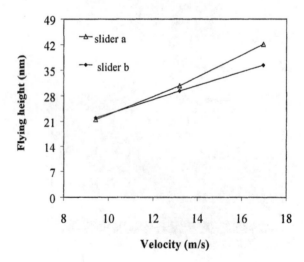

Figure 3 Flying height at point 'A.'

2.5 Hardness of DLC Overcoats on AlO$_3$-TiC Substrates

To minimize the effect of the substrate on the measured hardness for the 9.2 nm thick films, extremely small indentation loads were used. A nanoindenter made by Hysitron Inc. was used to create very small indentations with very small loads. The displacement of the indenter was simultaneously recorded using a force-displacement transducer. The indenter consists of a 3-plate capacitive force displacement transducer with electrostatic actuation [7]. To avoid substrate effects the contact depth needs to be less than 30% of the film thickness [8]. Hence, indentations with contact depths less than 3 nm were required to evaluate the 9.2 nm thick DLC samples. Figure 4 shows typical load-unload curves in DLC. The load-unload data were analyzed as described by Oliver and Pharr.[9].

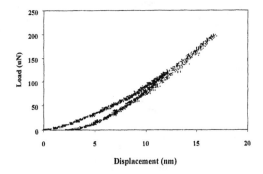

Figure. 4 Load unload curves in DLC.

The hardness of the DLC film was calculated as:

$$\text{Hardness} = \frac{Load}{Area} = \frac{P_{max}}{f(h_c)}, \qquad (1)$$

where $f(h_c)$ is the tip geometry function.

The slope of the initial part of the unloading curve gives the contact stiffness S. The reduced modulus E_r of the DLC is related to the contact stiffness S and the area of contact A as follows:

$$E_r = \left(\frac{dP}{dh}\right)\left(\frac{\pi}{A}\right)^{\frac{1}{2}} \text{ where } \frac{dP}{dh} = S \qquad (2)$$

and

$$\frac{1}{E_r} = \left(\frac{1-\gamma_{dlc}^2}{E_{dlc}}\right) + \left(\frac{1-\gamma_i^2}{E_i}\right) \qquad (3)$$

where $v_I = 0.07$ (Poisson's ratio of diamond), and $v_{DLC} = 0.2$ (Poisson's ratio of diamond like carbon). $E_{\text{fused quartz}}$ was taken to be 72 GPa, E_i = modulus of elasticity of diamond =1141 GPa and E_{DLC} = modulus of elasticity of the DLC film. P is the maximum indentation load and h is the displacement of the indenter. A power law was used to fit the unloading curve from 95% of the maximum load to 65% of the maximum load. The resulting function was analytically differentiated at maximum load to obtain S. The frame compliance of 6×10^{-6} m/N was subtracted from the total compliance calculated from the unloading curves. Figure 5a shows the tip geometry function, $f(h_c)$ obtained from the tip calibration procedure. The resulting hardness of fused quartz after calibration is shown in Fig. 5b. As seen in Fig. 5b there is negligible depth dependence of the hardness of fused quartz.

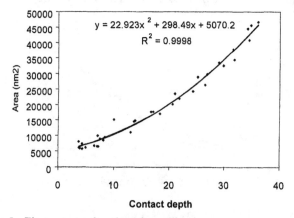

Figure 5a. Tip geometry function after calibration.

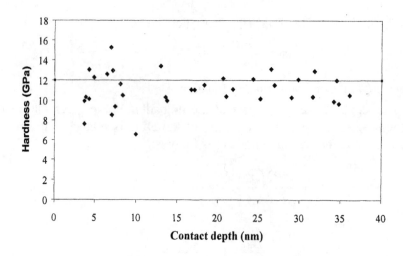

Figure 5b. Hardness of fused quartz after calibration.

3 RESULTS

3.1 Hardness of the DLC Overcoats

The hardness of the DLC overcoats evaluated as a function of indentation load is shown in Fig. 6. The hardness values showed more scatter at low load than at high loads.

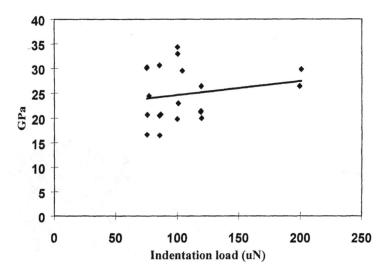

Figure 6. Hardness of DLC as a function of indentation load.

For an indentation load of 100 µN, the contact depth was measured to be between 4–6 nm, *i.e.*, data obtained at this load are influenced by both the hardness of the Al_2O_3-TiC substrate. From Fig. 6 we observe that at 100 µN, the DLC had a mean hardness of 27.8 GPa and a mean elastic modulus of 257.5 GPa.

3.2 Wear of Sliders after CSS Testing

Wear of the slider was less than 1 nm after 20k CSS cycles in tests against lubricated laser zone textured (LZT) disks. After take-off the sliders had no "interference" with the disk since the flying height was much larger than the glide avalanche height of the LZT disks. Hence negligible wear occurs during the dwell at maximum speed. However, CSS testing against unlubricated mechanically textured disks resulted in wear of the slider. An AFM scan of a typical "trench" near the poles of the slider after CSS testing against a mechanically textured disk is shown in Fig. 7a.

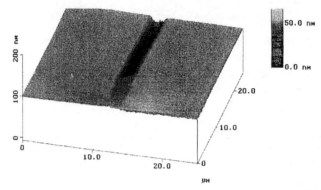

Figure 7a. "Trench" in the pole region.

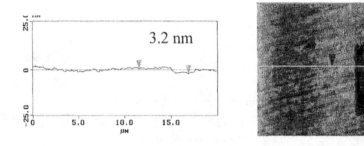

Figure 7b. "Trench" in the slider rail.

From Fig. 7a we note that the depth of the "trench" is not uniform after the test. Regions of the "trench" closer to the trailing edge of the slider exhibit a large decrease in depth. Figure 7b shows an AFM scan of a trench in the slider rail after 50k CSS cycles.

Figure 8a shows the wear depths near the "trenches" for sliders coated with DLC as a function of CSS cycles. The location of the points of wear measurement is marked in Fig. 8b.

P1 and P2 are "trenches" located in the pole region of the head while T1 and T2 are "trenches" in the air bearing surface of the slider. The interface failed after ~20k CSS cycles. Hence, the wear rates for DLC in Fig. 8 are presented only for the first 20k CSS cycles. The wear test was repeated with a second slider. The wear data for the second set of tests is labeled "slider2." As seen in Fig. 8a, the region near T1 has a higher wear rate than the region near T2 for all tests. Hence, it can be seen that the wear rate decreases with increasing distance from the trailing edge of the slider.

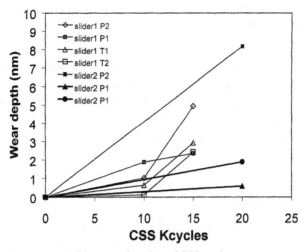

Figure 8a. Wear of sliders as a function of CSS cycles.

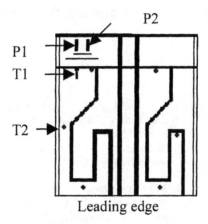

Figure 8b. Locations of the trenches.

4 DISCUSSION

The sliders used in our experiments had an average wear of 2 nm in the first 15k cycles. The sliders failed between 20k and 25k cycles. The locations of lowest head-disk spacing on the air-bearing surface of the sliders were the locations with the highest wear. None of the "trenches" in the air-bearing surface except "T1" which was near the trailing edge exhibited any decrease in depth. All regions of the air-bearing surface except the trailing edge clear the asperities of the disk due to the pitch of the slider at 9.4 m/s (160-170 nm). CSS tests with laser textured disks showed that without contact between the slider and the disk at maximum speed, negligible wear occurs during start and stop.

5 CONCLUSIONS

The direct method of measuring wear of sliders is useful to evaluate the wear of proximity recording sliders. For a target areal storage density of 100 Gb/in^2 the thickness of the protective wear overcoat on the head would be less than 3 nm. While measuring the hardness of DLC films thinner than 2 nm using the nanoindentation method would be a challenging task, direct wear measurements can be performed even on sliders with ultra-thin DLC overcoats.

ACKNOWLEDGEMENTS

The authors would like to acknowledge Thomas Wyrobek, Lance Kuhn and Ashok Kulkarni of Hysitron Inc. for their support and help with the experimental equipment. Thanks are also due to Victor Dunn of ReadRite Corp. for providing the DLC samples and the ion-milled heads for wear measurements.

REFERENCES

[1] Varanasi. S. S, Lauer. J. L. , Talke. F. E., 'Friction and wear studies of carbon overcoated thin film magnetic sliders: Application of Raman microspectroscopy', J. Tribology, Vol. 119, pp. 471-75, Jul 1997.
[2] Bogy. D. B. , Yun. X. , 'Enhancement of head-disk interface durability by use of diamond-like carbon overcoats on the slider rails.', IEEE Trans. Magnetics, Vol. 30, No. 2, pp. 369-374, 1994.
[3] Gupta, B.K., Bhushan. B, 'Micromechanical properties of amorphous carbon coatings deposited by different deposition techniques.' , Thin Solid Films, 270, 1995, 391-8.
[4] Tsui. T. Y, Pharr. G. M, Oliver. W. C, Bhatia. C. S, White. R. L, Anders. S, Anders. A, and Brown. I. G, 'Nanoindentation and nanoscratching of hard carbon coatings for magnetic disks.', Mater. Res. Soc. Symp. Proc. Vol. 383, 1995, 447-452.
[5] Kawakubo, Y. Ishihara, Tsutsumi. H and Shimizu. J., 'Spherical pin sliding test on coated magnetic recording disks.', Tribology and Mechanics of Magnetic Storage Systems, 1986, Vol. 3, SP-21, 118-24.
[6] Yiao-Tee Hsia; Rottmayer, R.; Donovan, M.J. , 'A new wear measurement technique for pseudo-contact magnetic recording heads.' , IEEE Transactions on Magnetics, Sept. 1996, .32, No. 5, pt.1, 3750-2.
[7] Bhushan. B. , Kulkarni. A. V. , Bonin. W. and Wyrobek. J. T. , 'Nanoindentation and picoindentation measurements using a capacitive transducer system in atomic force microscopy.', Philosophical Magazine A, . 74, No. 5, 1996, 1117-1128.
[8] Bhushan. B, Editor, Handbook of Micro/Nanotribology, (CRC Press, 1995), pp. 323.
[9] Oliver, W.C.; Pharr, G.M. , 'An improved technique for determining hardness and elastic modulus using load and displacement sensing indentation experiments.', J. Mater Res. , Jun 1992, . 7, No. 6, 1564-83.

Chapter 27

BONDING AND INTERPARTICLE INTERACTIONS OF SILICA NANOPARTICLES:
Probing Adhesion at Asperity-Asperity Contacts

James D. Batteas
Department of Chemistry
The City University of New York
College of Staten Island and the Graduate School, 2800 Victory Boulevard,
Staten Island, NY 10314
batteas@postbox.csi.cuny.edu

Marcus K. Weldon and Krishnan Raghavachari
Bell Laboratories-Lucent Technologies
Murray Hill, NJ 07974

Abstract Adhesion at asperity-asperity contacts plays a critical role in tribology. Studies of nanoparticle films with atomic force microscope (AFM) tips provide a means of probing adhesion at such nanoscale contacts. Here, the molecular details of adhesion at silica nanoparticle (~50 nm diameter) surfaces under various aqueous solution conditions have been investigated using a combination of atomic force microscopy and *ab initio* quantum chemical calculations. The adhesion between surface bound silica particles and the AFM tip mimics the colloidal interactions between silica particles in solution and depends on the intrinsic surface composition under the varying solution conditions. Depending on pH, the measured adhesion between the surfaces is a mixture of hydrogen bonding ($\equiv$Si-OH $\cdots$ HO-Si$\equiv$), anionic hydrogen bonding ($\equiv$Si=(OH)$_2^-$ $\cdots$ HO-Si$\equiv$) and formal covalent bonding, via formation of siloxane ($\equiv$Si-O-Si$\equiv$). In order to decouple the number density and energetics of each type of interaction within the nanospheric contacts formed between the surfaces, the adhesion measurements by AFM have been referenced to detailed *ab initio* quantum chemical calculations of the relevant interactions. This combined approach affords the description of the adhesion with molecular definition and the statistical distribution of interactions between particles can be evaluated.

1 INTRODUCTION

For many years colloidal particles on the order 5–1000 nm have been used to synthesize materials by controlling the aggregation or 'gelling' of the particles. The molecular interactions found between the colloidal particles as they contact each other in solution play the key role in the formation of bonds between the particles and control the ultimate nanoscale and macroscale organization of materials synthesized in this manner. Networks of bonded particles form during the sol-to-gel transformation resulting from the creation of siloxane ($\equiv$Si-O-Si$\equiv$) linkages between the colloidal particles in the sol [1]. The formation of these bonds depends on the silica surface chemistry, which varies with solution pH. At the isoelectric point (IEP) for silica (pH ~2–3), the surface is electrically neutral and terminated by surface silanol groups ($\equiv$Si-OH) [1]. Above this point there is a gradual increase in the surface charge, which we have recently shown to be dominated by the trapping of OH$^-$ by $\equiv$Si-OH, yielding a pentavalent $\equiv$Si=(OH)$_2^-$ anionic species [2]. With increasing pH a distribution of neutral and charged Si surface species is present at the surface, with the relative proportion of each dependent on the pH [1,2]. The formation of siloxane bonds, that is the essence of gelation, is optimal when an $\equiv$Si-OH and $\equiv$Si=(OH)$_2^-$ on opposing surfaces interact, with the highest rate of siloxane bond formation occurring at intermediate pH ranges (pH ~ 4–6) [1,3]. Above pH 7 surface charge density becomes sufficiently high that adhesion and inter-particle bonding is dramatically impeded.

The ability to directly probe the adhesive forces between nanoscopic particles is an essential step towards developing a complete understanding of the inter-particle interactions in these material systems. Several direct measurements of surface interactions related to oxides have been made in recent years using both the surface forces apparatus (SFA) [4-8] and the atomic force microscope (AFM) [9-17]. We have recently described the qualitative details of silica particle-particle adhesion under aqueous solution conditions using the AFM tip as a model of a silica nanoparticle [18]. In addition to probing the molecular interactions relevant to silica sol-gel systems, the sharply curved surfaces in contact at the AFM tip/particle junction provides a unique opportunity to explore adhesion and wear at nanometer scale asperity-asperity contacts, an important factor in tribology. Nanotribology studies have advanced significantly since the development of the AFM as detailed in a recent review by Carpick and Salmeron [19], with many studies focusing on utilizing the AFM probe tip as a model for a single asperity. We have applied a simple modification to this approach, by examining the interactions between AFM tips with nanoparticles (dispersed as surface films) to come one step closer to mimicking the conditions found at asperity-asperity contacts. This approach should be readily applicable to many systems.

As described above, an additional complication in studies of oxide surfaces such as silica in aqueous solutions is that the adhesion between surfaces may result from a combination of different interactions depending on the pH of the environment. In the case of silica, the adhesion between the surfaces can involve, hydrogen bonding, anionic hydrogen bonding and covalent bonding in various permutations depending on pH. In order to decouple the contributions to the adhesion measurements from these various interactions, detailed quantum chemical cluster calculations of the relevant interactions have been made.

Here we describe force microscopy studies designed to directly probe the interactions at nanoscale particle-particle interfaces of silica as a function of solution conditions. The particle-particle interactions in the sol-gel system are modeled as the interactions between colloidal silica particles deposited as a thin film on an atomically smooth Si(100) single crystal wafer with the hydrolyzed surface of the native oxide of Si_3N_4 AFM tips. The radius of curvature of the AFM tips are comparable to that of the particles being probed (~50 nm), such that the AFM tip is a good approximation of a second single silica "nanoparticle." Altering the solution pH modifies the surface chemistry and adhesion with the AFM probe tip is then measured *in situ*. Initial results of an interesting time-dependence on the adhesion measurements with tip-particle contact time are also described. The molecular details of the measured adhesion as a function of pH are only fully revealed when compared to quantum chemical cluster calculations of the relevant surface interactions.

2 METHODS

The details of preparing nanoparticle films and their measurement by AFM have been described previously [18]. Briefly, films of colloidal silica with an average diameter of 50 nm (OX-50, Degussa GmbH, Frankfurt, Germany) are dispersed in a sol and are spin coated (Headway Research, Garland, TX) onto a cleaned and oxidized silicon wafer. AFM images and force-distance measurements were made with a commercially available AFM (ThermoMicroscopes Explorer with ECU-plus electronics, Santa Clara, CA). All data are collected in aqueous solutions of high purity water (EASYpure RF, 18.2 MΩ·cm, Barnstead, Dubuque, IA) at room temperature (22±3 °C). The pH values of the solutions used were adjusted with HCl and NaOH (Aldrich, Milwaukee, WI). The experiments employed commercially available (ThermoMicroscopes, Santa Clara, CA) Si_3N_4 cantilevers. The normal force constants of the levers used were determined to be 0.36±0.08 N/m, based on measurements of the cantilever spring constant measured against a lever of known spring constant [20]. The proximal tip shape and radius of curvature was determined by imaging of a $SrTiO_3$(305) single crystal [21-23] in an environmental chamber under dry air (RH < 1%) conditions and tips were routinely found to exhibit a radius of curvature of

53±10 nm. In using Si_3N_4 based AFM tips to approximate a silica surface, the Si_3N_4 surface chemistry must be considered. Under aqueous solutions the Si_3N_4 surface has been shown to be oxidized and to expresses almost exclusively surface silanol groups, thus behaving chemically almost identically to silica surfaces [14,24-26].

The adhesion forces between the AFM tip and the surface bound particles were determined from force-distance measurements. Only the highest particulate asperities in the topographic images were selected for acquisition of force-distance data to eliminate interference from neighboring particles on a given particle-tip interaction. Each force-distance curve consists of 500 data points over a 100 nm range and is the average of 50 approach and retraction curves collected at a speed of 1 µm/s. Imaging of the surface following force-distance acquisition, indicated that the nanoparticles were not displaced by measurements and that piezo drift was insignificant. The influence of tip-particle contact time was examined by varying the approach and retraction rate between 0.1 – 4 µm/s.

3 RESULTS AND DISCUSSION

3.1 pH Dependence of Adhesion

The nanoparticle films formed on the Si wafers typically show randomly dispersed, spherically shaped particles (Figure 1). We have previously described the qualitative behavior of the adhesion with increasing pH [18].

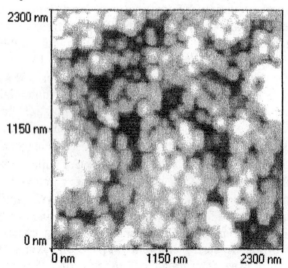

Figure 1: Topographic AFM image of a typical silica nanoparticle film on Si(100) used in this study collected under aqueous solution at pH 4.5.

The adhesion initially increases with pH, reaches a maximum between pH 4–5, and then decreases again (Figure 2). This trend is consistent with a previous SFA study of adhesion at silica surfaces as a function of pH, which also finds the maximum inter-particle adhesion between pH 4–5 [4].

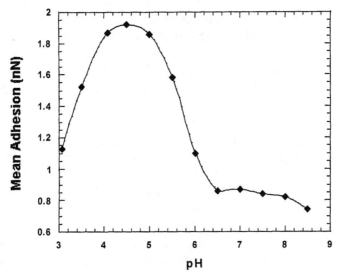

Figure 2. The pH dependence of the adhesion for individual silica particles with the AFM tip. Each data point represents the mean value collected from 40 different particles at a given pH, with 50 averaged measurements collected at each particle.

In aqueous environments at pH 2–3, the silica surface [1,27,28] is covered by silanol groups (≡Si-OH) with a surface density of ~4–5 silanol groups/nm^2 [1]. Under these conditions, the inter-particle adhesion is dominated by hydrogen bonding interactions between the silanol groups on opposing surfaces. As described in our previous work [18], the surface energy, contact area and interaction strength for ≡Si-OH ··· HO-Si≡ in water may be estimated using the contact mechanics model developed by Johnson, Kendall and Roberts (JKR) [29-36]. Using the JKR model the force of adhesion is related to the work of adhesion, W$_{adh}$, and the reduced radius, R, of the tip-particle contact,

$$F_{adh} = -\frac{3}{2}\pi R W_{adh}. \qquad (1)$$

The work of adhesion is a combination of the tip-particle (γ_{tp}), tip-solvent (γ_{ts}) and particle-solvent (γ_{ps}) interfacial energies, and for surfaces that have the same composition, the surface energy may be estimated as $1/2W_{adh}$. From our data, we find the surface energy of the silanol covered surface in water to be ~5.0–7.5 mJ/m^2, which is within the typical range of 4–9 mJ/m^2 found for alcohol terminated surfaces in contact with water [30,32].

Use of the JKR model also allows for estimation the effective contact area yielding adhesion at pull-off, making a determination of the number of silanol groups interacting within the contact at pull-off possible. Based on typical values for adhesion between the silanol-covered surfaces at pH 3 (0.8 – 1.2 nN), the critical contact area at pull-off in our experiments is ~1–2 nm^2. Thus, as the average surface density of silanol species on silica is approximately 4–5 ≡Si-OH groups/nm^2 [1], our measured adhesion at pH 3 results from only ~4–10 ≡Si-OH ··· HO-Si≡ molecular interactions. From this we can further estimate the strength of individual hydrogen bonding interactions in water to be 0.4±0.1 kcal/mol [37].

With increasing pH a maximum in adhesion appears between pH 4–5, with the mean adhesion nearly double that initially observed at pH 3. With increasing pH the surface charge increases via the formation of anionic centers in the form of ≡Si=(OH)$_2^-$ from trapping of OH$^-$ by a silanol group [1,2]. The presence of this species introduces anionic hydrogen bonding interactions (≡Si=(OH)$_2^-$ ··· HO-Si≡) to the adhesion in addition to the resident neutral hydrogen bonding. Due to the sharp curvature of the surfaces at the AFM tip/particle contact, with the estimated contact area at separation of only ~1–2 nm^2 – our adhesion measurements are very sensitive to the statistical distribution of surface species with increasing pH. As we move from particle to particle in the measurements, the adhesion we measure can be thought of as a 'snapshot' of the local chemistry within the nanospheric contact. Thus, examination of the statistical distribution of adhesion values at given pH must be done. At pH 3 the distribution is roughly symmetric as the tip-particle interaction is dominated almost exclusively by neutral ≡Si-OH ··· HO-Si≡ interactions (Fig. 3a). Increasing to pH 4.3 leads to an asymmetric broadening to high adhesion in the distribution of adhesion values (Fig. 3b), undoubtedly due to the addition of anionic interactions, but how much of a contribution is made by each?

Decoupling how each of these types of interactions influences the adhesion requires knowledge of the discrete energy contributed by each type of interaction. While the neutral hydrogen bonding interaction energy can be estimated experimentally as demonstrated above, measurement of isolated ≡Si=(OH)$_2^-$ ··· HO-Si≡ interactions is not possible in this experimental configuration due to the large contact area. If we ever hope to understand adhesion at such asperity-asperity contacts under increasingly realistic conditions however, the problem of deconvoluting numerous heterogeneous interactions contributing to adhesion must be tackled.

Our approach to unraveling the molecular details of the adhesion present at these oxide surfaces has been to utilize quantum chemical cluster calculations to quantify the relative strengths of the key interactions present between silica surfaces. To this end, the interaction energetics for ≡Si-OH ··· HO-Si≡ and ≡Si-(OH)$_2^-$ ··· HO-Si≡ have been determined from *ab initio* quantum cluster calculations, with the effects of solvation explicitly included. The details of these calculations are described elsewhere [3]. From these calculations we find that a single neutral ≡Si-OH ··· HO-Si≡ interaction in water has an interaction energy of ~0.25 kcal/mol, in good agreement with our experimentally determined value of 0.4±0.1 kcal/mol and from force-distance measurements of the discrete hydrogen bonding interaction on flat silica surfaces using ultrasharp tips [3]. The anionic hydrogen bond ≡Si=(OH)$_2^-$ ··· HO-Si≡ is found to be ~10 times that of the neutral case at ~2.6 kcal/mol.

Armed with these energies we can begin to examine the molecular nature of the adhesion in the mid-pH region. By pH 4–5 where pentavalent ≡Si=(OH)$_2^-$ species are present with a density of ~ 0.04 ≡Si=(OH)$_2^-$/nm^2 [1,27,28], the measured adhesion events fluctuate as the tip-particle contact probes various permutations ranging from exclusively ≡Si-OH ··· HO-Si≡ interactions in the contact to a mixture with ≡Si=(OH)$_2^-$ ·· HO-Si≡ interactions as well. While the measured adhesion distributions are of course inherently broadened by slight variations in silanol density from particle to particle, the observed adhesion depends on the specific number density of each type of interaction within the contact area at the time of the measurement. Looking at the statistical distribution of adhesion values for pH 3 and 4.3 collected using the same tip, it is clear that the distribution shifts from being symmetric at pH 3 (centered about ~0.8 nN) to being asymmetrically broadened to higher adhesion values by pH 4.3 (Figure 3a,b). When the composite of the distributions is plotted (Fig. 3c) it can be seen that at pH 4.3, in addition to interactions at low adhesion arising from neutral hydrogen bonds, a second distribution of values appears centered about ~1.3 nN. This is 0.5 nN above the neutral adhesion case and the shift of 0.5 nN is consistent with the 2.6 kcal/mol calculated for the addition of an anionic hydrogen bond. Thus, values in this range results from a mixture of neutral hydrogen bonds with a single anionic hydrogen bond. In addition, there is a noticeable spread out to even higher values of adhesion (>1.5 nN) that cannot be explained by the presence of anionic hydrogen bonds alone. However, we have recently shown that the ≡Si=(OH)$_2^-$ ··· HO-Si≡ species interacting between the tip and surface can further *react* to create siloxane

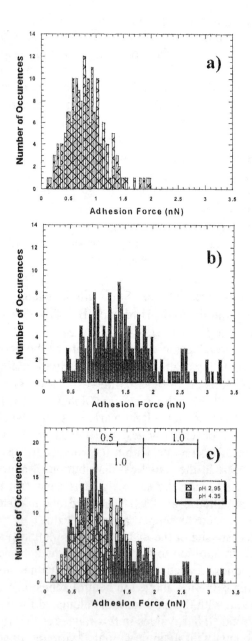

Figure 3. Statistical distribution of adhesion at a) pH 2.95, b) pH 4.35 and c) a composite of the distributions, collected from 200 particles with an average of 50 measurements per particle. On the composite, the buildup of an anionic hydrogen bond is observed ~ 0.5 nN above the neutral case (centered at 0.8 nN). Contributions from single siloxane bonds appear at 1.8 nN and 2.8 nN, at incremental increases of 1 nN above the neutral hydrogen bond and the first siloxane bond, respectively.

bonds between the surfaces. When this occurs, the measured adhesion will then contain a contribution from the energy needed to break this bond. Here again, quantum chemistry can help resolve the relevant interaction energetics. Cleavage of the covalent siloxane bond ($\equiv$Si-O-Si$\equiv$) which is ~25 kcal/mol in water, can also occur via a charged intermediate ($\equiv$Si-O-Si(OH)$^-\equiv$) which lowers the barrier to cleavage to ~6–8 kcal/mol, or equivalently ~1–1.5 nN. Based on this value, the highest adhesion values in the distributions must result from a combination of hydrogen bonding and a single siloxane bond. For example, the distinct component of the pH 4.3 distribution at 1.8 nN (see Figs. 3b, 3c) is consistent with the presence of a single covalent bond in addition to the background of H-bonds (*i.e*, 0.8 nN + 1 nN); similarly the values between 2.5–3 nN are consistent with the presence of two such bonds in the limit. These values becomes more apparent when ultrasharp AFM tips are used to further reduce the contact area whereby discrete values for single hydrogen bonds, single anionic hydrogen bonds and single covalent bonds have been resolved [3]. Above pH 7 the tip and particle surfaces become sufficiently negatively charged (~0.4 charges/nm^2) that there now is a significant probability of repulsive charge/charge interactions within the contact, which lowers the measured adhesion by screening the remaining attractive interactions.

3.2 Time Dependence of Adhesion

To investigate the influence of the contact time between the tip and surface bound particles on the adhesion, the force-distance measurements were also made with varying approach and retraction rates from 0.1–4 μm/s to vary the contact time from 10–400 ms (Fig. 4). At pH ~3 the measured adhesion increases with increasing tip-particle contact time. While at pH 4.5 the measured adhesion is essentially constant with increasing contact time. Again, at high pH (8.5) the adhesion is observed to increase with increasing contact time. Such time dependent adhesion of silica surfaces has been reported previously. An SFA study of adhesion of silica nanoparticle films found that under basic conditions (pH > 7) leaving the surfaces in the SFA together for longer periods increased the measured adhesion. The authors attributed this time dependence to the kinetics of siloxane bond formation [8]. However, under the various pH conditions, several mechanisms may be in play, including configurational ordering of the hydrogen bonding interactions with the water environment to adopt preferred conformations within the contact and/or diffusion of OH$^-$ into and out of the contact area, thus modifying the interactions within the contact as described above. Such loading rate variations on weak non-covalent interactions such as hydrogen bonding have also been described for

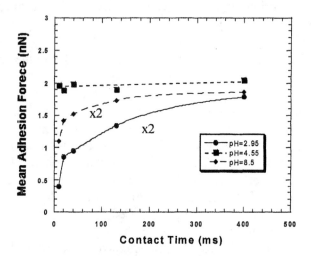

Figure 4. Plot of mean adhesion vs. contact time at pH 2.95, 4.55 and 8.50. Here the contact time is defined as the length of time the tip spends within 3 Å of the particle surface depending on the approach and retraction rate during our measurements. The curves for pH 2.95 and 8.50 are multiplied by a factor of 2 in the adhesion axis to better illustrate the observed change in adhesion.

biological systems such as biotin-streptavidin [38,39]. Interestingly, the same experiments conducted on a cleaned and oxidized Si(100) wafer show almost no time dependence on the adhesion at these pH's for the same approach and retraction rates. This difference may be due to the ability of charges to migrate more rapidly on the flat surface, versus the highly curved surface, and points to a need to investigate systems beyond ideally flat substrates. A thorough investigation of this phenomenon is currently underway.

4 SUMMARY

By combining experimental measurements of individual particle-particle interactions with *ab initio* quantum chemical cluster calculations, we have been able decouple the principal origins of the adhesion between an AFM tip and a silica surface under various pH conditions, including the formation of genuine chemical (Si-O-Si) bonds. The necessity of directly probing the interactions at such asperity-asperity contacts is clearly indicated by the contrasting time dependence of the adhesion measurements between the curved and flat oxide surfaces. Further, the small contact areas probed by studying such nanometer scale asperity-asperity contacts allows for the delineation of the discrete components in the distribution of interaction energies that exist under different conditions. Combined experimental and theoretical approaches such as these are needed if we are to ever draw molecular level conclusions for such heterogeneous systems.

ACKNOWLEDGMENTS

We gratefully acknowledge support of this work from the City University of New York-College of Staten Island and from Bell Labs-Lucent Technologies. JDB acknowledges support of this work from the American Chemical Society-Petroleum Research Fund (34792-G5), the PSC-CUNY Research Awards Program (61349-0030) and the National Science Foundation (DGE-9972892). We also acknowledge Dr. Robert W. Carpick, Department of Engineering Physics, University of Wisconsin-Madison, for many useful discussions on our data. JDB also acknowledges a collaboration agreement with ThermoMicroscopes.

REFERENCES

1. R.K. Iler, *The Chemistry of Silica* (Wiley-Interscience, New York, 1979).
2. M.K. Weldon, K.T. Queeney, K. Raghavachari and J.D. Batteas (*submitted to Science*).
3. M.K. Weldon, J.D. Batteas and K. Raghavachari (in preparation).
4. D. Atkins, P. Kékicheff and O. Spalla, J. Coll. Interface. Sci.188 (1997) 234.
5. G. Vigil, Z. Xu, S. Steinberg and J. Israelachvili, J. Coll. and Interface Sci. 165 (1994) 367.
6. J.-P. Chapel, J. Coll. and Interface. Sci. 162 (1994) 517.
7. L. Meagher, J. Coll. and Interface. Sci. 152 (1992) 293.
8. O. Spalla and P. Kékicheff, J. Coll. and Interface. Sci. 192 (1997) 43.
9. W.A. Ducker, T.J. Senden and R.M. Pashley, Nature 353 (1991) 239.
10. Y.Q. Li, N.J. Tao, J. Pan, A.A. Garcia and S.M. Lindsay, Langmuir 9 (1993) 637.
11. X.-Y. Lin, F. Cruzet and H. Arribart, J. Phys. Chem. 97 (1993) 7272.
12. W.A. Ducker, T.J. Senden and R.M. Pashley, Langmuir 8 (1992) 1831.
13. I. Larson, C.J. Drummond, D.Y.C. Chan and F. Grieser, Langmuir 13 (1997) 2109.
14. J.H. Hoh, J.P. Cleveland, C.B. Prater, J.-P. Revel and P.K. Hansma, J. Am. Chem. Soc. 114 (1992) 4917.
15. T. Miyatani, M. Horii, A. Rosa, M. Fujihira and O. Marti, Appl. Phys. Lett. 71 (1997) 2632.
16. A. Marti, G. Hähner and N.D. Spencer, Langmuir 11 (1995) 4632.
17. T.J. Senden, C.J. Drummond and P. Kékicheff, Langmuir 10 (1994) 358.
18. J.D. Batteas, X. Quan and M.K. Weldon, Tribol. Lett. 7 (1999) 121.
19. R.W. Carpick and M. Salmeron, Chemical Reviews 97 (1997) 1163.
20. M. Tortonese and M. Kirk, SPIE Proceedings 3009 (1997) 53.
21. S.S. Sheiko, M. Möller, E.M.C.M. Reuvekamp, E.M. and H.W. Zandbergen, Phys. Rev. B 48 (1993) 5675.
22. R.W. Carpick, N. Agrait, D.F. Ogletree and M. Salmeron, J. Vac. Sci. Tech. B 14 (1994) 1289.

23. D.F. Ogletree, R.W. Carpick and M. Salmeron, Rev. Sci. Instrum. 67 (1996) 3298.
24. L. Bousse and S. Mostarshed, J. Electroanal. Chem. 302 (1991) 269.
25. B.V. Zhmud, A. Meurk and L. Bergström, J. Coll. and Interface. Sci., 207 (1998) 332.
26. L.A. Wenzler, G.L. Moyes, L.G. Olsen, J.M. Harris and T.P. Beebe, Anal. Chem. 69 (1997) 2855.
27. G.H. Bolt, J. Chem. Soc. 61 (1957) 116.
28. P.W. Schlindler, B. Fürst, R. Dick and P. Wolf, J. Coll. Interface Sci. 55 (1976) 469.
29. K.L. Johnson, K. Kendall and A.D. Roberts, Proc. R. Soc. London A 324 (1971) 301.
30. J. Israelachvili, *Intermolecular and Surface Forces, 2nd ed.* (Academic Press, San Diego, 1992).
31. A. Noy, C.D. Frisbie, L.F. Rozsnyai, M.S. Wrighton and C.M. Lieber, J. Am. Chem. Soc. 117 (1995) 7943.
32. A. Noy, D.V. Vezenov and C.M. Lieber, Annu. Rev. Mater. Sci. 27 (1997) 381.
33. A. Noy, D.V. Vezenov, A. Noy, L.F. Rozsnyai and C.M. Lieber, J. Am. Chem. Soc. 119 (1997) 2006.
34. V. Tsukruk and V.N. Bliznyuk, Langmuir 14 (1998) 446.
35. S.K. Sinniah, A.B. Steel, C.J. Miller and J.E. Reutt-Robey, J. Am. Chem. Soc. 118 (1996) 8925.
36. E.W. van der Vegte and G. Hadziioannou, Langmuir 13 (1997) 4357.
37. Values used in JKR calculations: F_{adh} = 0.8 - 1.2 nN; $E_{Si_3N_4}$ = 290 GPa, E_{SiO_2} = 73 GPa, $v_{Si_3N_4}$ = 0.23, v_{SiO_2} = 0.17; this yields a K = 80.47 GPa. The values for Si_3N_4 are for LPCVD deposited silicon nitride. The reduced radius, R = 17 – 17.3 nm, using $R_{particle}$ = 25 nm and R_{tip} = 53 – 57 nm as determined from imaging of $SrTiO_3$ (305). The effective contact radius at separation, r_s, from the JKR model is given as:

$$r_s = \left(\frac{3\pi W_{adh} R^2}{2K}\right)^{1/3},$$

where K is the combined elastic moduli of the tip and particle, with the contact area at separation being $A_s = \pi r_s^2$. The principle source of error in this type of analysis is the determination of K, which is the reduced elastic modulus of the materials in the contact.38. R. Merkel, P. Nassoy, A. Leung, K. Ritchie and E. Evans, Nature 397 (1999) 50.
39. E. Evans and K. Ritchie, Biophys J. 72 (1997) 1541.

Chapter 28

MACROSCALE INSIGHT FROM NANOSCALE TESTING:
Manufacturing Applications of Single Asperity Plowing

Steven R. Schmid
Dept. Mechanical Engineering
University of Notre Dame
Notre Dame, IN

Louis G. Hector, Jr.
ALCOA Technical Center
ALCOA Center, PA.

Abstract With a very large number of applications, macroscale behavior is based upon hypotheses or guesses about material or process behavior, with no direct experimental evidence to confirm or deny the hypothesis. This is clearly the case with most tribological phenomenon, such as abrasion-based friction and wear, as well as boundary lubrication. Design rules such as the Archard wear law do not explain the function or suggest the value of a wear coefficient, for example, which in effect regulates tribologists to empiricism. Often, nanoscale simulations with only experimental or theoretical components are equally limited, since parameters such as physical dimensions and material properties are chosen to conveniently match the researcher's tools instead being correlated to a real problem.

This paper presents a summary of the research performed to date on single asperity plowing at depths and length scales that directly simulate real tribological contacts in manufacturing and machine design applications. Using a coupled experimental and theoretical approach, plausible and quantitative data is obtained to explain real-world, macroscale observations based on nanoscale phenomena. Applications emphasized in this paper are debris generation in metal forming, mechanisms of boundary lubrication, crystallographic anisotropy effects, and wear of polymers for orthopedic applications.

1 INTRODUCTION

There are numerous examples in engineering practice wherein a soft material surface is damaged by the plowing action of hard tool or mating surface asperities. Examples include metal rolling and extrusion, contact of gears of all types, rolling element bearings, artificial implants, grinding operations, *etc.* Asperity plowing of substrates can lead to elevated friction and wear levels since it is fundamentally a plastic deformation process. Excessive debris generation in the interface often results and the debris particles can transfer from one surface to another thereby degrading surface finish. Plowing can also occur from the action of loose particles in concentrated contacts, as is especially a concern in synovial contacts.

Engineering problems can be divided into two major classes, those where plowing is desirable or necessary (grinding, non-slip surfaces, ultrasonic machining, *etc.*) and those where plowing is undesirable (machine element contacts, extrusion, brake pads, *etc.*). Metal rolling is unusual in that plowing is both necessary so that sufficient friction can be generated to pull material through the rollers and also undesirable because of excessive friction and abrasive wear particles or smudge that result. Asperity plowing is a fundamental problem of enormous importance.

2 MATHEMATICAL APPROACHES: AZARKHIN-RICHMOND MODEL

A model system of plowing with a pyramidal indenter is shown in Fig. 1. A deformation field is posed wherein the plowed material moves

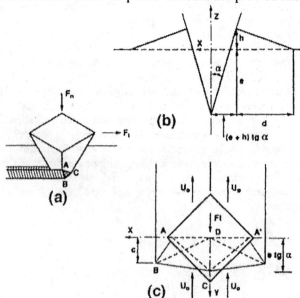

Figure 1. Geometry of the model program. (a) Oblique view, (b) edge view and (c) top view.

rigidly as it is sheared along planes BCD and ABD in response to the indentor motion. Furthermore, the material inside tetrahedron ABCD slides against the indentor along surface ACD. The rate of working, W, during the plowing process is the sum of the power dissipated in shear and the power dissipated in sliding against the indentor, or

$$W = 2|\vec{u}_0 - \vec{u}_1|\frac{\sigma_0}{\sqrt{3}}(<ABD> + <BCD>) + \\ + 2m\frac{\sigma_0}{\sqrt{3}}|\vec{u}_1| < ACD >, \quad (1)$$

where $\vec{u}_0$ is relative speed between the indentor and surrounding substrate (that is unaffected by plowing) and $\vec{u}_1$ is sliding velocity between the indentor and the material in the tetrahedron ABCD. The solution is obtained through the modified upper bound approach of Azarkhin & Richmond (1992), with typical results shown in Fig. 2. The important features of this model with respect to the remainder of this paper are that the ridge geometry is a function of the friction factor and that the penetration depth is a function of the material flow stress.

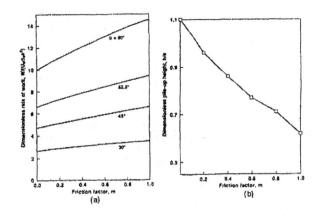

Figure 2. Typical results from the mathematical model. (a) Dimensionless power versus friction factor, and (b) ridge height versus friction factor.

3 EXPERIMENTAL METHOD

3.1 Atomic Force Microscope Operations

The instrumentation used for plowing experiments was a Digital Instruments Multimode AFM, which allows three principal operating modes: tapping mode, contact mode, and lithography mode. Tapping mode is best suited for imaging of surfaces because of the small contact forces, but this feature also makes this mode unsuitable for plowing operations. Contact mode has been used by some researchers to evaluate coating wear resistance (see, for example, Bhushan *et al.* 1994). However, this mode requires that indentors trace and then retrace a single track and hence during one of the traces, the blunt face of the indentor would be a leading edge. This geometry lies outside of the theoretical framework described above and is also not representative of most engineering applications where the geometry of the asperities does not change appreciably during plowing.

The operating mode used was lithography mode, which allowed considerable flexibility. Using C language based macros specific commands can be given to the piezoelectric scanner to control circuitry that executes these generally more austere tip-surface interactions.

3.2 Cantilever Developments

A wide variety of AFM cantilevers have been used by researchers. In order to obtain sufficient cantilever stiffness for plowing experiments and to utilize a hard and sharp cantilever, specially fabricated diamond-tipped stainless steel cantilevers were used in experiments. A typical SEM image of a diamond indentor is shown in Fig. 3. It should be noted that cantilevers with necessary sharpness were unavailable before this research encouraged development of new cantilever fabrication techniques, and these cantilevers are currently used in coatings and MEMS (Micro Electro-Mechanical Systems) research.

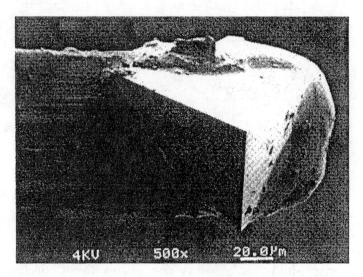

Figure 3. SEM image of diamond-tipped cantilever.

Determination of forces acting on an indentor in an atomic force microscope is a topic of considerable research as well as a fair amount of controversy. On one hand, some researchers recommend measurement of the cantilever dimensions in an SEM and then calculation of indentor forces from elastic cantilever beam theory and AFM-measured tip deflections. Alternatively, others suggest a purely experimental approach to determine cantilever stiffness (Ruan and Bhushan 1994).

All of these approaches are somewhat limited; cantilever beam theory cannot compensate for the contributions of the diamond and adhesive to the stiffness of the cantilever, and the experimental approaches require the use of a cantilever with known properties, which ultimately requires SEM measurements and cantilever theory application.

The approach of Hector & Schmid is strongly related to the experimental approach, with one major difference. The normal stiffness is directly measured using a precision scale (which can only be done with the subject extremely stiff cantilevers). This normal stiffness and the friction force mode investigations described by Ruan & Bhushan (1994) are used to obtain torsional stiffness. This approach then allows direct measurement of forces acting on the cantilever.

4 APPLICATIONS

4.1 Plowing of Aluminum Alloys

The main intention of the aluminum research to date was to investigate the reasons for the fundamentally different material responses to tooling and to explain the practical experiences of production personnel in the manufacture of aluminum. Three alloys, a 99.99% pure aluminum, a 5182 alloy and a 7150 alloy were prepared and plow tracks were generated at a variety of normal forces and plowing speeds.

Measurement of the plowed geometry and indentor forces allowed determination of the material flow stresses by applying the mathematical model described above. Typical results are shown in Figure 4, and it was noted that an indentation size effect (ISE) previously encountered in micro Vickers hardness testing also applied to plowing experiments. Figure 4 uses a curve fit of the form

$$H = ch^{m-2}, \qquad (2)$$

where H is measured hardness, c and m are experimentally determined values and h is the indentation depth. m is the indentation size effect (ISE) index, and has a value of 2 for a material whose hardness is independent of penetration depth.

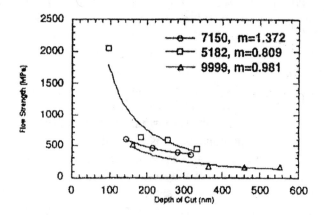

Figure 4. Typical flow stress results from AFM experiments.

From the experiments conducted, Schmid & Hector (1998) noted:

1. Pure aluminum was the best-behaved material, and always resulted in well defined plow tracks and shoulders. The plowing process reached steady state quickly and did not normally encounter fluctuations in indentor motion.

2. The 5182 material has magnesium as its main alloying element, and is recognized as a challenging material to form, having especially high friction. In the plowing experiments, 5182 displayed pronounced waviness in the ridge shoulders for many circumstances. Such circumstances occur in metal cutting applications when an unstable built up edge occurs on the tool nose, and this was verified to be the case here as well by light microscope observations of attached material after plowing. The pronounced ISE index from Figure 4 can be readily explained by a built up edge phenomenon as well – a built up edge makes the indentor appear blunter, requiring larger stresses to effect plowing.

3. 7150 is an aerospace alloy with high strength and good fracture toughness but less ductility than the other materials considered, using zinc as the major alloying element. The 7150 displayed necking and detachment of the ridges in plowing experiments, and an occasional ridge waviness, although far less severe than the 5182 material.

4.2 Plowing in the Presence of Boundary Additives

In the rolling and ironing of some aluminum alloys, especially 3004 (used extensively in the beverage container industry), small wear particles are often removed from the workpiece and adhere to the tooling. 3004 contains hard $Al_{12}(Mn|Fe)_3Si$ constituent particles which are harder than the surrounding media and are thought to aid in the removal of tooling debris through an abrasion process. This behavior can be clearly seen in Fig. 5(a), where a plow track crosses such a particle at a constant normal force. This cleaning of debris from the tooling is necessary, or else surface degradation quickly results in aesthetically unappealing products, but this debris then collects on the workpiece surface. These small loosely attached particles give the impression of a dirty workpiece, hence the practice of referring to this residue as smudge.

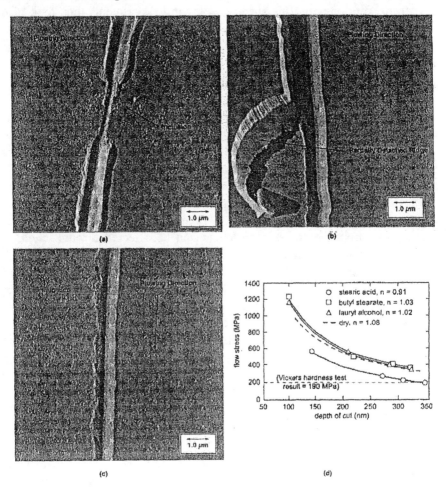

Figure 5. Results from plowing with boundary lubricants. (a) Effect of inclusion in plowing 3004 alloy, (b) plowing in the presence of butyl stearate, (c), plowing in the presence of stearic acid, and (d) resultant surface flow stress predictions.

Reich et al. (1994) found that smudge was directly related to abrasion processes, and it was known in the manufacturing environment that the lubricant additives used had a strong effect on the amount of smudge generated. Since all operations occur in the boundary regime, there is no apparent reason for this effect - asperity plowing takes place regardless of boundary additive used.

A number of 3004 alloy pucks were prepared and boundary films of selected additives (butyle stearate, stearic acid, and lauryl alcohol) were cast from solutions onto the surface. Plowing experiments then took place in a similar fashion to previous work on unlubricated surfaces. A clear distinction could be made, as seen in Figs. 5(b) and 5(c), where the ridges for the 3004 lubricated with butyl stearate are partially detached but still connected to the material, while the ridges for a stearic acid lubricated material are totally removed.

A more provocative finding (Opalka, et al., in press) was the ability to discern the mechanism through which a boundary lubricant acts. Most tribology texts will describe a boundary film mechanism, where a polar film strongly attaches to a surface and prevents adhesion with opposing asperities. A less often described mechanism is one of chemo-mechanical effects, to which the interested reader is directed to the excellent work of Westwood and Lockwood (1982). Basically, the thought is that the mechanical properties of surface layers can be affected through chemical surface interactions. Since the plowing experiments conducted on an AFM are not strongly affected by adhesion, a boundary lubricant which depends on a boundary film mechanism will not be readily discernable from dry plowing. On the other hand, if the surface layer mechanical properties are affected, plowing experiments will readily detect this mechanism.

This difference is clearly visible when comparing the flow stresses obtained from the mathematical model for the three boundary additives compared to dry plowing. As seen in Figure 5(d), the flow stress is totally unaffected by the presence of butyl stearate or lauryl alcohol, but is clearly weakened by the presence of stearic acid.

These experiments have demonstrated that these single asperity plowing experiments are useful for screening lubricant additives in terms of preventing smudge, and also for determining the lubrication mechanism for a given additive. As such, this is a very valuable diagnosis tool and also a method of performing fundamental research on lubrication theory.

4.3 Single Grain Plowing

The pucks used in experiments are annealed, with columnar equiaxed grains 1.0 to 5.0 millimeters in diameter (they are visible with the naked eye). This assures that plowing experiments do not occur on multiple grains or on grain boundaries, and is one reason for the well-behaved systems investigated to date. Another subtle point is that surface slip lines are visible on the surface of the puck (apparently from using pliers to remove the pucks from the adhesive in the fixture used in polishing). All plowing experiments conducted in the research described above were performed so that the plowing direction was parallel to visible surface slip lines. It was desired to determine the effect of grain orientation on the plowing process.

The crystallographic orientations of selected surface grains on the pucks were first determined with a scanning electron microscope-based technique known as orientation imaging microscopy (OIM). During OIM, the sample volume is inclined to the incident electron beam by 70° and subsequently stepped across a crystalline surface. Any sample volume with an undisturbed lattice will give rise to a characteristic diffraction pattern which is a two- dimensional projection of a three-dimensional unit cell. From this diffraction pattern, the three angles relating the crystal coordinate system with the sample system (that is the crystallographic orientation) are calculated using automated pattern recognition procedures. In this way, approximately 5000-10000 orientations (*i.e.*, data points) can be determined in one hour. The probe/sample interaction volume for aluminum determines the lateral resolution which is typically 0.5 μm. In addition to providing quantitative information about crystallographic texture, OIM can also provide insight into grain size, misorientation across grain boundaries, and local strain.

Plowing was conducted in the [111], [100], and [110] directions of the crystals, as well as on a (111) plane perpendicular to the initial [111] direction. Two sets of plow tracks were generated in the four crystallographic directions, each set containing four plow tracks generated at progressively increasing normal forces.

While a mathematical model incorporating anisotropic behavior of single crystals is in development, we applied the existing model to the experiments. Typical results are shown in Figure 6. The high purity alloy has the lowest flow stresses under all normal forces and directions, as expected. Low amounts of alloying elements result in substitutionally embedded atoms which act as obstacles to dislocation movement along glide planes, resulting in measurable increase in flow stress.

4.4 In-Situ Tension Test and Roughening Measurements

It is well-known that metals roughen when subjected to plastic deformation. It is also well-known that this is a very complex phenomenon, driven by the competition between Lüders band propagation to surfaces,

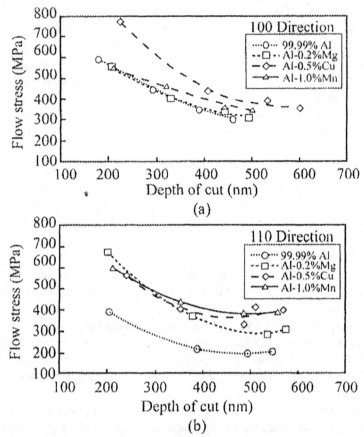

Figure 6. Flow stress as a function of indentation depth for know crystallographic orientation.

relative grain motion at the surface, and flattening of asperities by hard tooling. The understanding and control of surface roughening is essential for aerospace and automotive industries, and has been a topic of considerable macroscale research to date. Lüders bands will cause roughening if a slip system exists which can bring these slip bands to a surface; hence crystallographic orientation and crystal structure are primary contributing factors to roughening. This can be seen in Figure 7, where the evolution of roughness is seen in a grain at the surface of aluminum strip.

ACKNOWLEDGEMENTS

The author gratefully acknowledges the continued support of Dr. Jorn Larsen-Basse, and the National Science Foundation under CAREER grant

No. NSF-CMS96-23412 and GOALI grant No. NSF-CMS96-31384. Matt Thompson and Jeff Elings of Digital Equipment helped with the experimental method and cantilever developments, respectively, and Dr. Bharat Bhushan generously offered his advice on many aspects of the work.

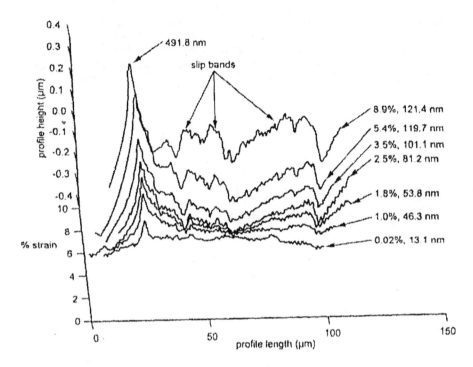

Figure 7. Evolution of surface roughness in aluminum under tensile strain.

REFERENCES

Azarkhin, A. and Richmond, O., 1989. A generalization of the upper bound method with particular reference to the problem of a ploughing indentor. *J. App. Mech.* 56: 10-14.

Azarkhin, A., and Richmond, O., 1990. On friction of ploughing by rigid asperities in the presence of straining. Upper bound method. *J. Trib.* 112: 324-329.

Azarkhin, A. and Richmond, O., 1992. A model of ploughing by a pyramidal indentor ñ upper bound method for stress-free surfaces. *Wear* 157: 409-418.

Bhushan, B., Koinkar, V.N., & Ruan, J. 1994. Microtribology of magnetic media. *Proc. Inst. Mech. Eng. Ser. J: J. Eng. Trib.* 116 (1994): 17-29.

Hector, L.G., and Schmid, S.R. 1998. Simulation of asperity plowing in an atomic force microscope. Part I: experimental and theoretical methods. *Wear* 215: 247-256.

Opalka, S.M., Hector, L.G., Schmid, S.R., Reich, R.A., and Epp, J.A., in press. Boundary additive effect on abrasive wear during single asperity plowing of a 3004 aluminum alloy. To appear in *J. Trib.*

Reich, R.A., Epp, J.A., and Gantzer, D.E. 1994. A mechanism for generating aluminum debris in the roll bite and its partitioning between the surface of the work roll and the surface of the sheet as smudge. *Trib. Trans.* 39: 23-32.

Ruan, J., and Bhushan, B. 1994. Atomic-scale friction measurements using friction-force microscopy: Part I. General principals and new measurement techniques. *J. Trib.* 116: 378-388.

Schmid, S.R., and Hector, L.G., 1998. Simulation of asperity plowing in an atomic force microscope. Part II: plowing of aluminum alloys. *Wear* 215: 257-266.

Westwood, A.R.C, and Lockwood, F.E., 1982. Chemomechanical effects in Lubrication. In Georges, J.M., ed., Microscopic Aspects of Adhesion and Lubrication. New York: Elsevier

Chapter 29

SURFACE FORCES AND FRICTION BETWEEN CELLULOSE SURFACES IN AQUEOUS MEDIA

Stefan Zauscher
State University of New York-College of Environmental Science and Forestry
Faculty of Paper Science and Engineering
Syracuse, NY 13210
zauscher@esf.edu

Daniel J. Klingenberg
University of Wisconsin-Madison
Department of Chemical Engineering and Rheology Research Center
Madison, WI 53706

Abstract Colloidal probe microscopy was employed to study normal forces and sliding friction between model cellulose surfaces in aqueous solutions. Hydrodynamic interactions must be accounted for in data analysis. Long-range interactions are governed by double layer forces, and once surfaces contact, by osmotic repulsive forces and visco-elasticity. Increasing the ionic strength decreases surface potentials and increases adhesive forces. Polyelectrolytes cause strong steric repulsion at high surface coverage, where interactions are sensitive to probe velocity. Polymer bridging occurs at low coverage.
Regardless of scan size, friction exhibits irregular stick-slip behavior related to surface roughness. At small scan sizes (on the order of 10 nm) the lateral force decreases with increasing load. Above a critical scan size of about 100 nm corresponding to the average size of asperities on one of the model surfaces–lateral forces are independent of scan size, but depend on the load. Hydrodynamic forces contribute little to friction. Small amounts of high molecular weight, water-soluble polymers significantly decrease sliding friction between cellulose surfaces.

1 INTRODUCTION

In papermaking, paper converting, and end-use, a great number of colloidal, hydrodynamic, and nano-mechanical interactions are important. For example, cellulosic fibers and fiber fragments, polymers, filler particles, water, and air interact in a complex way while paper is formed on the paper machine. Nanometer-sized cellulose particles and mineral fillers interact with polymers and cellulose surfaces dynamically in fractions of seconds and are retained in the fiber network during drainage [1, 2]. In practice the paper maker is interested in increasing the efficiency of particle deposition, and maximizing the filler retention without causing negative effects on paper formation by fiber flocculation. Aside from particle size and hydrodynamic conditions [1], polymer adsorption kinetics, extent of the adsorbed layer, and solvent conditions are important for particle-fiber, and fiber-fiber interactions [2]. Very little is known about friction between cellulosic surfaces in the presence of polyelectrolytes. Direct measurement of these interactions on the particle can thus provide insight to particle retention mechanisms, particle adhesion, and ultimately to suspension structure.

Although our paper's focus is on interactions between cellulose surfaces, this study should have bearing on other areas of colloid science, bio-interfaces, and nanotechnology, where the direct measurement of surface and friction forces between polymeric surfaces in aqueous environments is important.

Forces acting between slowly approaching cellulose surfaces were first measured by Neuman et al. [3] on regenerated, spin-casted cellulose films, and more recently by Holmberg et al. [4] on Langmuir-Blodgett (LB) cellulose films, using the surface forces apparatus (SFA, [5]). Cellulose-cellulose interactions were also measured between spherical cellulose particles with colloidal probe microscopy (CPM, [6, 7]) [8, 9].

Normal forces between surfaces in aqueous solutions of uncharged polymers [10-13] and polyelectrolytes have been studied extensively with the SFA [3, 14-19] and CPM [20, 21]. The surfaces employed were typically mica, silica, or alumina. Despite the importance of polymer-mediated interactions in papermaking, there have been very few studies in which these interactions were measured directly [22-24].

Recent simulations have shown that the tendency of fibers to form persistent flocs is a direct consequence of inter-fiber friction [25, 26]. Few friction measurements for cellulosic fibers in liquid media are available [27, 28]. In view of the limited data, there is need for a more fundamental understanding of sliding friction between cellulose surfaces in aqueous systems, particularly in presence of water-soluble polymers.

Force measurements between well-characterized model surfaces provide a means to better understand the complex phenomena occurring in papermaking suspensions. In this paper we summarize results from normal and lateral force measurements between model cellulose surfaces in aqueous media and how these forces are affected by water-soluble polymers.

To the author's knowledge the present work discusses for the first time friction measurements between polymeric surfaces in aqueous systems using colloidal probe microscopy. Our findings may therefore be relevant for tribological phenomena between hydrophilic polymeric surfaces in general.

2 MATERIALS AND METHODS

2.1 Chemicals

Sodium chloride (ACS grade) was used as received. Filtered MQ-water ("Milli-Q;" Resistivity: 18.2 MΩ/cm, pH 5.5, Millipore, Bedford, MA) was used for all aqueous solutions and for rinsing. Anionic sodium carboxymethyl cellulose was used as the polyelectrolyte (Na-CMC 7H4-F, Mw: $(5.7\pm1.0)\times10^5$ g/mol, Rg: 104±8 nm, Hercules, Wilmington, DE).

2.2 Cellulose Surfaces

We encountered serious problems in preparing flat cellulose surfaces that would sustain the shear forces in lateral force measurements without delamination from the sample holder. Ultrathin cellulose films (order 70 nm) that were spin-casted from an organic solvent onto etched (70% H_2SO_4, 30% H_2O_2) SiO_2 surfaces and subsequently regenerated proved good substrates for normal force measurements, but were unsuitable for lateral force measurements. We finally chose regenerated cellulose membranes that were mounted dry. The demands on the adhesive bond between the film and the sample holder are severe because the film swells considerably when exposed to water. Typical two-component epoxy resins proved to be unsuitable for sample mounting because of the risk of sample contamination by uncured monomers. XPS analyses on two-component epoxy mounted films showed higher fractions of C-C bonds on the sample surface than for the unmounted control. Single component UV-curing adhesives performed extremely well with respect to both sample contamination as well as maintaining a bond after water exposure.

Flat cellulose surfaces were prepared from solvent-extracted, regenerated cellulose films (Spectra/Por 4, Spectrum, Laguna Hills, CA). Spherical, regenerated cellulose beads (Cellufine GC-15(uf), diameter 15–25 μm, Chisso Corp., Tokyo, Japan) were used for cantilever modification. Beads were washed in acetone, and after drying, dialyzed in several changes of MQ-water. Colloidal probes were prepared by attaching a cellulose bead to a silicon nitride AFM cantilever with a one-component, solvent-free, epoxy adhesive (NOA 81, Norland Products, New Brunswick, NJ) [29]. We verified the cellulosic nature and the cleanliness of the regenerated cellulose surfaces with x-ray photoelectron spectroscopy by evaluating the chemical surface composition and the carbon bonding environment by C1s peak deconvolution [30].

Adsorption of Na-CMC on regenerated cellulose surfaces was measured by depletion. Adsorption isotherms for Na-CMC adsorption onto cellulose films and beads are plotted in Fig. 1. Na-CMC adsorption onto negatively charged cellulose surfaces occurs in spite of the negative charge of the polyelectrolyte. This is partly explained by the high electrolyte concentration (0.01 M NaCl) that lowers the electrostatic barrier to adsorption by screening surface charges, as well as polymer charges. Na-CMC has a rigid backbone, and in the case where non-electrostatic interactions drive adsorption, adsorbs in a flat conformation, *i.e.*, mostly in trains, with some chain ends extending into solution [31].

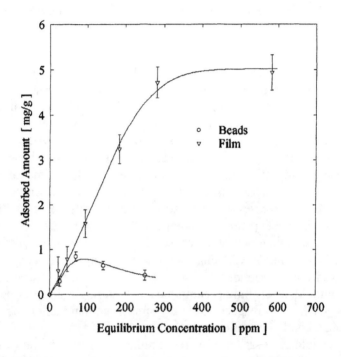

Figure 1. Adsorbed amount as a function of polymer equilibrium concentration for Na-CMC (in 0.01 M NaCl), adsorbed onto regenerated cellulose beads and films.

The charge densities of cellulose films and beads were determined by conductometric titration [32]. Charge number densities were 19 and 5 mmol/kg for the regenerated cellulose films and beads, respectively. Titration plots indicated that charge arises from weak acids.

Surface topography and roughness of cellulose films and beads in MQ-water were determined from flattened, but otherwise unfiltered AFM images. AFM scans on numerous films revealed a surface composed of domains on the order of 100 nm. For a 1×1 μm scan, the root-mean-squared (rms) roughness was between 3 and 4 nm. Cellulose beads were rougher, with an rms roughness typically on the order of 35 nm for a 5×5 μm scan.

2.3 Colloidal Probe Microscopy

2.3.1 Cantilever Calibration

Normal and lateral forces were measured with standard, 100 µm long, wide-legged, triangular, silicon nitride cantilevers (Model NP, Digital Instruments, Santa Barbara, CA). Bending and twisting stiffness was determined analytically from measured cantilever dimensions assuming a modulus of elasticity of 150 GPa and a Poisson's ratio of 0.25 [33, 34], taking into account cantilever inclination [35] and the attached colloidal probe [36].

2.3.2 Force Measurements and Cross-Correlation

We used a MultiMode AFM with a Nanoscope IIIa controller (Digital Instruments, Santa Barbara, CA) [36]. All aqueous solutions were filtered, silicone hoses and fittings were rinsed in several changes of MQ water and boiled for at least 1 hr prior to an experiment. All experiments were performed at room temperature. Mechanical and thermal drift were minimized by equilibrating the system at least 2 hrs prior to measurements. The typical wait period after a change of the liquid in the fluid cell was between 30 min and 1 hr.

The normal force zero was determined by centering the approach and retraction traces at sufficiently large surface separations about the zero deflection voltage. Surface separation and force magnitude were determined from the constant compliance regime [7, 36-39].

Friction was measured with lateral force microscopy [40, 41]. Friction experiments were performed under controlled normal forces in line-scan mode (*i.e.*, the y-scan direction was disabled). The torsion signal from the photodetector (PSD) was corrected for tilt and an average torsion signal was calculated with

$$\langle \Delta U_t \rangle = \frac{\langle \Delta U_t^{(+)} \rangle - \langle \Delta U_t^{(-)} \rangle}{2}, \quad (1)$$

where $\langle \Delta U_t^{(+)} \rangle$ and $\langle \Delta U_t^{(-)} \rangle$ are the average torsion voltages for sliding in the positive and negative x-directions, respectively.

The subtraction in Equation 1 removes signal contributions due to PSD misalignment [40] and also cancels some topographic influence that is common in both scan directions [42, 43]. End effects due to static friction before the onset of sliding were excluded.

The torsional sensitivity of the PSD was calibrated with two colloidal probe modified cantilevers and with an unmodified lever using the method by O'Shea [44]. We recorded friction loops over a few nm displacement in linescan mode at loads on the order of 10 nN. The initial slope in the lateral force versus lateral displacement plot was assumed to represent the total lateral stiffness,

$$\frac{dF_{lat.}}{dx} = k_{lat.}^{tot.} = \left(\frac{1}{k_{lat.}^{lev.}} + \frac{1}{k_{lat.}^{cont.}}\right)^{-1}, \quad (2)$$

where $k_{lat.}^{lev.}$ is the lateral stiffness of the lever/colloidal probe assembly, and $k_{lat.}^{cont.}$ is the lateral stiffness of the contact. Assuming a single asperity contact, then the lateral contact stiffness depends on the reduced surface shear modulus of the contacting materials, and on the contact radius [45]. The contact radius is a function of load, asperity size, and the reduced surface elastic modulus of the interacting materials [45, 46]. A calculation–assuming an elastic modulus of wet cellulose of 0.1 GPa–shows that $k_{lat.}^{lev.}$ and $k_{lat.}^{cont.}$ likely are of the same order [45]. Since we have not performed lateral stiffness measurements at different loads, and have not attempted to separate lateral lever stiffness from contact stiffness, there is an uncertainty associated with the one-point PSD calibration. This results in an uncertainty in the magnitude of the lateral force but does not affect relative differences in lateral force when comparing different experiments. The lateral force is not necessarily equal to the friction force because of the convolution of lateral stiffness and load-dependent contact stiffness, and in this paper we report only lateral force. Other implications of the convolution of lateral stiffness and contact stiffness will be considered throughout the paper.

During friction measurements, the lateral force and height signal were recorded simultaneously. The height signal represents a convolution of the topographies of the sphere and the film surfaces and contains a contribution due to the twist in the cantilever because normal and lateral cantilever motion are coupled. We found that at low applied loads there was a significant spatial correlation between the lateral force signal and the height signal. This prompted us to quantify this correlation with the cross-correlation function (CCF), which is a time-averaged measure of shared signal properties. The CCF is useful when comparing two signals because it retains information about the relative phases of the two signals which it is comparing [47]. The cross-correlation between the two signals x and y sampled with a sampling interval T is

$$r_{xy}(k) = \lim_{N \to \infty} \frac{1}{(2N+1)} \sum_{m=-N}^{N} x_m y_{m+k}, \quad (3)$$

where x_m and y_{m+k} represent two sample values separated by kT seconds, and the summation has $(2N+1)$ terms [47]. The CCF is thus a function of the imposed time-shift (in our case expressed as a spatial shift or lag) between the two signals. Since we shifted two non-periodic, discrete signals relative to each other, end effects were accounted for by "zero-padding" the data [48].

2.4 Data Modeling

2.4.1 Equilibrium Forces

Surface forces in aqueous electrolyte solutions were modeled with DLVO theory [49, 50] using the linearized Derjaguin approximation for a sphere interacting with a plane [51]. Electrostatic forces were calculated analytically in the limit of small potentials using an improved solution for the constant charge boundary condition [52].

The Hamaker constant for the interaction of two surfaces in an aqueous medium was calculated from optical and dielectric properties of the system using Lifshitz theory, taking ion screening into account [51]. For silica glass the Hamaker constant ranged from $(0.9–1.2) \times 10^{-20}$ J in agreement with Refs. [7, 53]. For cellulose the values ranged from $(0.64–0.88) \times 10^{-20}$ J, in agreement with Refs. [4, 54].

2.4.2 Hydrodynamic Forces

Hydrodynamic forces arise in CPM from the motion of the colloidal probe and cantilever relative to the surrounding fluid. When the separation between the probe and film surfaces is small, the flow in the small gap produces the dominant contribution to hydrodynamic force on the probe/cantilever assembly (hydrodynamic lubrication). Despite their importance, little attention has been devoted to these forces in CPM [35, 55].

During a typical CPM normal force experiment, the separation between the probe and film surfaces varies from a few hundred nanometers to near zero, while the distance between the film surface and cantilever remains roughly constant at approximately the probe diameter [order of 10 μm]. Thus the contribution of the hydrodynamic lubrication between the probe and film surfaces to the total hydrodynamic force on the probe/cantilever assembly varies rapidly with separation; however, the contribution of the hydrodynamic drag on the cantilever to the total hydrodynamic force is essentially independent of surface separation.

We model the hydrodynamic force on the probe/cantilever assembly as the sum of the separation-dependent hydrodynamic force on the probe plus the hydrodynamic drag on the cantilever. The hydrodynamic force on the probe is described by the lubrication approximation for the approach of a smooth sphere toward a smooth plane [56-58], neglecting all surface deformations [56]. Thus, for a surface separation $h(t)$, the total hydrodynamic force on the probe/cantilever assembly normal to the surface is

$$F_{hyd} = -6\pi\eta a \frac{dh(t)}{dt}\left[\frac{a}{h(t)} - \frac{1}{5}\ln\left(\frac{a}{h(t)}\right) + 0.971264\right] - A\eta\frac{dh(t)}{dt}, \quad (4)$$

where η is the fluid viscosity, a is the sphere radius, $h(t) = x_1(t) - x_2(t) - a$ is the surface separation, $x_1(t)$ is the bead center-of-mass position, $x_2(t)$ is the surface position, and A is a geometric prefactor describing the drag on the cantilever.

DLVO and hydrodynamic forces between two surfaces depend sensitively on surface separation and particle geometry. In reality, most particles are not smooth. The effects of these asperities on interactions become noticeable when the surfaces interact at distances on the order of the asperity size. The local radius of curvature of the largest asperity will then become the relevant length scale [59]. The magnitude of observed interaction forces is often less than that predicted by calculations using the overall particle dimension.

2.4.3 Forces Due to Polymers

In the presence of polymers, repulsive steric forces, attractive bridging forces, and attractive depletion forces can arise. The magnitude and extent of these forces depend on polymer concentration, surface affinity, solvent quality, ionic strength, and on the dynamics of the measurement [10, 11, 60]. Our normal force measurements were performed on time scales that were likely too short for thermodynamic equilibrium of the adsorbed polymer to be reached [10]; thus we did not model these forces.

2.4.4 Contact Forces

The contact area between the cellulose surfaces is not well defined because of their surface roughnesses, however, in case of a single asperity, the normal contact stiffness is directly proportional to the contact radius (Hertz case). Estimates for the reduced elastic modulus for the wet cellulose sphere-film system are on the order of 0.1 GPa [36]. Using Hertz's theory, assuming a contact size of 100 nm and a load of 10 nN, the contact stiffness will only be about a factor 10 larger than the cantilever stiffness [46]. This means that some of the deformation in a normal force experiment will occur in the contact.

Carpick et al. [45] have shown that the shear strength of a contact can be calculated from measurements of total lateral stiffness (measured by sinusoidal, angstrom-sized oscillations to avoid slip) and from lateral force data, without assuming a particular interaction model. For a sphere-plane contact the lateral contact stiffness is $k_{lat.}^{cont.} = 8Gr$, where G is the reduced shear modulus of the sphere-film system, and r is the contact radius, which is independent of the tip-sample interaction forces. Carpick et al. [45] also showed that the total lateral stiffness can be a function of load; therefore, in order to accurately separate the effects of lateral contact stiffness from friction, the total lateral stiffness needs to be determined in conjunction with each friction experiment. This was not done in the present study.

3 RESULTS AND DISCUSSION

3.1 Normal Force Measurements

3.1.1 Hydrodynamic Interactions

To study the effect of hydrodynamic interactions, we measured the forces between two silica-glass surfaces in 0.1 M NaCl at several approach speeds (Fig. 2). Data at each speed are averaged over five consecutive approaches. The solid curves in Figure 2 are the best fits of the linear superposition of the hydrodynamic force (Eq. 4) and the screened van der Waals force, with the particle radius and the constant A as fitting parameters. Electrostatic repulsion is negligible in 0.1 M NaCl for separations larger than a few nanometers. The fitted sphere radius (8.6 ± 0.2 μm) was independent of approach speed, and matched the optically determined radius (8.6±0.3 μm). The parameter A obtained from the fit was $4.9 \times 10^{-3} \pm 0.1 \times 10^{-3}$ m.

This analysis shows that hydrodynamic forces must be considered at large approach speeds, and that hydrodynamic drag accounts for a small, speed-dependent force offset at large separations. The excellent agreement of the fitted and the measured sphere radii confirm our cantilever calibration.

For cellulose normal force experiments we therefore measured normal forces at approach speeds that were low enough so that contributions to hydrodynamic forces could be neglected. However, in data modeling we took into account the hydrodynamic drag term.

3.1.2 Normal Forces Between Cellulose Surfaces in Electrolyte Solutions

In most normal force measurements between cellulose surfaces, a speed-dependent hysteresis occurred between approach and retraction in the constant compliance regime. This hysteresis likely arises from load-dependent friction forces between the sphere and the surface [35].

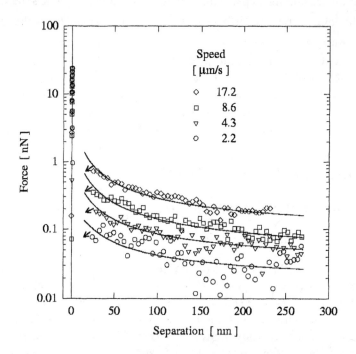

Figure 2. Force as a function of separation between two approaching silica-glass surfaces in 0.1 M NaCl. The solid curves are the predictions from linear superposition of hydrodynamic and van der Waals forces. Arrows indicate the onset of cantilever instability.

Figure 3 shows the interaction between two approaching cellulose surfaces in aqueous NaCl solutions as a function of apparent surface separation. The data represent the average of five consecutive approach-retraction cycles, each at an approach speed of 0.9 μm/s.

In contrast to quasi-static force measurements that showed that cellulose conformations relax slowly upon first compression [3, 4], we did not notice any difference in the range of the interaction between the first and subsequent approaches. Our force-separation profiles thus likely represent a transient rather than a thermodynamic equilibrium state.

Two monotonically repulsive interaction regimes are apparent in Figure 3(a). In Regime I, at separations larger than about 10 nm, interactions decay almost exponentially and their range decreases with increasing electrolyte concentration, in agreement with predictions of DLVO theory. Predicted surface potentials are low, and agree with zeta-potentials reported for cellulose [61, 62], and a value reported by Rutland *et al.* [8] using CPM. Low potentials also reflect the low charge densities found by charge titration. In Regime II, below about 10 nm, the interactions decrease more steeply with surface separation. The transition between Regime I and II was associated with the first contact between the approaching surfaces.

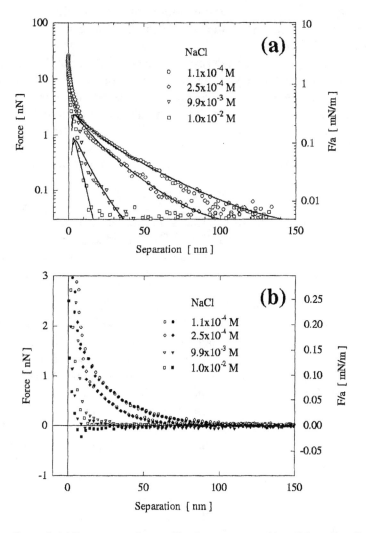

Figure 3. (a) Force-separation profiles for two approaching (0.9 μm/s) cellulose surfaces in aqueous NaCl solutions. The solid curves are the fits to DLVO theory, using the linearized Derjaguin approximation with constant charge boundary conditions, including hydrodynamic forces. Corresponding surface potentials (Ψ) and Debye lengths (κ^{-1}) are: in 1.1×10^{-4} M NaCl, $\Psi = -21$ mV, $\kappa^{-1} = 31$ nm; in 2.5×10^{-4} M NaCl, $\Psi = -18$ mV, $\kappa^{-1} = 20$ nm; in 9.9×10^{-4} M NaCl, $\Psi = -11$ mV, $\kappa^{-1} = 10$ nm; and in 1.0×10^{-2} M NaCl, $\Psi =$ uncertain, $\kappa^{-1} = 4$ nm. (b) Force upon approach (open symbols) and retraction (filled symbols) as a function of surface separation for the data from Figure 3(a).

To describe the force behavior in Regime II, we assume that the cellulose-solution interface is diffuse. When the surfaces are brought together, they contact at discrete contact spots, and both the polymer configuration and the charge distribution in the gel matrix accommodate the applied pressure. Upon compression, more volume is excluded and osmotic repulsive forces increase, in addition to the elastic restoring force of the compressed gel network.

Adhesion and adhesion hysteresis between cellulose surfaces interacting in MQ water occurred rarely, in agreement with observations by Holmberg et al. [4] for the interaction after several approach-retraction cycles. At high ionic strengths a small secondary minimum upon approach and an adhesive minimum upon retraction developed. A small pull-off force was required to separate the surfaces (Fig. 3(b)). Furthermore, hysteresis between approach and retraction increased with increasing electrolyte concentration.

3.1.3 Forces Between Cellulose Surfaces in Polyelectrolyte Solutions

Forces between cellulose surfaces in polyelectrolyte solutions were always repulsive at large separations. The reproducible interaction in the non-contact, repulsive regime resembles the non-equilibrium (un-relaxed) interaction found for adsorbed, uncharged polymers after repeated approaches [10, 11, 13].

Figure 4(a) shows force-separation profiles for interacting cellulose surfaces at different Na-CMC (high MW) concentrations in 0.01 M NaCl. Best fit parameters for a DLVO constant charge fit for no Na-CMC are $\Psi = -9$ mV and $\kappa^{-1} = 8.2$ nm. The data are averaged over five consecutive approach-retraction cycles, at a constant speed of 0.9 μm/s. To avoid clutter, only every fifth data point is plotted. The equilibration time between changes in polymer concentration was about 1 hr, with the surfaces separated approximately 50 μm.

Forces at high polymer concentrations (98–493 ppm) are monotonically repulsive, and their range increases with increasing polymer concentration. Force onset upon compression appears in the separation range 200–400 nm (Fig. 4(b)). At small polymer concentration (21 ppm), the interaction is significantly different. The repulsive force increases slowly with compression at large separations, but increases steeply below about 50 nm. The interaction matches that of two bare cellulose surfaces in electrolyte, but is offset by about 15 nm to a larger separation.

The force decay lengths at large polymer concentrations are significantly larger than the Debye screening length. This likely arises from osmotic interactions between compressed adsorbed polymer layers in contact [10, 11]. Adhesive forces were absent at high polymer concentration. Although the approach speed was constant, hydrodynamic-

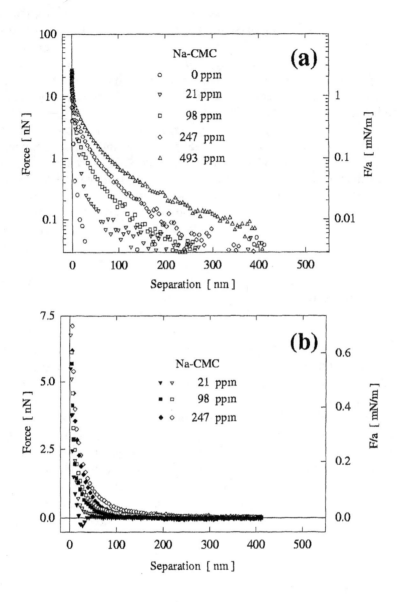

Figure 4. (a) Forces between two cellulose surfaces as a function of surface separation in 0.01 M NaCl and at four different Na-CMC concentrations upon approach. (b) Force upon approach and retraction as a function of surface separation for three of the polymer concentrations from Figure 4(a). Open and filled symbols are for approach and retraction, respectively.

like offset forces increased with polymer concentration. We attribute this to an increased hydrodynamic force caused by increased solution viscosity.

The extent of the steric interaction is on the order of 1–2 Rg per surface and increases with increasing polymer concentration. For adsorbed, uncharged homopolymers, equilibrium interaction distances of up to 4 Rg per surface have been reported [10, 11]. The lack of electrostatically controlled interactions in the outer region of the adsorbed polymers is attributed to ionic screening, and to a diffuse adsorbed region due to the broad molecular weight distribution.

Conformational information is contained in the short range data. Figure 4(a) shows that between 21 and 98 ppm, the interaction changes from double-layer controlled to sterically dominated. This concentration regime also corresponds to the transition from partial to full bead surface coverage. These results suggest also that a transition occurs in this concentration range from an essentially flat to a more expanded adsorbed polymer conformation.

The small range of the repulsive force upon approach, and the corresponding adhesive force minimum upon retraction at low polymer concentration indicate polymer bridging (Fig. 4(b)). Polymer bridging occurs if surfaces are sufficiently close and adsorption sites are available on the opposing surface, *i.e.*, at partial surface coverage [2, 63]. We did not observe attraction on approach, likely because the approach times were shorter than those necessary for Na-CMC to form a bridge. Repeated contact at polymer concentrations that promote bridging lowered the adhesive force magnitude upon separation. At small separation speeds (about 0.1 μm/s) and small polymer concentrations, adhesive force minima were more noticeable. Similar trends for polyelectrolytes adsorbed onto hard surfaces have been observed [20, 21]. Small speeds also imply long surface contact times in the constant compliance regime. The longer the contact time, the higher the probability for polymer inter-diffusion; at polymer concentrations that promote bridging, more contact time will favor the formation of polymer bridges. Adhesive force minima upon separation are consistent with disentanglement of individual polymer chains that are tethered to the two separating surfaces [64, 65]. Adhesive minima were absent at large polymer concentrations (>98 ppm).

If polymer is adsorbed on dynamically interacting surfaces, then non-equilibrium forces depend on polymer chain dynamics and viscous effects. Force hysteresis occurred at large polymer concentrations and large speed (0.9 μm/s); hysteresis was absent when the speed was reduced to 0.09 μm/s. This speed dependence of the force hysteresis between approach and retraction can be explained by hydrodynamic forces [60], and a non-equilibrium state of the polymer segment density in the gap [10]. Hydrodynamic forces that arise from viscous drag of the solvent on the immobilized polymer network decrease with decreasing speed [60].

Klein and Luckham [10] attributed force hysteresis to a larger, non-equilibrium polymer surface segment density after compression. Upon decompression, the polymer concentration in the gap is momentarily lower, and a lower disjoining pressure results. Although their analysis was applied to explain the difference between relaxed and un-relaxed force profiles [11], an analogous mechanism may also explain differences in our un-relaxed force profiles.

Experiments with medium and low molecular weight Na-CMC showed that the extent of the interaction depends on molecular weight, and increases linearly with increasing molecular weight at small and intermediate polymer concentrations. At large polymer concentrations (250 and 500 ppm) the extent of interaction depends only weakly on polymer concentration, which suggests that both surfaces reached complete polymer coverage.

Interactions in solutions containing a cationic polyelectrolyte (cationic polyacrylamide) were similar to those found for Na-CMC; however, due to molecular differences between the polymers, the extent of the steric layer was larger for C-PAM at comparable concentrations [36, 66].

3.2 Friction Force Measurements

Friction behavior depends on scan size. In regime I (< 100 nm), the friction (lateral) force increases rapidly with scan size, and in regime II (> 100 nm) it is nearly independent of scan size. A likely explanation for the change in lateral force behavior at a scan size of about 100 nm is the change in surface topography near 100 nm. Friction at scan sizes below about 100 nm is likely associated with sliding within 100 nm domains. The scan size independent lateral force in regime II suggests that bead asperities interact with multiple film domains during a sliding cycle.

3.2.1 Friction Force in Scan Size Regime I

The lateral force pattern for sliding in 0.01 M NaCl at 20 nm/s is discussed below and is representative of that at smaller sliding speeds, for sliding in MQ water, and for scan sizes < 40 nm. At smaller sliding speeds, signal quality deteriorated because of signal drift and vibrational disturbances, and at larger scan sizes, the surface roughness contribution to lateral force became more dominant.

At small applied loads, sliding was associated with a stick-slip pattern, where cantilever vibrational noise was superimposed on larger stick segments (Fig. 5(a)). In contrast to the behavior in scan size regime II, the average lateral force did not increase linearly with increasing normal force for scan sizes < 40 nm. The lateral force approaches a constant at large normal forces. The characteristic stick-slip pattern of the lateral force, observed at very low applied loads (Figs. 5(a) and 5(b)), vanished with increasing load (Figs. 5(c) and 5(d)).

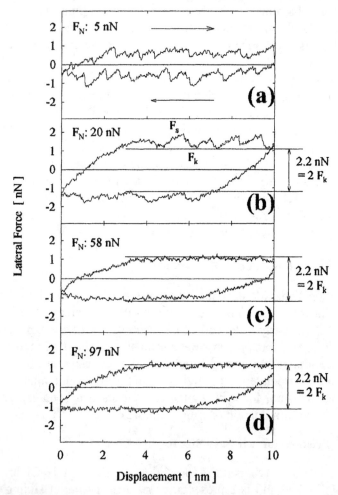

Figure 5. Lateral force for two sliding cellulose surfaces as a function of lateral displacement at four applied load levels. (a) 5 nN, (b) 20 nN, (c) 58 nN, and (d) 97 nN (speed: 20 nm/s; 0.01 M NaCl).

This behavior may be caused by flattening of load carrying asperities at sufficiently large loads, which in turn facilitates sliding because topographical features do not further obstruct sliding. This explanation is consistent with Figures 5(b-d) which reveal that the lateral force magnitude at the minimum of the stick-slip cycles, labeled F_k in Figure 5(b), is equal to the lateral force necessary to sustain sliding at larger applied loads (Figs. 5(c) and 5(d)). The approach to steady sliding at large applied loads is more gradual and large scale stick-slip events are entirely absent. The observations at large applied loads suggest a load-independent friction law, and the measured lateral force may therefore reflect an intrinsic resistance to sliding of the water-swollen cellulose surfaces. Similar friction-load behavior was reported for sliding friction between gellan hydrogels by Gong et al. [67].

We note, however, that the measured lateral force contains contributions from the load dependent shear strength (contact stiffness), and is therefore not equal to friction force. More work is necessary to deconvolute such contributions as contact deformation and sliding friction to the lateral force [45].

3.2.2 Friction Forces in Scan Size Regime II

The lateral force and the corresponding height profile for two cellulose surfaces in 0.01 M NaCl, sliding past each other at 10 μm/s with an applied normal force of 4.5 nN are plotted as functions of lateral displacement in Figures 6(a) and 6(b), respectively.

The friction behavior displayed in Figure 6(a) is typical for experiments in scan size regime II. After a steep initial rise in lateral force surfaces clearly began to slide. During steady-state sliding, lateral force data showed irregular transitions between two or more dynamic states. This "stick-slip" behavior is typical for surfaces with non-periodic roughness [68-71].

Although surface tracking was good at small normal forces a small lateral shift is apparent upon reversal of the scan direction (Fig. 6(b)). In contrast, accurate tracking without lateral shift was obtained with a bare cantilever tip under similar scan conditions. This suggests that the shift is associated with the cellulose sphere attached to the end of the cantilever, and is not caused by hysteresis effects of the piezoelectric tube scanner. The shift is likely caused by the reversal of the bead and cantilever tilt when sliding direction is reversed (see inset of Fig. 6(b)).

Lateral Force-Height Cross-Correlation. The lateral force-height cross-correlation is plotted as a function of lateral distance (lag) for four different normal loads in Figure 7 (sliding in 0.01 M NaCl at 10 μm/s). For a normal load of 4.5 nN, the cross-correlation has a pronounced peak at zero lag, indicating a strong correlation between height and lateral force signal. As the normal load increases, the magnitude of the cross-correlation at zero lag decreases (Fig. 7(a-d)). This likely results from flattening of surface asperities, as well as the formation of multiple simultaneous contacts–both of which distribute the lateral force over a larger area and thus decrease the spatial correlation.

The Effect of Sliding Speed. Lateral force increased only slightly with increasing sliding speed (2 μm/s–200 μm/s). The weak dependence was universally observed in all experiments.

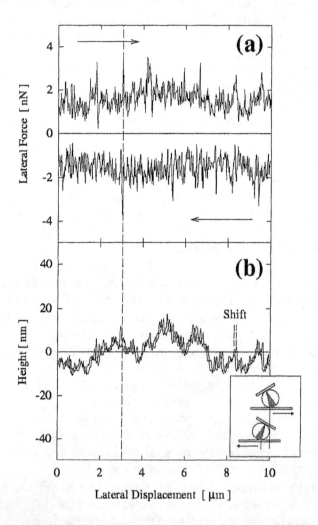

Figure 6. (a) Lateral force (friction loop), and (b) corresponding topographical profile as functions of lateral displacement for two sliding cellulose surfaces (load: 4.5 nN; speed: 10 μm/s; 0.01 M NaCl). Inset: schematic of probe rolling upon reversal of the scan direction.

The Effect of Polyelectrolytes. The lateral force is plotted as a function of normal force for cellulose surfaces in differently concentrated Na-CMC solutions in Figure 8(a) (speed: 20 μm/s; 0.01 M NaCl).

Na-CMC decreases the lateral force at a given normal force. Also plotted in Figure 8(b) is the normal force (without sliding) as a function of surface separation for these cellulose surfaces in different Na-CMC solutions. This data illustrates that the friction measurements correspond to relatively large normal forces where the adsorbed polymer layers are highly compressed (labels A, B, C, and D in Figure 8) [66].

At small polymer concentrations, the lateral force increases roughly linearly with the normal force (Fig. 8(a)). At large polymer concentrations,

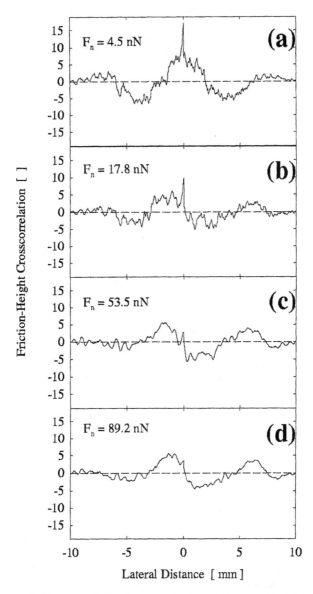

Figure 7. Cross-correlation between the lateral force and height profile as a function of lateral distance for four different loads. (a) 4.5 nN, (b) 17.8 nN, (c) 53.5 nN, and (d) 89.2 nN (speed: 10 μm/s; 0.01 M NaCl).

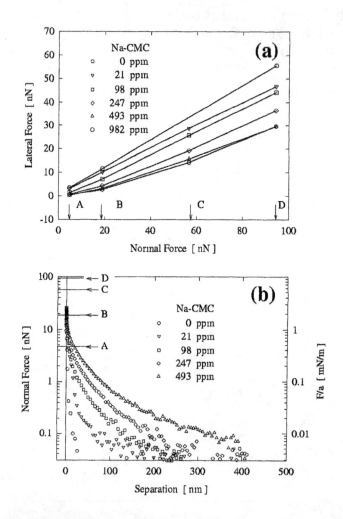

Figure 8. (a) Lateral force as a function of normal force for different Na-CMC concentrations (speed: 20 μm/s; 0.01 M NaCl). (b) Normal force as function of surface separation (without sliding) for different Na-CMC solution concentrations (0.01 M NaCl). The labels A, B, C, and D indicate the normal forces applied in friction experiments (Fig. 8(a)) and correspond to normal forces in Fig. 8(b)).

the lateral force becomes a nonlinear function of the normal force, with the slope increasing with increasing load. The cross-correlations of the lateral force and height profile are plotted as functions of lateral distance for four polymer concentrations in Figure 9. Without polymer, there is a strong cross-correlation at zero lag, indicating a correlation between the height and lateral force signal. With increasing polymer concentration (0–247 ppm), the cross-correlation at zero lag decreases, much like an increase in normal force decreases the correlation (Fig. 9(a-c)). Above 247 ppm, cross-correlation does not change significantly with increasing polymer concentration (Fig. 9(d)). This suggests a connection to the adsorbed amount, which also does not change significantly above polymer concentrations of about 300 ppm (Fig. 1).

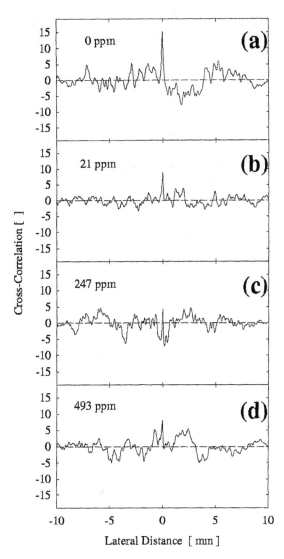

Figure 9. Cross-correlation between lateral force and height profile as a function of lateral distance for four different Na-CMC concentrations. (a) 0 ppm, (b) 21 ppm, (c) 247 ppm, and (d) 493 ppm (speed: 20 μm/s; load: 56.6 nN; 0.01 M NaCl).

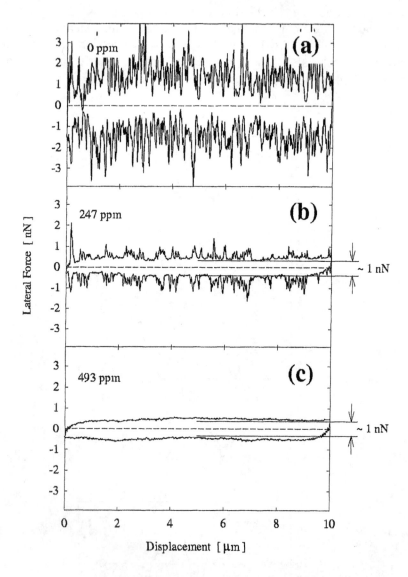

Figure 10. Lateral force for two sliding cellulose surfaces as a function of lateral displacement at three polymer concentrations. (a) 0 ppm, (b) 247 ppm, and (c) 493 ppm (load: 4.7 nN; speed: 203 µm/s; Na-CMC).

Figure 10 illustrates that the adsorbed polymer also alters the stick-slip-like patterns observed in the lateral force-displacement loops. For small loads in the absence of adsorbed polymer, the lateral force signal appears to be composed of chaotic stick-slip events (Fig. 10(a)). With increasing load, the lateral force magnitude becomes more uniform. Similar behavior is observed for small polymer concentrations (< 100 ppm). At a larger polymer concentration (247 ppm), the behavior at small loads (< 5 nN) is different. Here the lateral force is relatively smooth, punctuated by less frequent stick-slip events (Fig. 10(b)). With a further increase in polymer concentration to 493 ppm, sliding proceeds smoothly without stick-slip (Fig. 10(c)). The lateral force is equal to the minimum lateral force to sustain sliding at 247 ppm. The stick-slip pattern reappears for large polymer concentrations under large loads.

The cross-correlation data and the friction behavior at low loads suggest that the adsorbed polymer may reduce friction by "smoothing" the surfaces–perhaps by simply maintaining the surfaces at a larger separation for a given normal load, thus reducing the intimacy of contact.

We define the coefficient of friction (COF) as the ratio of lateral to normal force. It is important to note, however, that the lateral force contains contributions of the load dependent, lateral contact stiffness that we have not separated in this study [45]. The effect of polymer on lateral force becomes more apparent by plotting the COF as a function of polymer concentration (Fig. 11). The COF drops rapidly with increasing polymer concentration and appears to level off at larger polymer concentrations. The reduction in COF with polymer addition is largest at low normal forces.

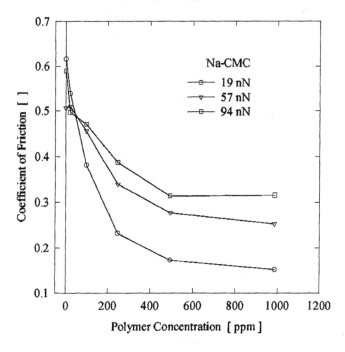

Figure 11. Coefficient of friction as a function of polymer concentration (Na-CMC, 20 μm/s) at three different normal forces.

This behavior is consistent with normal force measurements and with polymer adsorption data (Fig. 1). The extent of steric repulsion increases significantly with increasing polymer concentration in normal force experiments.

The adsorption isotherms in Figure 1 show that monolayer coverage is reached at about 300 ppm. This concentration corresponds roughly to the concentration at which the COF begins to level off (Fig. 11). This suggests that a polymer's effectiveness to reduce sliding friction is a function of the adsorbed amount.

The effect of high molecular weight cationic polyelectrolyte on lateral force in scan size regime II is quite similar to that of Na-CMC [72]. Differences in the behavior between the polymers can be explained by their molecular differences and differences in adsorption onto cellulose surfaces that lead to different surface conformation.

4 CONCLUSIONS

Colloidal probe microscopy was used to study normal and friction interactions between model cellulose surfaces. CPM is a dynamic force measurement technique, and therefore non-equilibrium (*e.g.* hydrodynamic) forces must be accounted for in the data interpretation. Normal forces between regenerated cellulose surfaces in electrolyte are governed by double layer forces. The fitted surface potential is low, in agreement with the low measured charge density, and zeta potential measurements reported in the literature [9, 61].

Normal forces between cellulose surfaces in polyelectrolyte solutions depend on the polymer concentration and on the compression rate. Force distance profiles above monolayer coverage (large polymer concentrations) were highly reproducible at a given compression rate. However, the interaction reflects a transient rather than an equilibrium state.

The extent of the steric interaction at high polymer concentrations increased approximately linearly with polymer molecular weight. At low surface coverage (small polymer concentrations) polymer bridging was observed, apparent by an adhesive force minimum upon surface separation. Bridging interactions occurred below monolayer coverage for cellulose beads. Electric double layer interactions were only observed at partial surface coverage.

For Na-CMC at low surface coverage, the molecules adsorbed relatively flat on cellulose surfaces. With increasing polymer concentration, the layer became more expanded, with a layer thickness of 1–3 R_g, depending on polymer concentration.

The scan size dependent lateral force between cellulose model surfaces in aqueous solutions is generally characterized by irregular stick-slip behavior. While lateral forces at scan sizes well below 100 nm appear to become load independent, lateral forces at scan sizes above 100 nm increase with increasing loads.

The critical scan size of about 100 nm corresponds to the average size of imaged surface asperities on the cellulose films. These asperities thus appear to be the dominant topographical feature responsible for friction at larger scan sizes. The loss of lateral force-height cross-correlation at zero lag with increasing load suggests the formation of multi-asperity contacts with increasing load. Lateral forces depend only weakly on sliding speed and solution viscosity which suggests that hydrodynamic forces contribute little to the lateral force.

Lateral forces are largest at large electrolyte concentrations and at small polymer concentrations, and decrease significantly with increasing polymer concentration. This behavior is consistent with normal force measurements at zero and small polymer concentrations where adhesive forces and polymer bridging forces were measurable. These adhesive and bridging forces effectively increase the applied load.

The reduction in lateral force with increasing polymer concentration levels off as complete monolayer coverage is approached. This is yet another indicator that adsorbed layer thickness plays an important role in reducing friction between sliding cellulose surfaces. A possible mechanism, consistent with our observations, is one where adsorbed polymers mask surface asperities, thus allowing sliding to proceed with less resistance. The effectiveness of a polymer to mask surface asperities depends on the adsorbed layer thickness, which is a function of polymer conformation on the surface, polymer molecular weight, the adsorbed amount, and the applied load.

5 FUTURE RESEARCH

The study of friction between "soft" polymeric surfaces in aqueous media at nano-newton loads and nanometer scales needs further exploration. Challenges will be to accurately describe the contact mechanics so that a reasonable deconvolution of cantilever stiffness and lateral contact stiffness will be possible.

While polyelectrolyte adsorption onto polymeric surfaces can be obtained by bulk measurements, polymer adsorption kinetics and conformation are more difficult to measure. Adsorption kinetics and polymer conformation, however, play an important role in a wide variety of colloidal phenomena ranging from micro-particle flocculation to protein binding to biosensors. We are currently working on combining solid state surface plasmon resonance (SPR) and colloidal probe microscopy to gather adsorption kinetic data, spatial extent of the adsorbed polymer layer, and force information directly and simultaneously. Particularly interesting is the aspect of evaluating changes in polymer conformation and interaction strength as a function of changes in the solvent conditions. This is important for protein binding and screening of bio-adhesive polymers. The increased spatial resolution raises the potential of mapping bio-compatibility on heterogeneous surfaces.

Available lateral speeds in CPM are scan-size dependent, and range form nanometers/second to micrometers/second. In real systems the relative speeds between particles may be much higher, on the order of meters/second. It is therefore important to measure friction at speeds where the sliding times over molecular length-scales (*e.g.* order of 100 nm for polyelectrolytes) approach or exceed the relaxation times for the backbone motion of the interacting polymers.

ACKNOWLEDGEMENTS

This research was supported by NSF (grant No. CTS-9502276).

REFERENCES

[1] T. Van de Ven. Physicochemical and hydrodynamic aspects of fines and fillers retention. In C. F. Baker and V. W. Punton, editors, *Fundamentals of Papermaking*, pages 471–494. Mechanical Engineering Publishers, Ltd., London, UK, 1989.

[2] T. Lindström. Some fundamental chemical aspects on paper forming. In C. F. Baker and V. W. Punton, editors, *Fundamentals of Papermaking*, pages 311–412. Mechanical Engineering Publishers, Ltd., London, UK, 1989.

[3] R. D. Neuman, J. M. Berg, and P. M. Claesson. Direct measurement of surface forces in papermaking and paper coating systems. *Nordic Pulp and Paper Research Journal*, 8(1):96–104, 1993.

[4] M. Holmberg, J. Berg, S. Stemme, L. Ödberg, J. Rasmusson, and P. Claesson. Surface force studies of Langmuir-Blodgett cellulose films. *Journal of Colloid and Interface Science*, 186:369–381, 1997.

[5] J. N. Israelachvili and G. E. Adams. Measurement of forces between two mica surfaces in aqueous electrolyte solutions in the range 0–100 nm. *Journal of the Chemical Society, Faraday Transactions I*, 74:975–1001, 1978.

[6] W. A. Ducker, T. J. Senden, and R. M. Pashley. Direct measurement of colloidal forces using an atomic force microscope. *Nature*, 353(2):239–241, 1991.

[7] W. A. Ducker, T. J. Senden, and R. M. Pashley. Measurement of forces in liquids using a force microscope. *Langmuir*, 8:1831–1836, 1992.

[8] M. W. Rutland, A. Carambassis, G. A. Willing, and R. D. Neuman. Surface force measurements between cellulose surfaces using scanning probe microscopy. *Colloids and Surfaces A: Physicochemical and Engineering Aspects*, 123-124:369–374, 1997.

[9] A. Carambassis and M. W. Rutland. Interactions of cellulose surfaces: Effect of electrolyte. *Langmuir*, 15:5584–5590, 1999.
[10] J. Klein and P. F. Luckham. Forces between two adsorbed poly(ethylene oxide) layers in a good aqueous solvent in the range 0-150 nm. *Macromolecules*, 17(5):1041–1048, 1984.
[11] P. F. Luckham and J. Klein. Interactions between smooth solid surfaces in solutions of adsorbing and non-adsorbing polymers in good solvent conditions. *Macromolecules*, 18(4):721–728, 1985.
[12] P. F. Luckham. Measurement of the interaction between adsorbed polymer layers: The steric effect. *Advances in Colloid and Interface Science*, 34:191–215, 1991.
[13] G. J. C. Braithwaite, A. Howe, and P. F. Luckham. Interactions between poly(ethylene oxide) layers adsorbed to glass surfaces probed by using a modified atomic force microscope. *Langmuir*, 12:4224–4237, 1996.
[14] P. M. Claesson, M. A. G. Dahlgren, and L. Eriksson. Forces between polyelectrolyte-coated surfaces: Relations between surface interaction and floc properties. *Colloids and Surfaces A: Physicochemical and Engineering Aspects*, 93:293–303, 1994.
[15] P. M. Claesson and B. W. Ninham. pH dependent interactions between adsorbed chitosan. *Langmuir*, 8:1406–1412, 1992.
[16] M. A. G. Dahlgren, P. M. Claesson, and R. Audebert. Interaction and adsorption of polyelectroytes on mica. *Nordic Pulp and Paper Research Journal*, 8:62–67, 1993.
[17] M. A. G. Dahlgren, Å. Waltermo, E. Blomberg, P. M. Claesson, L. Sjöström, T. Åkesson, and B. Jönsson. Salt effects on the interaction between adsorbed cationic polyelectrolyte layers - theory and experiment. *Journal of Physical Chemistry*, 97:11769–11775, 1993.
[18] Y. Kamiyama and J. Israelachvili. Effect of pH and salt on the adsorption and interactions of an amphoteric polyelectrolyte. *Macromolecules*, 25:5081–5088, 1992.
[19] J. Marra and M. L. Hair. Forces between two poly(2-vinylpyridine)-covered surfaces as a function of ionic strength and polymer charge. *Journal of Physical Chemistry*, 92:6044–6051, 1988.
[20] S. Biggs. Steric and bridging forces between surfaces bearing adsorbed polymer: an atomic force microscopy study. *Langmuir*, 11:156–162, 1995.
[21] S. Biggs and A. D. Proud. Forces between silica surfaces in aqueous solutions of a weak polyelectrolyte. *Langmuir*, 13:7202–7210, 1997.
[22] M. Holmberg, R. Wigren, R. Erlandson, and P. M. Claesson. Interactions between cellulose and colloidal silica in the presence of polyelectrolytes. *Colloids and Surfaces A; Physicochemical and Engineering Aspects*, 129-130:175–183, 1997.
[23] I. V. Shchukin, E. D. Nad Videnskii, E. A. Amelina, A. I. Bessonov, A. M. Parfenova, G. Aranovich, and M. Donokhi. Adhesion of cellulose fibers in liquid media: 2. measurement of contact force of attraction. *Colloid Journal*, 60(5):541–543, 1998.
[24] E. Popotoshev, M. W. Rutland, and P. M. Claesson. Surface forces in aqueous polyvinylamine solutions. 2. Interactions between glass and cellulose. *Langmuir*, 16:1987–1992, 2000.
[25] C. F. Schmid, L. H. Switzer, and D. J. Klingenberg. Simulations of fiber flocculation: Effects of fiber properties and interfiber forces. Accepted: *Journal of Rheology*, 1999.
[26] C. F. Schmid and D. J. Klingenberg. Mechanical flocculation in flowing fiber suspensions. *Physical Review Letters*, 84(2):290–293, 2000.
[27] E. A. Amelina, E. D. Shchukin, A. M. Parfenova, A. I. Bessonov, and I. V. Videnskii. Adhesion of cellulose fibers in liquid media: 1. measurement of the contact friction force. *Colloid Journal*, 60(5):537–540, 1998.
[28] S. R. Andersson and A. Rasmuson. Dry and wet friction of single pulp and synthetic fibres. *Journal of Pulp and Paper Science*, 23(1):J5–J11, 1997.

[29] S. Zauscher. Putting a sphere on an atomic force microscope cantilever. *Microscopy Today*, (10):6, December 1997.
[30] G. M. Dorris and D. G. Gray. The surface analysis of paper and wood fibres by ESCA (electron spectroscopy for chemical analysis). *Cellulose Chemistry and Technology*, 12:9–23, 1978.
[31] C. W. Hoogendam, A. De Keizer, M. A. Cohen Stuart, B. H. Bijsterbosch, J. G., Batelaan, and P. M. Van der Horst. Adsorption mechanisms of carboxymethyl cellulose on mineral surfaces. *Langmuir*, 14:3825–3839, 1998.
[32] S. Katz, R. P. Beatson, and A. M. Scallan. The determination of strong and weak acidic groups in sulfite pulps. *Svensk Papperstidning*, 87(6):R48–R53, 1984.
[33] J. A. Neumeister and W. A. Ducker. Lateral, normal, and longitudinal spring constants of atomic force microscopy cantilevers. *Review of Scientific Instruments*, 65(8):2527–2531, 1994.
[34] D. Sarid. *Scanning Force Microscopy*. Oxford University Press, 2nd edition, 1994.
[35] P. Attard, A. Carambassis, and M. W. Rutland. Dynamic surface force measurement. 2. Friction and the atomic force microscope. *Langmuir*, 15:553–563, 1999.
[36] S. Zauscher. Polymer Mediated Surface Interactions in Pulpfiber Suspension Rheology. PhD thesis, University of Wisconsin–Madison, 1999.
[37] H.-J. Butt and M. Jaschke. Calculation of thermal noise in atomic force microscopy. *Nanotechnology*, 6:1–7, 1995.
[38] C. J. Drummond and T. J. Senden. Examination of the geometry of long-range tip-sample interaction in atomic force microscopy. *Colloids and Surfaces A: Physicochemical and Engineering Aspects,*, 87:217–234, 1994.
[39] Digital Instruments. *Interpreting Force Image Data*. Support Note No. 227, Rev. A. Digital Instruments, Santa Barbara, CA, 1996.
[40] B. Bhushan. Introduction: Friction measurement methods. In B. Bhushan, editor, *Handbook of Micro/Nanotribology*, pages 44–52. CRC Press, Inc., Boca Raton, FL, 1995.
[41] G. Meyer and N. M. Amer. Simultaneous measurement of lateral and normal forces with an optical-beam-deflection atomic force microscope. *Applied Physics Letters*, 57(20):2089–2091, 1990.
[42] M. Labardi, M. Allegrini, C. Ascoli, C. Frediani, and M. Salerno. Normal and lateral forces in friction force microscopy. In H.-J. Güntherodt, D. Anselmetti, and E. Meyer, editors, *Forces in Scanning Probe Methods*, pages 319–324. Kluwer Academic Publishers, Dordrecht, NL, 1995.
[43] E. Meyer, R. Lüthi, L. Howald, and H.-J. Güntherodt. Friction force microscopy. In H.-J. Güntherodt, D. Anselmeti, and E. Meyer, editors, *Forces in Scanning Probe Methods*, pages 285–306. Kluwer Academic Publishers, Dordrecht, NL, 1995.
[44] S. J. O'Shea, M. E. Welland, and T. Rayment. Atomic force microscope study of boundary layer lubrication. *Applied Physics Letters*, 61(18):2240, 1992.
[45] R. W. Carpick, D. F. Ogletree, and M. Salmeron. Lateral stiffness: A new nanomechanical measurement for the determination of shear strengths with friction force microscopy. *Applied Physics Letters*, 70(12):1548–1550, 1997.
[46] K. L. Johnson, K. Kendall, and A. D. Roberts. Surface energy and contact of elastic solids. *Proceedings of the Royal Society, London*, 324:301–313, 1971.
[47] P. A. Lynn. *An Introduction to the Analysis and Processing of Signals*. Hemisphere Publishing Corporation, New York, 1989.
[48] W. H. Press, B. P. Flannery, S. A. Teukolsky, and W. T. Vetterling. *Numerical Recipes in Pascal*. Cambridge University Press, New York, 1989.
[49] R. J. Hunter. *Foundations of Colloid Science*, volume I. Clarendon Press, Oxford, UK, 1989.

[50] W. B. Russel, D. A. Saville, and W. R. Schowalter. *Colloidal Dispersions.* Cambridge University Press, Cambridge, UK, 1989.
[51] J. Israelachvili. *Intermolecular and Surface Forces.* Academic Press Inc., San Diego, 2nd edition, 1992.
[52] J. Gregory. Approximate expression for the interaction of diffuse electrical double layers at constant charge. *Journal of the Chemical Society. Faraday Transactions*, 69(2):1723–1728, 1973.
[54] P. G. Hartley, I. Larson, and P. J. Scales. Electrokinetic and direct force measurements between silica and mica surfaces in dilute electrolyte solutions. *Langmuir*, 13:2207, 1997.
[54] J. Visser. On Hamaker constants: A comparison between Hamaker constants and Lifshitz-van der Waals constants. *Advances in Colloid and Interface Science*, 3:331–363, 1972.
[55] P. Attard, J. C. Schultz, and M. W. Rutland. Dynamic surface force measurement. 1. van der Waals collisions. *Review of Scientific Instruments*, 69(11):3852–3866, 1998.
[56] D. Y. C. Chan and R. G. Horn. The drainage of thin liquid films between solid surfaces. *Journal of Chemical Physics*, 83:5311–5324, 1985.
[57] R. G. Cox and H. Brenner. The slow motion of a sphere through a viscous fluid towards a plane surface – II, small gap widths, including inertial effects. *Chemical Engineering Science*, 22:1753–1777, 1967.
[58] S. Kim and S. J. Karrila. *Microhydrodynamics-Principles and Selected Applications.* Butterworth-Heinemann, Stoneham, MA, 1991.
[59] S. Bhattacharjee, C. Ko, and M. Elimelech. DLVO interactions between rough surfaces. *Langmuir*, 14:3365–3375, 1998.
[60] J.Klein. Shear, friction, and lubrication forces between polymer-bearing surfaces. *Annual Reviews of Material Science*, 26:581–612, 1996.
[61] J. Brandrup, E. H. Immergut, and E. A. Grulke, editors. *Polymer Handbook.* Wiley & Sons, Inc., New York, NY, 4th edition, 1999.
[62] R. Pelton. A model of the external surface of wood pulp fibers. *Nordic Pulp and Paper Research Journal*, 8(1):113–119, 1993.
[63] V. K. La Mer and T. W. Healey. Adsorption-flocculation reactions of macromolecules at the solid-liquid interface. *Review of Pure and Applied Chemistry*, 13:112–133, 1963.
[64] C. Ortiz and G. Hadziioannou. Entropic elasticity of single polymer chains of poly(methacrylic acid) measured by atomic force microscopy. *Macromolecules*, 32(3):780–787, 1999.
[65] M. Rief, P. Schulz-Vanheyden, and H. E. Gaub. Single molecule force spectroscopy by AFM reveals details of polymer structure. In N. Garcia, M. Nieto-Vesperinas, and H. Rohrer, editors, *Nanoscale Science and Technology*, pages 41–47. Kluwer Academic Publishers, Dordrecht, Netherlands, 1998.
[66] S. Zauscher and D. J. Klingenberg. Normal forces between cellulose surfaces measured with colloidal probe microscopy. Submitted: *Journal of Colloid and Interface Science*, 1999.
[67] J. Gong, M. Higa, Y. Iwasaki, Y. Katsuyama, and Y. Osada. Friction of gels. *Journal of Physical Chemistry, B*, 101:5487–5489, 1997.
[68] A. D. Berman, W. A. Ducker, and J. N. Israelachvili. Origin and characterization of different stick-slip friction mechanisms. *Langmuir*, 12:4559–4563, 1996.
[69] P. J. Blau. *Friction Science and Technology.* Marcel Dekker, Inc., New York, NY, 1996.
[70] E. Rabinowicz. *Friction and Wear of Materials.* John Wiley and Sons, New York, NY, 1965.
[71] H. Yoshizawa and J. Israelachvili. Fundamental mechanisms of interfacial friction. 2. stick-slip friction of spherical and chain molecules. *Journal of Physical Chemistry*, 7:11300–11313, 1993.

[72] S. Zauscher and D. J. Klingenberg. Friction forces between cellulose surfaces measured with colloidal probe microscopy. Submitted: *Colloids and Surfaces A; Physicochemical and Engineering Aspects*, 1999.

INDEX

accelerator, 165, 177
actuator, 139, 165, 221, 239
adhesion, 79, 139, 327, 387
air bearing, 79
atomic force microscopy, 109, 115, 139, 387, 399, 411

bearing, 197
bonding, 327
buckyball, 95
buckytube, 55, 95, 109

cantilever, 115, 123, 399
ceramics, 271
continuum approach, 29, 45, 63, 197

degradation, 291
deposition, 271
diamond, 55, 377
diffusion, 291

elasticity, 29
electro-discharge machining, 247
electrostatics, 139, 177, 221
etching, 197

fabrication, 197, 247, 259
film, 123, 185, 271, 327
flow, 63
focused ion beam, 247
friction, 15, 29, 115, 123, 139, 185, 291, 347, 411

friction force microscopy, 123, 139
fullerene, 95
funding, 15, 23

hydrocarbon, 55
hydrodynamics, 63

hardness, 377

injector, 249
instrumentation, 109, 115, 123

laser, 247
lubrication, 79, 95, 139, 197, 291, 327, 347, 361, 377, 399

magnetic storage, 79, 291, 377
manufacturing, 177
materials, 165, 221, 259, 327
mesomanufacturing, 259, 271
metal, 395
metrology, 259
MEMS, 15, 23, 123, 139, 165, 177, 185, 197, 221, 239, 259, 347
micrometer 165, 239
milling, 247
mirror, 139
molecular dynamics, 29, 45, 55
motor, 139

nanotechnology, 15
nanoparticle, 383
nanotube, 55, 95, 109

overview, 1, 15, 23
oxidation 327

phase transition, 29
plasma, 271
pressure, 165
pump, 239
reliability, 165
rotation, 197
rotor, 197

safety 185,
self-assembled monolayer, 139, 221, 343,
scanning probe microscopy, 109, 115, 123, 139, 361, 387, 399, 411
seal, 271
sensor, 165
silane, 123
silicon, 139
speed, 197
stress, 165
stiction, 139, 177
surface, 123, 139, 177, 411
surface tension, 29, 239
switch, 239

theory, 29, 45, 55, 63, 79, 383
tribology, 15, 29, 115, 123, 139, 185, 291, 347, 411

vacuum, 361
viscosity, 29

weapon, 185
wear, 45, 123, 139, 185, 327, 347, 377
wetting, 239

CPSIA information can be obtained
at www.ICGtesting.com
Printed in the USA
LVHW081236160220
647086LV00013B/779